Editorial Board Members' Collection Series: Pollutants and Climate Change, Runoff, Behaviour and Adverse Effects in Aquatic Ecosystems

Editorial Board Members' Collection Series: Pollutants and Climate Change, Runoff, Behaviour and Adverse Effects in Aquatic Ecosystems

Editors

Mauro Marini

Tomás Ángel DelValls

MDPI

Basel • Beijing • Wuhan • Barcelona • Belgrade • Novi Sad • Cluj • Manchester

Editors

Mauro Marini	Tomás Ángel DelValls
National Research Council,	Santa Cecília University
CNR-IRBIM	(UNISANTA)
Ancona	Boqueirão
Italy	Brazil

Editorial Office
MDPI
St. Alban-Anlage 66
4052 Basel, Switzerland

This is a reprint of articles from the Special Issue published online in the open access journal *Applied Sciences* (ISSN 2076-3417) (available at: https://www.mdpi.com/journal/applsci/special_issues/J7HCK84575).

For citation purposes, cite each article independently as indicated on the article page online and as indicated below:

Lastname, A.A.; Lastname, B.B. Article Title. *Journal Name* **Year**, *Volume Number*, Page Range.

ISBN 978-3-7258-0943-1 (Hbk)
ISBN 978-3-7258-0944-8 (PDF)
doi.org/10.3390/books978-3-7258-0944-8

Cover image courtesy of Mauro Marini

Contents

About the Editors

Mauro Marini

After obtaining his degree in Agriculture from the University of Perugia, Italy, in 1990, he has worked with the CNR for more than 35 years, studying the chemical processes of soil, groundwater, and seawater organic and inorganic compounds. Since June 1997, he has focused on processes that occur in the sea. He has been in charge of numerous oceanographic cruises organized to collect physical and chemical samples throughout the Adriatic and the Mediterranean Seas. He is especially interested in biogeochemical processes that take place in the water column and at the water–sediment interface, as well as in pollutant and xenobiotic interactions in marine sediments and organisms. He is the author or co-author of more than 300 scientific publications. He was section head of the CNR ISMAR (Ancona, Italy) from September 2009 until it became the CNR IRBIM in September 2018. Currently, he holds the position of Research Director at the CNR Institute for Marine Biological Resources and Biotechnology.

Tomás Ángel DelValls

Dr. DelValls holds a PhD in science (chemical engineering). He has held various research and academic positions at a number of international institutions: CSIC, Spain; University of Cádiz, Spain; Scripps Institution of Oceanography at U.C., San Diego, USA; Universidade Santa Cecilia, Brazil; and others. He is Senior Scientist in charge of ecotoxicology and environmental science and technology at UNISANTA (Brazil) and is Chief Scientific Officer at Water Challenge, S.L. (Spain). Dr. DelValls held the position of coordinator of the UNESCO UNITWIN WiCop (UNESCO Chair program—IOC/UNESCO). He has more than 250 peer-reviewed research publications included in JCR/SCI and has coordinated more than 50 research and academic projects. With more than 30 years of experience in coordinating international and national research programs across a number of scientific disciplines, he has supervised more than 60 Master's students, 30 Doctoral students, and 15 Post-doctoral fellows from more than 20 different countries. Dr. DelValls was awarded the first National Research Prize of the Ministry of the Presidency (Spain, 2006) and was also given an award by the Ministry of Environment (2008), both of which were related to environmental quality in marine ecosystems.

applied sciences

MDPI

Editorial

Pollutants and Climate Change, Runoff, Behavior and Adverse Effects in Aquatic Ecosystems

Mauro Marini [1,2,*] and T. Ángel DelValls [3,4,*]

1 Institute for Biological Resources and Marine Biotechnologies (IRBIM), National Research Council of Italy (CNR), 60125 Ancona, Italy
2 Fano Marine Center, The Inter-Institute Center for Research on Marine Biodiversity, Resources and Biotechnologies, 61032 Fano, Italy
3 Environmental Science and Technology Department, University of Santa Cecilia, Santos 11045-907, Brazil
4 Water Challenge, S.L., 28043 Madrid, Spain
* Correspondence: mauro.marini@cnr.it (M.M.); delvalls@unisanta.br (T.Á.D.)

1. Introduction

This Special Issue addresses topics related to the sources of pollutants and their diffusion in ecosystems and aquatic organisms. Aspects of how to improve the monitoring of pollutants in the environment and within the food chain are also covered. The eleven research articles and two reviews included in this Special Issue improve the knowledge of the adverse effects on aquatic ecosystems induced by climate change, as well as the resilience potential of marine organisms within polluted environments.

The geographical areas of particular interest considered in this Special Issue are as follows:

(i) The Adriatic Sea, a sea particularly conditioned by the supply of fresh water [1–3], shallower to the north and with greater bathymetry towards the south [4,5], where the waters encounter the rest of the Mediterranean Sea [6,7]. The circulation is characterized by hydrological processes typical of the northern hemisphere [8], where there are important bottom currents that transport suspended particles, including contaminants [9,10]. Various matrices from the Adriatic Sea were studied and treated, such as sediment, water, and organisms, coming from different areas: some with restricted circulation, such as in the inlets of Kotor [11–13], some along the coastal strip [2,14], and some in the open sea [15,16].

(ii) The Gulf of Gabès, in the southeastern Mediterranean Sea: a shallow sea rich in nutrients [17] and biodiversity [18], strongly influenced by the main currents of the Mediterranean Sea coming from the west [19–21], that is of particular importance for fishing activity [22], albeit highly anthropized [23,24].

(iii) The eastern boundary current, off the Chilean coast, flows along the west coast of South America and is one of the largest ocean current systems, known as upwelling, where cold, nutrient-rich water promotes high primary and secondary productivity [25]. For this reason, the water masses off the coast of Chile are among the best areas for fishing in the world [26,27]. The coastal waters in the South Pacific Ocean, especially along the Atacama Region of Northern Chile, are affected by intense anthropic activities. The rapid increase in industrial, agricultural, and fishing activities, together with the development of ports, desalination plants, and tourism, have strongly influenced the quality of the surrounding environment [28].

The mining sector has grown expeditiously in this geographical area, with the extraction of gold, silver, molybdenum, iron, and lithium, but above all, copper.

Furthermore, two drainage basins of important rivers, one in Spain and one in Brazil, were also studied to investigate metal/loid contamination and the associated adverse

Citation: Marini, M.; DelValls, T.Á. Pollutants and Climate Change, Runoff, Behavior and Adverse Effects in Aquatic Ecosystems. *Appl. Sci.* **2024**, *14*, 2823. https://doi.org/10.3390/app14072823

Received: 14 March 2024
Revised: 24 March 2024
Accepted: 25 March 2024
Published: 27 March 2024

effects on organisms. Finally, a novel and disruptive technology has been described in this Special Issue, demonstrating its validity for brine decontaminations and extrapolating it to a new paradigm related to the circular economy and its application to other sectors with contaminated waters or fluids. An overview of published articles is shown in the next paragraphs.

2. Environmental Contamination in Sediments of Aquatic Systems

Environmental contamination by different anthropogenic activities has been analyzed in sediments in different aquatic environments, such as river basins and marine sediment, both near shore and offshore. The tolerance level of some target benthic organisms was also evaluated. The areas surveyed in this research were: the Adriatic Sea, the southern Pacific Ocean, and two river basins: one in the southwest of Spain and one in Brazil. Mandić J. et al. (Contribution 1) showed the observed concentrations of polycyclic aromatic hydrocarbons (PAH) and their sources in the Central Adriatic Sea, both near the eastern coast and offshore. Pyrogenic sources were the main contributors to the total concentration of PAH in the coastal area. At the same time, in the mid-Adriatic offshore sediments, PAHs originated mainly from petrogenic sources. Bonnail E. et al. (Contribution 2) studied the state of ecological health in the northern coastal area (26° S) of the Atacama Region (Chile, Pacific Ocean), including a marine protected area, with sub-areas for tourism and benthic management, where there are many important industrial activities. Results of heavy metal pollution levels in sediments, including toxicity levels, measured on target organisms were presented. Riba et al. (Contribution 3) studied a segment of the Guadalquivir River (SW Spain), through an integrative assessment of sediment quality. The study was perfected in laboratory conditions with a biological test, using different organisms exposed to sediments to determine their toxicological risk. Additionally, Riba et al. (Contribution 4) presented a study on the São Francisco River (Brazil) with a weight-of-evidence approach to identify sources of pollution in the river and in the city of Três Marias (Minas Gerais, Brazil), based on the structure of the benthic community and the physicochemical characterization of the sediments. Three different sentinel species have been identified to improve sediment quality assessments in river areas.

3. Contaminant Accumulation in the Organisms

Aquatic organisms are excellent indicators of the presence of contaminating substances in the environment. In these studies, it was possible to see how consolidated knowledge of target organisms is useful for defining in-depth contamination evaluation criteria, as seen for heavy metals, as well as for new organisms, such as some invasive species. Martinovic et al. (Contribution 5) determined the content of trace metals in the soft tissues, shells, and shell demineralization products of the Mediterranean mussel *Mytilus galloprovincialis* L., in the Bay of Kotor (Adriatic Sea). The study aimed to investigate accumulation patterns of trace metals in the shell and soft tissues to assess pollution sources in coastal marine ecosystems. The study by Dghim et al. (Contribution 6) was based on demonstrating the accumulation capacity of the invasive blue crab species, *Portunus segnis*, by measuring the concentrations of heavy metals in the gills and hepatopancreas, together with its usefulness as a sentinel species for the Gulf of Gabès (southeastern Mediterranean Sea).

4. Effect of Climate Change

Even studies on the effects of climate change, such as ocean acidification, can compromise the vitality of some organisms, such as bivalves, which are widely used for monitoring studies in coastal areas as useful bioindicators of persistent pollutants. These results should be taken into consideration when environmental monitoring approaches are carried out, particularly in areas most affected by these changes, like in tropical marine areas where recorded pH variations interfere with the health status of organisms and their metabolic functions, as described in the papers by da Silva Souza et al. (Contribution 7) and Cunha Passarelli et al. (Contribution 8). Over the past three decades, the ecology of benthic foraminifera has been intensively studied to improve the potential application of these marine organisms as a proxy for the effects of climate change and other global change phenomena, as well as the ability to retain persistent pollutants. An interesting study was made by Rostami et al. (Contribution 9) in a large stretch of the Adriatic Sea, where climatic variations were also observed, further verifying how the fluvial water regime has changed in the last century, carrying higher loads of nutrients that contribute to eutrophication, making the Adriatic Sea one of the most productive areas of the Mediterranean Sea (study presented by Marini and Grilli, Contribution 10).

5. Other Topics: Applications and Technology

Desalination of aquifers and marine water represents a viable choice to meet global domestic and industrial water needs. With the rise in sea levels and saltwater intrusion into groundwater, desalination becomes increasingly necessary. Removing dissolved salts from seawater to obtain fresh water of sufficient quality for many purposes generates highly concentrated salt waste, known as brine. Bonnail et al. (Contribution 11) described a clean, low-impact technology for producing drinking water without producing pollution. Rampazzo et al. (Contribution 12) used very advanced technologies to define the decomposition of plastic materials in the marine environment, in particular bioplastics, such as starch-based compostable shopping bags. This leads to useful considerations for defining the use of increasingly rapidly degrading materials in the marine environment. Bellucci and Giuliani (Contribution 13) consider coupling scientific evidence from sediment cores with historical information as an effective way to reconstruct and quantify anthropogenic impacts in transitional and marine-coastal areas and thus establish the responsibilities for pollution that may also be used in the legal field. Good practices for selecting sampling sites and specific investigation techniques are key to understanding pollution histories and dynamics, along with reliable dating methods and analytical procedures. Examples are shown of how the results have been used in preliminary investigations or judicial debates to attribute responsibility for environmental pollution to past or present industrial management.

6. Conclusions and Perspectives on the Future of Research

The work presented in this Special Issue contains concrete progress and new insights into the study of coastal and river waters in different geographical areas.

The presented studies offer innovative ideas for the correct monitoring and evaluation of the negative effects of pollutants on aquatic ecosystems. From monitoring itself to remediation technologies, it has been demonstrated that this type of study could potentially help to improve our knowledge and applications in different contexts, such as in the legal field. It was thus highlighted that the evaluation of the effects of pollution requires interdisciplinary and multidisciplinary efforts, as well as the application of new investigative methodologies.

Author Contributions: M.M. and T.Á.D. both conceived the idea and wrote the manuscript. All authors have read and agreed to the published version of the manuscript.

Funding: This research received no external funding.

Institutional Review Board Statement: Not applicable.

Informed Consent Statement: Not applicable.

Data Availability Statement: Not applicable.

Conflicts of Interest: The authors declare no conflicts of interest.

List of Contributions

1. Application of Positive Matrix Factorization for Source Apportionment of Polycyclic Aromatic Hydrocarbons (PAH) in the Adriatic Sea, and the Evaluation of PAH-Related Carcinogenic Risks
2. An Integrated Approach for the Environmental Characterization of a Coastal Area in the Southern Atacama Desert
3. Sediment Contamination and Toxicity in the Guadalquivir River (Southwest, Spain)
4. Integrative Assessment of Sediment Quality in the São Francisco River (Mina Gerais, Brazil)
5. Shell Organic Matrix (Conchix) of the Mediterranean Mussel *Mytilus galloprovincialis* L. as the Medium for Assessment of Trace Metals in the Boka Kotorska Bay
6. Assessment of the Accumulation of Trace Metals and Oxidative Stress Response Biomarkers in the Portunid Portunus segnis
7. Integrated Assessment of CO_2-Induced Acidification Lethal and Sub-Lethal Effects on Tropical Mussels Perna perna
8. Risk Assessment of a Coastal Ecosystem from SW Spain Exposed to CO_2 Enrichment Conditions
9. Understanding the Distributions of Benthic Foraminifera in the Adriatic Sea with Gradient Forest and Structural Equation Models
10. The Role of Nitrogen and Phosphorus in Eutrophication of the Northern Adriatic Sea: History and Future Scenarios
11. A new disruptive technology for zero brine discharge: towards the paradigm shift
12. An FTIR and EA-IRMS Application to the Degradation Study of Compostable Plastic Bags in the Natural Marine Environment
13. The Role of Sediment Records in Environmental Forensic Studies: Two Examples from Italy of Research Approaches Developed to Address Responsibilities and Management Options

References

1. Campanelli, A.; Fornasiero, P.; Marini, M. Physical and Chemical Characterization of the water Column in the Piceno Coastal Area (Adriatic Sea). *Fresenius Environ. Bull.* **2004**, *13*, 430–435.
2. Cozzi, S.; Giani, M. River water and nutrient discharges in the Northern Adriatic Sea: Current importance and long-term changes. *Cont. Shelf Res.* **2011**, *31*, 1881–1893. [CrossRef]
3. Marini, M.; Campanelli, A.; Sanxhaku, M.; Kljajić, Z.; Betti, M.; Grilli, F. Late Spring Characterization of Different Coastal Areas of the Adriatic Sea. *Acta Adriat.* **2015**, *56*, 27–46.
4. Vilibić, I. An analysis of dense water production on the North Adriatic shelf. *Estuar. Coast. Shelf Sci.* **2003**, *56*, 697–707. [CrossRef]
5. Turchetto, M.; Boldrin, A.; Langone, L.; Tesi, T.; Foglini, F. Particle transport in the Bari Canyon (southern Adriatic Sea). *Mar. Geol.* **2007**, *246*, 231–247. [CrossRef]
6. Kovacevic, V.; Gacic, M.; Poulain, P. Eulerian current measurements in the Strait of Otranto and in the Southern Adriatic. *J. Mar. Syst.* **1999**, *20*, 255–278. [CrossRef]
7. Civitarese, G.; Gačić, M.; Lipizer, M.; Eusebi Borzelli, G.L. On the impact of the Bimodal Oscillating System (BiOS) on the biogeochemistry and biology of the Adriatic and Ionian Seas (Eastern Mediterranean). *Biogeosciences* **2010**, *7*, 3987–3997. [CrossRef]
8. Artegiani, A.; Paschini, E.; Russo, A.; Bregant, D.; Raicich, F.; Pinardi, N. The Adriatic Sea General Circulation. Part II: Baroclinic Circulation Structure. *J. Phys. Oceanogr.* **1997**, *27*, 1515–1532. [CrossRef]
9. Tesi, T.; Miserocchi, S.; Goñi, M.A.; Langone, L.; Boldrin, A.; Turchetto, M. Organic matter origin and distribution in suspended particulate materials and surficial sediments from the western Adriatic Sea (Italy). *Estuar. Coast. Shelf Sci.* **2007**, *73*, 431–446. [CrossRef]
10. Spagnoli, F.; De Marco, R.; Dinelli, E.; Frapiccini, E.; Frontalini, F.; Giordano, P. Sources and Metal Pollution of Sediments from a Coastal Area of the Central Western Adriatic Sea (Southern Marche Region, Italy). *Appl. Sci.* **2021**, *11*, 1118. [CrossRef]
11. Milanovic, S. Hydrogeological characteristics of some deep siphonal springs in Serbia and. Montenegro karst. *Environ. Geol.* **2007**, *51*, 755–759. [CrossRef]
12. Campanelli, A.; Bulatović, A.; Cabrini, M.; Kljajić, Z.; Mosetti, R.; Paschini, E.; Penna, P.; Marini, M. Spatial distribution of physical, chemical and biological oceanographic properties, phytoplankton, nutrients and coloured dissolved organic matter (CDOM) in the Boka Kotorska Bay (Adriatic Sea). *Geofizika* **2009**, *26*, 215–228.
13. Bellafiore, D.; Guarnieri, A.; Grilli, F.; Penna, P.; Bortoluzzi, G.; Giglio, F.; Pinardi, N. Study of the hydrodynamical processes in the Boka Kotorska Bay with a finite element model. *Dyn. Atmos. Oceans* **2011**, *52*, 298–321. [CrossRef]
14. Marini, M.; Fornasiero, P.; Artegiani, A. Variations of Hydrochemical Features in the Coastal Waters of Monte Conero: 1982–1990. *Mar. Ecol.* **2002**, *23*, 258–271. [CrossRef]
15. Janeković, I.; Mihanović, H.; Vilibić, I.; Tudor, M. Extreme cooling and dense water formation estimates in open and coastal regions of the Adriatic Sea during the winter of 2012. *J. Geophys. Res.* **2014**, *119*, 3200–3218. [CrossRef]
16. Maric, D.; Kraus, R.; Godrijan, J.; Supi'c, N.; Djakovac, T.; Precali, R. Phytoplankton response to climatic and anthropogenic influences in the north-eastern Adriatic during the last four decades. *Estuar. Coast. Shelf Sci.* **2012**, *115*, 98–112. [CrossRef]

17. Bel Hassen, M.; Drira, Z.; Hamza, A.; Ayadi, H.; Akrout, F.; Messaoudi, S.; Issaoui, H.; Lotfi, A.; Bouaïn, A. Phytoplankton dynamics related to water mass properties in the Gulf of Gabes: Ecological implications. *J. Mar. Syst.* **2009**, *75*, 216–226. [CrossRef]
18. Crocetta, F.; Agius, D.; Balestrei, P.; Barichei, M.; Bayhan, Y.K.; Çakir, M.; Ciriaco, S.; Corsini-Foka, M.; Deidun, A.; El Zrelli, R.; et al. New mediterranean biodiversity records (October 2015). *Mediterr. Mar. Sci.* **2015**, *16*, 682–702. [CrossRef]
19. Roussenov, V.; Stanev, E.; Artale, V.; Pinardi, N. A seasonal model of the Mediterranean Sea general circulation. *J. Geophys. Res.* **1995**, *100*, 13515–13538. [CrossRef]
20. Tonani, M.; Pinardi, N.; Dobricic, S.; Pujol, I.; Fratianni, C. A high-resolution free-surface model of the Mediterranean Sea. *Ocean Sci.* **2007**, *4*, 213–244. [CrossRef]
21. Malanotte-Rizzoli, P.; Artale, V.; Borzelli-Eusebi, G.L.; Brenner, S.; Crise, A.; Gacic, M.; Kress, N.; Marullo, S.; Ribera d'Alcalà, M.; Sofianos, S.; et al. Physical forcing and physical/biochemical variability of the Mediterranean Sea: A review of unresolved issues and directions for future research. *Ocean Sci.* **2014**, *10*, 281–322. [CrossRef]
22. Colloca, F.; Scarcella, G.; Libralato, S. Recent trends and impacts of fisheries exploitation on Mediterranean stocks and ecosystems. *Front. Mar. Sci.* **2017**, *4*, 244. [CrossRef]
23. Smaoui-Damak, W.; Hamza-Chaffai, A.; Berthet, B.; Amiard, J.C. Preliminary study of the clam Ruditapes decussatus exposed in situ to metal contamination and originating from the Gulf of Gabes, Tunisia. *Bull. Environ. Contam. Toxicol.* **2003**, *7*, 961–970. [CrossRef]
24. El Zrelli, R.; Courjault-Radé, P.; Castet, S.; Michel, S.; Bejaoui, N. Heavy metal contamination and ecological risk assessment in the surface sediments of the coastal area surrounding the industrial complex of Gabes city, Gulf of Gabes, SE Tunisia. *Mar. Pol. Bull.* **2015**, *101*, 922–929. [CrossRef]
25. Thiel, M.; Macaya, E.C.; Acuña, E.; Vasquez, J.A.; Vega, J.M.A. The Humboldt Current System of northern and central Chile—Oceanographic processes, ecological interactions and socioeconomic feedback. *Ocean. Mar. Biol.* **2007**, *45*, 195–344.
26. Alheit, J.; Bernal, P.A. Effects of physical and biological changes on the biomass yield of the Hurnboldt Current System. In *Large Marine Ecosystems*; Sherman, K., Alexander, L.M., Gold, B.D., Eds.; American Assoclation for the Advancement of Science Press: Washington, DC, USA, 1993; pp. 53–68.
27. Daneri, G.; Dellarossa, V.; Quiñones, R.; Jacob, B.; Montero, P.; Ulloa, O. Primary production and community respiration in the Humboldt Current System off Chile and associated oceanic areas. *Mar. Ecol. Prog. Ser.* **2000**, *197*, 41–49. [CrossRef]
28. Bonnail, E.; Diaz, A.; García, A.; Cruces, E.; Borrero-Santiago, A. Coastal uses and contaminants spread in the Desert coastal region of Atacama. *Chemosphere* **2022**, *288*, 132519. [CrossRef]

applied sciences

MDPI

Article

A New Disruptive Technology for Zero-Brine Discharge: Towards a Paradigm Shift

Estefanía Bonnail [1,2], Sebastián Vera [2,*] and T. Ángel DelValls [2,3]

1 Centro de Investigaciones Costeras, Universidad de Atacama (CIC-UDA), Avenida Copayapu 485, Copiapó 1530000, Chile
2 Water Challenge S.L., Avda. Papa Negro, 63, 28043 Madrid, Spain; delvalls@unisanta.br
3 Environmental Science and Technology Department, University of Santa Cecilia, Santos 11045-907, Brazil
* Correspondence: svera@waterchallenge.eu

Abstract: The desalination of aquifers and seawaters is a viable choice to meet primarily domestic and industrial global water requirements. It removes salts from seawater to obtain freshwater with sufficient quality for different purposes, as well as a highly salt-concentrated waste stream known as brine. This residue is usually returned to the ocean, provoking, among other impacts, changes in temperature, salinity and oxygen and overall local aquatic ecosystem stress, as well as social rejection. Desalination in inland aquifers is more complicated because brine disposal is complicated or impossible. The current study presents a new zero-brine discharge technology able to achieve ecological liquid purification through distillation for the separation of the dissolved solids as crystallized salts (Adiabatic Sonic Evaporation and Crystallization, ASE&C). This new technology was used with seawater and three types of brine to test how it would work when coupled with reverse osmosis desalination plants. Analysis of the byproducts after treatment of the seawater and the different brines are presented here. A basic economic approach to calculating potential revenues is also presented. The results of the analyses revealed a complete depuration of water as distilled water, and crystallized solids with highly concentrated commercial salts (with different composition depending on their origin). The estimated economic value of annual revenue (taking into account only seven element recoveries and treatment of a volume of 1000 m^3/d) for three types of brines ranged between 1 and 11 million euros, compared to between 3.6 and 9.3 million euros when ASE&C is employed with seawater. The treatment of greater volumes for seawater desalination would increase these numbers significantly. ASE&C supposes a solution coupled (or not) to desalination plants to reduce the ecological impacts associated with brine discharges to zero, obtaining two significant commercial byproducts: (seawater: freshwater and commercial elements Br, Ca, Cs, Cl, NaOH, Mg, N, K, Rb, Na, Sr, Li, U, B, Sr, Ga, etc.; aquifers: a larger list than for saltwater, depending on the nature of the water body). It can solve environmental issues associated with brine discharge, with null CO_2 emissions (renewable energy) and profitable (i.e., with no costly pretreatment) technology.

Keywords: desalination; seawater treatment; aquifer desalination; ASE&C; brine mining optimization; circular economy

Citation: Bonnail, E.; Vera, S.; DelValls, T.Á. A New Disruptive Technology for Zero-Brine Discharge: Towards a Paradigm Shift. *Appl. Sci.* **2023**, *13*, 13092. https://doi.org/10.3390/app132413092

Academic Editor: Dae Sung Lee

Received: 15 September 2023
Revised: 10 October 2023
Accepted: 28 November 2023
Published: 8 December 2023

1. Introduction

Despite being called the Blue Planet because two thirds of it is covered by water, 97% of the water of the Earth is seawater (Figure 1a) and only 3% is freshwater, in the form of glaciers (2.5%) and other water bodies of groundwater (aquifers), superficial water (lakes and rivers), atmosphere (water vapor), and even organisms [1,2]. A combination of the exponentially rising population, the advance of drought and desertification as consequences of global warming and climate change, and the scarcity and unequal global distribution of water resources and other associated problems such as pollution is endangering, aggravating, and accelerating the loss of water available for human consumption, hydric stress, and the water crisis. This water stress is based on freshwater availability and affects different levels (ecology, biology, economy, social, etc.) depending on the geographic location. Around 4 billion people live in areas that suffer from severe physical water scarcity for at least one month per year [3]; these regions are mostly located between the tropics and the Northern Hemisphere (Figure 1b, [4]). It is estimated that more than 1.8 billion people will live in regions without water, and two-thirds of the global population will live under hydric stress. Unfortunately, the scarcity of water and water quality deterioration go hand in hand with famine, low development of agriculture, and low social development.

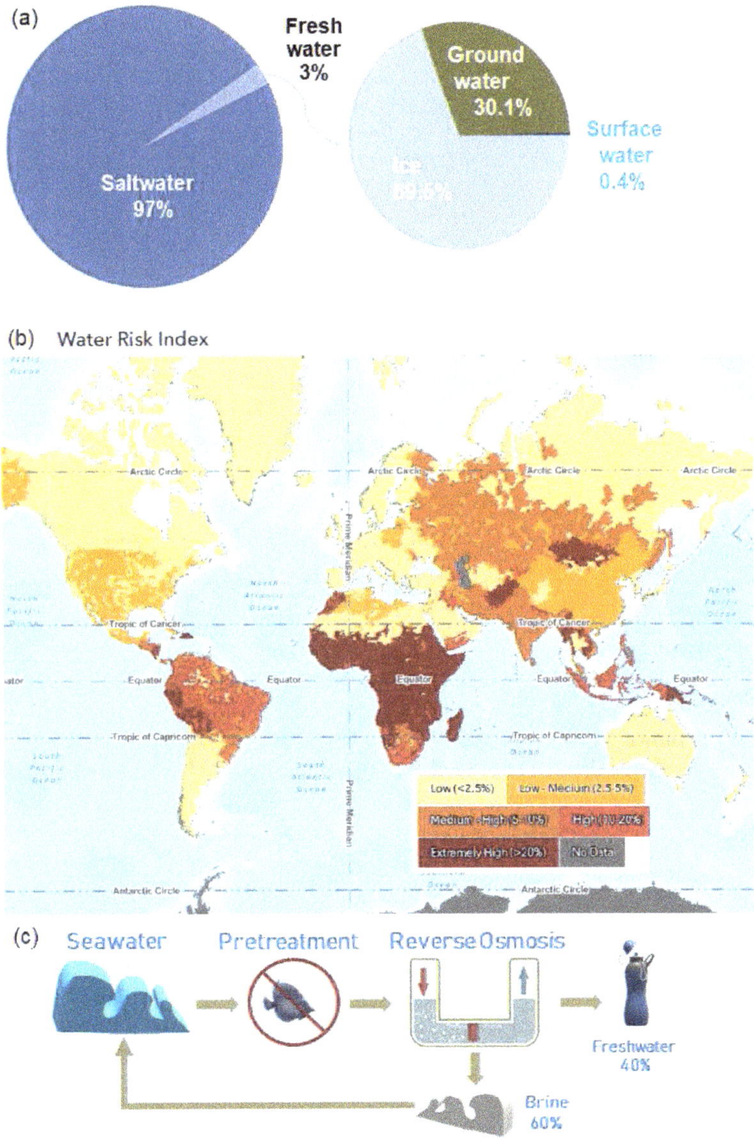

Figure 1. (**a**) Distribution of the total water on Earth in percentages (Source: [5]). (**b**) Water risk Index Map (Sources: [6,7]). (**c**) Desalination process.

According to the United Nations World Water Development Report [8], global water demand is expected to continue increasing at a similar rate until 2050, accounting for an increase of 20 to 30% above the current level of water use, mainly due to rising demand in the industrial, agricultural, and domestic sectors. Water demand is increasing, while supplies are shrinking (due to superficial and aquifer contamination, glacier melting, droughts, increasing urbanization, etc.). To cover this demand, the water industry must make better and more efficient use of available water resources, by means of water purification, water desalination, and waste water treatment and in industrial activities that require the use of water.

Due to the huge proportion of seawater (Figure 1a) and the fact that most countries have easy access to it (Figure 1b), one of the most feasible ways to obtain freshwater is through desalination. This water treatment is based on liquid–solid separation techniques such as evaporation and reverse osmosis (RO), but these techniques are not completely efficient (between 60–98%) due to the generation of a waste product known as brine (Figure 1c). A desalination plant operating using RO commonly recovers 40%, i.e., 0.40 m^3 of purified water is produced from each 1 m^3 of seawater, and 0.60 m^3 is transformed into brine to be discharged back into the ocean. The state of desalinization and brine discharges has been recently and thoroughly reviewed [9]. It is estimated that more than 15 thousand operational desalination plants produce 95 million m^3 of desalinated water per day, and more than 140 million m^3 of brine per day. Most of these plants are installed in the Middle East and North Africa (coinciding with areas under hydric stress, Figure 1b); although half of them are destined for industrial purposes, more than 62% of desalinated water is destined for human consumption. RO is by far the dominant treatment technology (84% of the operational plants).

The downsides of desalination are the intake and pretreatment strategies, use of land, elevated power costs, membrane separation saturations, chemical addition and biofouling, ecological and environmental problems associated with the brine returning into the sea, and social rejection [10–13].

The most abundant freshwater source is groundwater (Figure 1a), but continental brackish water occupies about 11% of aquifer volume and is located in the upland areas of sedimentary basins [14]. Desalination of geothermal and hypersaline water contained in aquifers is a possible alternative for municipal/agricultural supplies [14,15], but is also strongly linked to the extraction of minerals and elements with commercial value.

To achieve the objectives of sustainable development, there is a global challenge to obtain economically feasible high-quality water by means of "green technologies" to supply current and future hydric requirements. The main objective of this study is to show and discuss the results obtained using a new system (Adiabatic Sonic Evaporation and Crystallization, ASE&C) to separate water from solutes and solids associated with different saline and hypersaline samples (seawater and brines), obtaining a purified fluid (freshwater) and crystallized solids (with high potential commercial value). Also, this study analyses some performance tradeoffs to demonstrate the ASE&C system's feasibility in the treatment of seawater, brine, and other hypersaline fluids.

2. Materials and Methods

2.1. ASE&C Technology

A new innovative technology, termed Adiabatic Sonic Evaporation and Crystallization (ASE&C), was designed and patented (World International Patent Organization: EP3135635) for contaminated fluid purification using a physical approach, distillation for the separation of liquids and dissolved solids by promoting adiabatic changes and acceleration (Figure 2) [16]. In this case, it completely separated a seawater sample into freshwater and crystallized salts with a cost of energy between 0 and 20 kWh/m^3.

The specific energy consumption calculated by the ASE&C system is the total amount of power required for internal pumping and the heat exchanger. The required energy quantity for the ASE&C system depends on the complementary energy sources available.

The system consumes around 19.2 kWh/m^3 without any complementary energy source. If a residual heat source is available, the energy required to boost the water in the circuit is between 3–4 kWh/m^3. However, the energy consumption reaches 0 kWh/m^3 when the ASE&C technology is coupled to a hybrid solar thermal–photovoltaic system (for 24-h operation).

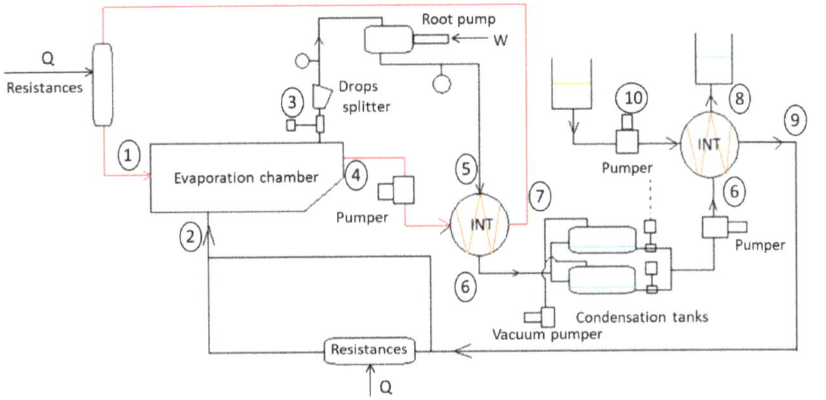

Figure 2. ASE&C™ flowchart diagram with energy feedback through vapor compression. 1. Inlet of the heat transfer fluid to the evaporator. 2. Inlet of the fluid to be processed (raw water) to the evaporator. 3. Output of the processed fluid (vapor) from the evaporator. 4. Heat-carrying fluid outlet from the root compressor. 5. Processed fluid (vapor) outlet from the root compressor to the heat exchanger. 6. Processed fluid (condensed) outlet from exchanger; after the fluid is condensed in the heat exchanger, it is deposited in two condensate tanks, and then taken to the condensate heat exchanger. 7. Exit of the heat transfer fluid after being heated by the exchanger. 8. Outlet of the processed fluid from the condensate heat exchanger. 9. Outlet of the fluid to be processed (raw water) from the condensate heat exchanger. 10. Input of the fluid to be processed to the condensate heat exchanger.

The ASE&C must be designed strategically to optimize the performance of the entire (zero-liquid discharge) ZLD system while depurating any contaminated fluid [17–20]. An improper design can reduce the overall efficiency or increase the operating costs.

2.2. Seawater Treatment

A volume of 79 L of seawater was supplied to the ASE&C technology; 77 L were ASE&C-treated and collected at the output of the systems (+1.3 L were used for the industrial set-up), and 0.7 L were collected for analytical purposes. For this particular case, the energy consumption was assumed to be 3–5 kWh/m^3.

2.3. Analytical Procedures

Aliquots from the liquid samples were collected in sterile bottles from the input and output for water quality analyses. Cations were determined by means of ICP-OES, anions by means of ionic chromatography/HPLC, carbonates and bicarbonates by means of Metrohm titration, and P with ICP-MS in certified laboratories as detailed in [17–20].

3. Results and Discussion

3.1. Seawater Treatment Results

Fluids collected before and after the ASE&C treatment were analyzed. Table 1 summarizes water parameters at the input and output of the ASE&C system. All the dissolved elements were almost completely removed (most of them below the analytical detection limit) to obtain distilled water with pH modified from 7.9 to 5.2, EC from 56 mS/cm to 142 µS/cm, and the Langelier index from 0.79 to −5.85.

Table 1. Physical-chemical analysis of water samples at the input (seawater) and the output (treated fluid).

Parameter		Input Sample (Seawater)	Output Sample (Distilled Water)
Langelier Index		0.79	−5.85
EC	µS/cm	53,327	142
pH		7.9	5.2
SS	mg/L	10	<3.0
COD	mg/L	1.1	0.6
Bicarbonates	mg/L	144.0	<5.0
Carbonates	mg/L	<0.2	<0.2
Nitrates	mg/L	<0.2	<0.2
Sulfates	mg/L	2524	<5.0
K	mg/L	403.7	<2.0
F	mg/L	1.4	<1.0
Br	mg/L	<0.1	<0.1
Ca	mg/L	421.2	<2.0
Mg	mg/L	1318.4	3.1
Na	mg/L	11,009.8	16.9
Cl	mg/L	18,200	30.9
B	mg/L	5.03	0.55

The values obtained for the crystallized salts formed are given in Table 2. Microbiological presence was not detected in the solid residue. Most of the newly formed conglomerates were chlorides (53%); 80.50% of the dried residue composition was NaCl, together with sulfates (4.80%), and Mg (3.34%).

Table 2. Crystallized solid salts' analysis after seawater treatment with the ASE&C system.

Microbiological Analysis		
Clostridium perfringens	absence/25 g	n.d.
Listeria monocytogenes	c.f.u./g	<20
Total coliforms	absence/g	n.d.
Intestinal enterococcus	absence/g	n.d.
Escherichia coli	absence/g	n.d.
Staphylococcus coagulase positives	absence/g	n.d
Mesophile aerobic microorganisms	c.f.u./g	<10
Mold and leaven (25 °C)	c.f.u./g	<10
Chemical analysis		
Fe	mg/kg	3.79
Ca	%	0.43
Humidity	%	<1
Insoluble residue	%	1.0
Sulfates	%	4.80
NaCl (over dried matter)	%	85
Mg (MgO)	%	5.54
B	mg/L	81.2
As	µg/kg	<25
Cd	µg/kg	<10
Zn	µg/kg	10,178
Co	µg/kg	<10
Cu	µg/kg	477
Mn	µg/kg	679
Hg	µg/kg	<5
Pb	µg/kg	132
K	mg/kg	8187

n.d.: not detected.

3.2. ASE&C Technology Coupled with RO

The ASE&C system can be coupled to the RO process to improve the efficiency of desalination. Applying the ASE&C depuration to brine (Figure 3) can improve the RO efficiency up to 100% freshwater recovery, so brine discharge into the ocean would be zero (ZLD), and a byproduct composed of crystallized commercial salts would be obtained (Figure 3a). Since brine from groundwater and inland reservoirs have different compositions, the pre-treatment and treatment are different, but the final products would be the same—complete freshwater recovery and dried crystallized salts with different compositions. The compositional chemistry of the Atlantic Ocean, the Mediterranean Sea, and brackish groundwater was studied in depth by [21]. This has an essential transcendence in the technoeconomic assessment of the efficiency process with the ASE&C system (material resistance and life cycle), the potential recovery process, and the revenue generated from the byproducts.

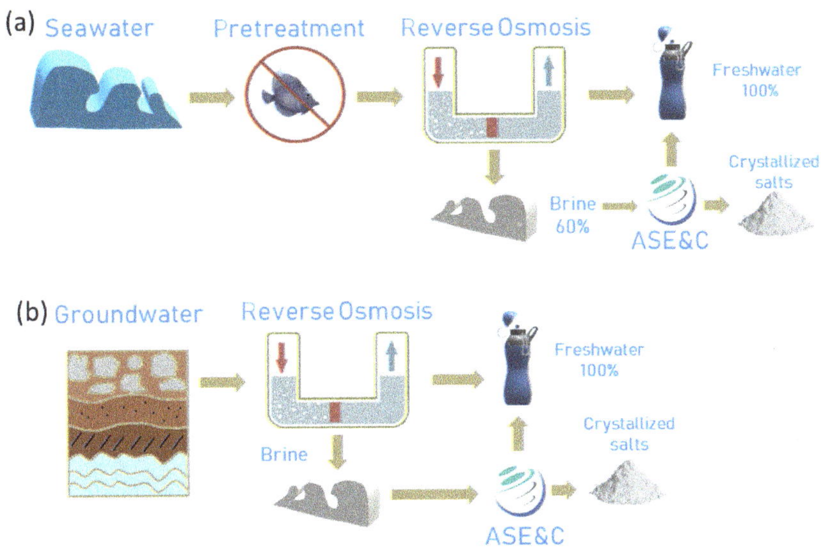

Figure 3. ASE&C coupled to the reserve osmosis cycle process for (**a**) the seawater desalination process and (**b**) groundwater brine treatment.

Some preliminary tests were adapted and run with different types of brine with different compositional and physicochemical characteristics (seawater and aquifer desalination) (Table 3). Three types of brine with different ranges of electrical conductivity were successfully separated into distilled water and crystallized solids (humidity < 1%). Brine A had an EC of 122 mS/cm and 78 g/L of TDS; it was oversaturated brine (hypersaline). Brine B was average seawater brine from desalinization (70 mS/cm and 44 g/L TDS). A third brine (Brine C), originally from brackish groundwater, had 28 mS/cm and 18 g/L TDS. Therefore, the ASE&C technology was adapted for the necessities of different types of fluids. The resulting products had similar volumes to the input, with chemical composition similar to distilled water (EC < 90 mS/cm, TDS < 60 mg/L) and removal of sulfates, chlorides and most of the dissolved elements (Table 3). The most interesting feature of the technology is the complete recovery of freshwater that can be used for other purposes with very low energy cost, but also the useful chemical compounds contained in the second byproduct as crystallized salts (with different compositions depending on the original fluid), with potential commercial value in the raw material market.

Table 3. Physicochemical characteristics of different brines (Brine A: hypersaline; Brine B: desalination brine; Brine C: groundwater brine) before (I) and after (O) treatment with ASE&C technology.

	Brine A		Brine B		Brine C	
	I	O	I	O	I	O
EC (mS/cm)	122,000	89	70,000	91	28,000	52
Eh (mV)	296	653	301	660	298	651
TDS (mg/L)	78,080	56	44,830	57	17,930	32
Sulfates (mg/L)	76,670	<5	4860	<5	3870	<5
Chlorides (mg/L)	67,060	<10	38,550	<10	14,569	<10
Na (mg/L)	35,990	<10	20,650	<10	6458	<10
Ca (mg/L)	1830	<10	1050	<10	986	<10
Mg (mg/L)	4880	<10	2810	<10	1860	<10
K (mg/L)	1463	<5	843	<5	453	<5
Bicarbonates (mg/L)	611	<5	351	<5	654	<5
Br (mg/L)	416	<0.1	240	<0.1	44.8	<0.1
Ga (mg/L)	650	<0.01	589	<0.01		<0.01
Sr (mg/L)	52	<0.01	29	<0.01	14.4	<0.01
Li (mg/L)	2.7	<0.01	1.6	<0.01		<0.01
B (mg/L)	17	<0.05	9.8	<0.05		<0.05
Rb (mg/L)	0.52	<0.01	0.28	<0.01		<0.01

Recovering metals, metalloids and rare earth elements as a byproduct of depuration activities is challenging, but some important concentrations are found in brines, such as lithium, rubidium, and potassium. Thus, extraction from brines and other waste streams is becoming economically viable, as previously reported in [16] with mining lixiviates depurated with the ASE&C technology.

Desalination processes require physical and chemical pre-treatment (reduction of algae and corrosive material, etc.) [22] (Figure 1c), but the cleaning processes of membranes increase the chemical components in the disposal brine. However, lately, the challenge in water desalination has involved two byproducts: freshwater and solid salts [23]. Therefore, in the race to obtain better water quality, brine may be an excellent potential source. Brine mining is commonly used to extract numerous materials from the same deposit (e.g., lithium carbonate, magnesium chloride, potassium chloride), regardless of brine origin. New technologies for the separation and recovery of sources from seawater and brine are appearing [23]. Sodium chloride from salt extraction is the most ancient resource exploited from seawater, but some others, such as Au, Mg and Br, have also been extracted throughout history [24]. The new trends in mineral market prices and critical raw materials highlight the interest in the search for new sources such as brines [21,25]. Na, Ca and Mg are elements that can be found in greater concentrations in oceans and brackish groundwater [21]. Some elements, although low in concentration, play an important role in the world market due to their technological applications; among them are Li, Rb and Sr, as well as trace elements such as rare earth elements.

Ref. [21] analyzed the economic potential of brine mining from desalination rejects containing B, Ca, Sr, Mg, Na, Li, Rb and Ga; they concluded that the economic value might range from EUR 13 billion to 29 billion annually. However, annual estimations (gross profit) for the complete recovery of elements with the studied brines (Table 3), applying the same prices to a production capacity of 1000 m^3/day, ranged from EUR 1 to 11 million for Brine A and Brine B, and from EUR 10 to 21 thousand for Brine C (Table 4). More than 96% of the revenues of Brine A and B come from gallium, due to its elevated market price. Gallium also has specialized uses in small quantities, e.g., optoelectronic devices, with economically viable recovery from seawater and other highly concentrated sources [26]. There would also be an important recovery of rubidium from Brine B. This rare alkali metal has high economic value for its emerging industrial applications (biomedicine, solar cells, electronics, etc.); it is extracted from land-based mineral ore sources (as a fraction of ores) [27]. However, [27] other alternative sustainable sources such as seawater brine or salt lake brines, with less energy extraction consumption and greater concentration than land ores, have been suggested. Some hydrometallurgical techniques can effectively separate Rb as rubidium carbonate (98%) from brines [28] and other minerals and brines [29].

Table 4. Potential economic range value (minimum–maximum in thousands of EUR) of the elements recovered from the waste of brines calculated for an annual depuration capacity of 1000 m^3/d and considering analytical data and market prices from [21].

	Seawater	Brine A	Brine B	Brine C
Na	2692–8117	24.11–72.7	13.8–41.7	4.33–13.0
Ca	66–291	0.79–3.46	0.5–2.0	0.43–1.9
Mg	148	15.03	8.7	5.7
Ga		1104–11,340	1000–10,275	
Sr		0.52–2.94	0.29–1.64	0.14–0.82
Li	703–755	1.03–1.11	0.61–0.66	
B	22–28	0.013–0.02	7–9	
Rb		4.91–6.31	2642–3396	
Total (thousands of EUR)	3631–9338	1150–11,441	1026–10,333	10.62–21.45

Almost all the revenue from Brine C comes from a combination of Na and Ca due to their high concentrations (there are missing data for the other elements). Calcium removal from seawater/brine has dual benefits: it improves the RO (secondary scaling) [25] and market potential (0.43–1.89 EUR/kg, [21]), thereby minimizing desalination costs. Calcium is sold as calcium carbonate pellets when produced from brackish water, and calcium sulfate when produced from seawater [25]. Sodium, as different sodium compounds (sodium chloride, sodium sulfate), is sold for many purposes (the food industry, cosmetics, and ice-melters, among others). Despite its low market price in comparison to other elements (0.67–2.02 EUR/kg, [21]), its elevated concentrations in seawater and brine make this element economically interesting.

The approximation was also calculated for seawater with the available data. The annual benefits obtained ranged (for Na, Ca, Mg, Li and B recovery) from EUR 3 million to 9 million, based on 75–87% Na recovery and 8–20% Li recovery. The strong uncertainty regarding the availability of Li and the pegmatite ores has raised interest in brine/seawater as a new potential source [25]. Although Li concentration in these materials seems very low (average 1.24 g/m^3, [21]), it can increase significantly after treatment with RO or other methodologies such as ASE&C, reaching economically feasible extraction from brines and solids (in the case of the ASE&C). Li extraction from brines, such as electrochemical extraction, is more developed from the point of view of environmental friendliness and competitiveness [30].

3.3. ASE&C Technology Competitive Advantages

(i) The ASE&C system is a cost-effective water quality treatment which does not need sample pre-treatment (it does not use chemical reagents). The resultant fluid from the treated seawater achieves 100% removal of most of the elements (Tables 1 and 5).

(ii) It is an energetically versatile low-cost technology (0–20 kWh/ m^3) when coupled to solar thermal energy, photovoltaic, residual heat, biomass, etc., considering that it requires electricity or heat for the operation.

(iii) The final two potential commercial byproducts (clean freshwater and commercial crystallized salts) could offset operational costs as a promising secondary source. It has been demonstrated that 100% distilled water can be produced from contaminated effluents. Volumes of 100 m^3 per day were treated and tested under controlled conditions for several weeks, demonstrating the production of 100 m^3/d distilled water (pilot plant). This is also scalable up to the treatment of 600 m^3/d to recover the same amount of water (a 40-foot plant); however, it is scalable to higher volumes, as replicates of this approach will not have limits in their design. Regarding the production of dry conglomerates, it obtains 95–99% of the total solutes included in the original input fluid.

(iv) The CO_2 footprint of the ASEC&C technology is estimated at 0.2 tons (for production of 600 m^3/d and 5 kWh/m^3), so these emissions would be approximately null if using photovoltaic and thermo-solar panels as a source of energy. Thus, considering a consumption of 5 kWh of electric energy, it would emit 0.5 kg/CO_2 eq m^3.

(v) This technology can be adapted to other fluid characteristics for a properly efficient process, and indeed has already been tested on some of them [16]. ASE&C is a portable module occupying 40 feet of container space and may be placed anywhere, but could also be designed as a full plant for higher volumes of fluid treatments.

(vi) It complies with the legislation of the EU Water Framework Directive [31–33] and the UN Sustainable Development Goals (SDGs) to achieve good ecological status of water bodies. The ASE&C effluent produces distilled water that can be reused for different purposes, such as industrial processes and human consumption; it might be injected in wells to fill and purify aquifers; after remineralization, it is suitable for drinking; it is suitable for irrigation after mixture with aquifer water; and it could also be dumped in superficial water, achieving good ecological status.

Table 5. ASE&C vs. other competing technologies (reverse osmosis: RO, evaporation and crystallization).

Feature	RO	Evaporation	Crystallization	Evap + Cryst 2 Steps	ASEC
Water/solid separation	45%	80%	80–100%	90–100%	100%
Energy consumption (kWh/m^3)	3–5	80	250–500	130–230	0–20 *
Use of chemicals	Yes	Yes	Yes	Yes	No
CAPEX (€/m^3)	700–2000	15,000	20,000–40,000	30,000	5000–15,000
Environmental concerns	Brine discharges	Brine discharges	High CO_2 emissions	Brine discharges	ZLD Low CO_2-missions

* variation depending on energy sources available, see Section 2.1.

One study [13] recognized several impact mitigation strategies, associated with (i) source water intake, (ii) brine discharge (water recovery, energy recovery, extraction of precious materials, beneficial use of brine), (iii) energy use, (iv) site selection, and (v) impact on the environment. The ASE&C can completely solve three of them and the others to a partial extent.

Other technologies for freshwater recovery of mineral extraction from brines are less efficient, imply high energy consumption, and require the addition of products and chemicals (Table 5). Although some of them have lower CAPEX values, they also include several environmental concerns (especially brine discharges). RO and ion exchange units are commonly used in lithium recovery from geothermal brines [34]. Direct lithium extraction employs significant volumes of water and chemicals, has high energy consumption, and generates wastes [35]. In contrast, other evaporitic technologies employed in brine mining (such as lithium extraction) have intensive water use, protracted duration, and particular use over continental brines [35]. Ultrafiltration combined with RO and/or nanofiltration have recently been used [36]. Some other zero-discharge desalinations use hybrid desalination with RO and nanofiltration or electrodialysis, metathesis and evaporation ponds, with significant recovery of water (97%) and fractions of multivalent ions (60%), but also with substantial energy consumption (0.77 kWh per g TDS/L^{-1}) [37] and potentially detrimental effects due to the costly use of membranes and chemicals.

4. Conclusions

The ASE&C technology can treat brackish water and seawater, including their brine waste from RO desalination, with complete separation of water as distilled water and crystallized salts (humidity < 1%), using very little energy (0–20 kWh/m^3). This study presents the results of the efficiency of separation in terms of physicochemical analysis. Laboratory results showed that the two byproducts derived (freshwater and crystallized salts) have commercial potential, and point to a paradigm shift in desalination processes, because of the zero-brine discharge converted to commercial salts, the implementation of the circular economy, and the avoidance of environmental concerns related to brine discharge back into the ocean.

The preliminary results demonstrated that the technology can be adapted successfully to different types of fluids with a wide range of electrical conductivity. Thus, the ASE&C technology coupled with the existing RO treatment plants might help to solve environmental problems associated with brine discharges, but it also supposes an alternative method of producing raw materials for brine mining. Annual economic estimations based on the studied samples were calculated: they averaged EUR 6.5 million when ASE&C is coupled with a seawater RO desalination plant (Na recovery was the greatest contributor, followed by Li). Other brine recoveries provided greater variable benefits (average benefits of EUR 4 million, and maximum benefits of EUR 11 million) due to composition disparity, with important recovery of Ga, Mg, and Na.

The innovative ASE&C technology transforms brines and improves water solutions to help reduce costs and waste volume, while increasing byproduct values with a low carbon footprint.

Author Contributions: Conceptualization, E.B., S.V. and T.Á.D.; methodology: E.B., S.V. and T.Á.D.; software: E.B., S.V. and T.Á.D.; validation: all authors; formal analysis: E.B., S.V. and T.Á.D.; investigation: E.B., S.V. and T.Á.D.; resources: all authors; data curation: E.B., S.V. and T.Á.D.; writing—original draft preparation, E.B. and T.Á.D.; writing—review and editing, all authors; visualization, E.B., supervision: S.V. and T.Á.D.; project administration, S.V. and T.Á.D.; funding acquisition, S.V. and T.Á.D. All authors have read and agreed to the published version of the manuscript.

Funding: This research was funded by ASEC Grant #190180588 through the EU's EIC/Accelerator Horizon Europe 2022 call.

Institutional Review Board Statement: Not applicable.

Informed Consent Statement: Not applicable.

Data Availability Statement: Data are contained within the article.

Acknowledgments: The authors thank the Water Challenge S.L. technical team for all the activities conducted during the study in the ASEC plant, and the technical and laboratory staff of the ICMAN-CSIC.

Conflicts of Interest: The authors declare that this study received funding from the European Innovation Council (EIC). The funder was not involved in the study design, collection, analysis, interpretation of data, the writing of this article or the decision to submit it for publication. Authors Estefanía Bonnail, Sebastián Vera and T. Ángel DelValls were employed by the company Water Challenge S.L. The authors declare that the research was conducted in the absence of any commercial or financial relationships that could be construed as a potential conflict of interest.

References

1. United Stated Geological Survey (USGS). The Distribution of Water on, in, and above the Earth. Water Science School. 2019. Available online: https://www.usgs.gov/media/images/distribution-water-and-above-earth#:~:text=About%252071%2520percent%2520of%2520the,in%2520you%2520and%2520your%2520dog (accessed on 28 November 2022).
2. UNESCO. *"Water for People, Water for Life" United Nations World Water Development Report, Part II: A Look at the World's Freshwater Resources*; 2003; UNESDOC Digital Library, Call Number: hnd 551.48 UNE2003; 576p, ISBN 978-92-3-103881-5. ISSN1 92-3-103881-8. ISSN2 1-57181-628-3. Available online: www.unesco.org (accessed on 27 November 2023).
3. Mekonnen, M.M.; Hoekstra, A.Y. Four billion people facing severe water scarcity. *Sci. Adv.* **2016**, *2*, e1500323. [CrossRef]

4. Vörösmarty, C.; McIntyre, P.; Gessner, M.; Dudgeon, D.; Prusevich, A.; Green, P.; Glidden, S.; Bunn, S.E.; Sullivan, C.A.; Liermann, C.R.; et al. Global threats to human water security and river biodiversity. *Nature* **2010**, *467*, 555–561. [CrossRef]
5. NASA/PennState. Earth in Future. Distribution of Water on the Earth´s Surface. Available online: https://www.e-education.psu.edu/earth103/node/701#:~:text=The%2520distribution%2520of%2520water%2520on,lakes%252C%2520rivers%252C%2520and%2520swamps (accessed on 27 November 2023).
6. National Geographic, MapMaker, Water Risk Index. 2023. Available online: https://www.arcgis.com/apps/instant/atlas/index.html?appid=0cd1cdee853c413a84bfe4b9a6931f0d&webmap=3871d636011b4101abfcbc75bef46802 (accessed on 27 November 2023).
7. Hofste, R.; Kuzma, S.; Walker, S.; Sutanudjaja, E.H.; Bierkens, M.F.; Kuijper, M.J.; Sanchez, M.F.; Van Beek, R.; Wada, Y.; Rodríguez, S.G.; et al. *Aqueduct 3.0: Updated Decision-Relevant Global Water Risk Indicators*; Technical Note; World Resources Institute: Washington, DC, USA, 2019. Available online: https://www.wri.org/publication/aqueduct-30 (accessed on 27 November 2023).
8. UNWWD. United Nations World Water Development Report. Valuing Water. 2021. Available online: https://unesdoc.unesco.org/ark:/48223/pf0000375751 (accessed on 28 November 2022).
9. Jones, E.; Qadir, M.; van Vliet, M.T.H.; Smakhtin, V.; Kang, S.-M. The State of desalination and brine production: A global outlook. *Sci. Total Environ.* **2021**, *657*, 1343–1356. [CrossRef]
10. Liyanaarachchi, S.; Shu, L.; Muthukumaran, S.; Jegatheesan, V.; Baskaran, K. Problems in seawater industrial desalination processes and potential sustainable solutions: A review. *Rev. Environ. Sci. Biotechnol.* **2014**, *13*, 203–214. [CrossRef]
11. Cambridge, M.L.; Zavala-Perez, A.; Cawthray, G.R.; Statton, J.; Mondon, J.; Kendrick, G.A. Effects of desalination brine and seawater with the same elevated salinity on growth, physiology and seedling development of the seagrass *Posidonia australis*. *Mar. Pollut. Bull.* **2019**, *140*, 462–471. [CrossRef]
12. Zapata-Sierra, A.; Cascajares, M.; Alcayde, A.; Manzano-Agugliaro, F. Worldwide research trends on desalination. *Desalination* **2021**, *519*, 115305. [CrossRef]
13. Ihsanullah, I.; Atieh, M.A.; Sajid, M.; Nazal, M.K. Desalination and environment: A critical analysis of impacts, mitigation strategies, and greener desalination technologies. *Sci. Total Environ.* **2021**, *780*, 146585. [CrossRef]
14. Person, M.A.; Sazeed, N. Continental Brackish Groundwater Resources. In *Unconventional Water Resources*; Qadir, M., Smakhtin, V., Koo-Oshima, S., Guenther, E., Eds.; Springer: Cham, Switzerland, 2022. [CrossRef]
15. Aparicio, J.; Candela, L.; Alfranca, O.; García-Aróstegui, J.L. Economic evaluation of small desalination plants from brackish aquifers. Application to Campo de Cartagena (SE Spain). *Desalination* **2017**, *411*, 38–44. [CrossRef]
16. Bonnail, E.; Vera, S.; Blasco, J.; Conradi, M.; DelValls, T.Á. Metal pollution and mining in the Iberian Pyrite Belt: New remediation technologies to improve the ecosystem services of the river basins. *Water* **2023**, *15*, 1302. [CrossRef]
17. Bonnail, E.; Vera, S.; Blasco, J.; DelValls, T.Á. Towards a cleaner textile industry: The liquid residue treatment challenge. *Water* **2023**. *under review*.
18. ICMAN/CSIC. *Scientific Report on ASEC Efficiency in the Purification of Contaminated Mining Waters*; Technical Report; ICMAN/CSIC: Cádiz, Spain, 2022.
19. ICMAN/CSIC. *Scientific Report on ASEC Efficiency in the Purification of Swine Slurries*; Technical Report; ICMAN/CSIC: Cádiz, Spain, 2022.
20. ICMAN/CSIC. *Scientific Report on ASEC Efficiency in the Purification of Contaminated Waters from Textile Industry*; Technical Report; ICMAN/CSIC: Cádiz, Spain, 2022.
21. Del Villar, A.; Melgarejo, J.; García-López, M.; Fernández-Aracil, P. The economic value of the extracted elements from brine concentrates of Spanish desalination plants. *Desalination* **2023**, *560*, 116678. [CrossRef]
22. Mavukkanfy, M.; Chabib, C.M.; Mustafa, I.; Al Ghaferi, A. Brine management in desalination industry: From waste to resources generation. *Desalination* **2019**, *472*, 114187. [CrossRef]
23. Panagopoulos, A.; Haralambous, K.-J.; Loizidou, M. Desalination brine disposal methods and treatment technologies—A review. *Sci. Total Environ.* **2019**, *693*, 133545. [CrossRef]
24. Thys, A. Sustainability and impact aspects of exploitation of marine salt, magnesium and bromine. *J. Coast. Res.* **2003**, *19*, 912–918.
25. Shahmansouri, A.; Min, J.; Jin, L.; Bellona, C. Feasibility of extracting valuable minerals from desalination concentrate: A comprehensive literature review. *J. Clean. Prod.* **2015**, *100*, 4–16. [CrossRef]
26. DuChanois, R.M.; Cooper, N.J.; Lee, B.; Patel, S.K.; Mazurowski, L.; Graedel, T.E.; Elimelech, M. Prospects of metal recovery from wastewater and brine. *Nat. Water* **2023**, *1*, 37–46. [CrossRef]
27. Sharma, S.K.; Truong, D.Q.; Guo, J.; An, A.K.; Naidu, G.; Deka, B.J. Recovery of rubidium from brine sources utilizing diverse separation technologies. *Desalination* **2023**, *556*, 116578. [CrossRef]
28. Chen, W.-S.; Lee, C.-H.; Chung, Y.-F.; Tien, K.-W.; Chen, Y.-J.; Chen, Y.-A. Recovery of rubidium and cesium resources from brine of desalination through t-BAMBP Extraction. *Metals* **2020**, *10*, 607. [CrossRef]
29. Xing, P.; Wang, C.; Chen, Y.; Ma, B. Rubidium extraction from mineral and brine resources: A review. *Hydrometallurgy* **2021**, *203*, 105644. [CrossRef]
30. Zhao, X.; Yang, H.; Wang, Y.; Sha, Z. Review on the electrochemical extraction of lithium from seawater/brine. *J. Electroanal. Chem.* **2019**, *850*, 113389. [CrossRef]

31. Directive 2000/60/EC of the European Parliament and of the Council of 23 October 2000 Establishing a Framework for Community Action in the Field of Water Policy. Available online: https://www.fao.org/faolex/results/details/en/c/LEX-FAOC023005/#:~:text=The%2520purpose%2520of%2520this%2520Directive,based%2520on%2520a%2520long%252Dterm (accessed on 27 November 2023).
32. Directive 2008/105/EC of the European Parliament and of the Council of 16 December 2008 on Environmental Quality Standards in the Field of Water Policy, Amending and Subsequently Repealing Council Directives 82/176/EEC, 83/513/EEC, 84/156/EEC, 84/491/EEC, 86/280/EEC and amending Directive 2000/60/EC of the European Parliament and of the Council. Available online: https://eur-lex.europa.eu/legal-content/EN/TXT/PDF/?uri=CELEX:02008L0105-20130913&from=PT (accessed on 27 November 2023).
33. Directive 2013/39/EU of the European Parliament and the Council of 12 August 2013 Amending Directives 2000/60/EC and 2008/105/EC as Regards Priority Substances in the Field of Water Policy. Available online: https://eur-lex.europa.eu/LexUriServ/LexUriServ.do?uri=OJ:L:2013:226:0001:0017:en:PDF (accessed on 27 November 2023).
34. SAMCOTECH. 2023. Available online: https://samcotech.com/how-is-brine-mining-used-for-lithium-recovery/ (accessed on 27 November 2023).
35. Vera, M.L.; Torres, W.R.; Galli, C.I.; Chagnes, A.; Flexer, V. Environmental impact of direct lithium extraction from brines. *Nat. Rev. Earth Environ.* **2023**, *4*, 149–165. [CrossRef]
36. Song, W.; Li, N.; Ding, S.; Wang, X.; Li, H.; Zhang, Y.; Feng, X.; Lu, J.; Ding, J. Nanofiltration desalination of reverse osmosis concentrate pretreated by advanced oxidation with ultrafiltration: Response surface optimization and exploration of membrane fouling. *J. Environ. Chem. Eng.* **2021**, *9*, 106340. [CrossRef]
37. Cappelle, M.; Walker, W.S.; Davis, T.A. Improving desalination recovery using zero discharge desalination (ZDD): A process model for evaluating technical feasibility. *Ind. Eng. Chem. Res.* **2017**, *56*, 10448–10460. [CrossRef]

applied sciences MDPI

Article

An FTIR and EA-IRMS Application to the Degradation Study of Compostable Plastic Bags in the Natural Marine Environment

Federico Rampazzo [1], Nicoletta Calace [2], Malgorzata Formalewicz [1], Seta Noventa [1], Claudia Gion [1], Lucia Bongiorni [3], Amelia De Lazzari [3], Valerio Causin [4] and Daniela Berto [1,*]

1 Istituto Superiore per la Protezione e la Ricerca Ambientale (ISPRA), Brondolo 5, 30015 Chioggia, Italy; federico.rampazzo@isprambiente.it (F.R.); malgorzata.formalewicz@isprambiente.it (M.F.); seta.noventa@isprambiente.it (S.N.); claudia.gion@isprambiente.it (C.G.)
2 Istituto Superiore per la Protezione e la Ricerca Ambientale (ISPRA), Via Vitaliano Brancati 48, 00144 Rome, Italy; nicoletta.calace@isprambiente.it
3 Consiglio Nazionale delle Ricerche, Istituto di Scienze Marine, Tesa 104–Arsenale, Castello 2737/F, 30122 Venezia, Italy; lucia.bongiorni@ve.ismar.cnr.it (L.B.); amelia.delazzari@ve.ismar.cnr.it (A.D.L.)
4 Dipartimento di Scienze Chimiche, Università di Padova, Via Marzolo 1, 35131 Padova, Italy; valerio.causin@unipd.it
* Correspondence: daniela.berto@isprambiente.it

Abstract: The present study aims to explore the degradation process of compostable, starch-based (i.e., Mater-Bi®) shopping bags in the marine environment using isotope ratio mass spectrometry and Fourier-transform infrared spectroscopy (FTIR). The mixing model applied to the isotopic data suggested that the compostable shopping bags had a mixed composition with a higher percentage of polyesters (61% to 72%). Changes in the isotopic composition over a 73-day period of marine water immersion showed a decrease in the corn starch constituent (of 14% to 13%), with a similar rate in both types of bags in accordance with the evidence derived from the infrared spectra. The time required for complete degradation of the starch fraction was estimated by an isotopic approach from 124 to 180 days, following zero-order kinetics. The coupled application of these two analytical methodologies promises to (i) show complementary evidence regarding the time-course degradability of different polymers via FTIR changes, and (ii) infer potential degradation mechanisms via carbon isotope analyzes. We encourage the use of this information to support advancements in the development of more sustainable-by-design plastic materials.

Keywords: biopolymers; shopping bags; carbon isotopes; IRMS; FTIR; biodegradation

Citation: Rampazzo, F.; Calace, N.; Formalewicz, M.; Noventa, S.; Gion, C.; Bongiorni, L.; De Lazzari, A.; Causin, V.; Berto, D. An FTIR and EA-IRMS Application to the Degradation Study of Compostable Plastic Bags in the Natural Marine Environment. *Appl. Sci.* **2023**, *13*, 10851. https://doi.org/10.3390/app131910851

Academic Editors: Jordi Puiggali and Mónica Calero de Hoces

Received: 5 July 2023
Revised: 31 August 2023
Accepted: 27 September 2023
Published: 29 September 2023

1. Introduction

Plastics have gained great importance in modern society due to their durability, versatility, lightness, ease of processing and low production costs as well as a broad range of polymer- and formulation-specific properties such as water repellency, resistance to corrosion and electrical, mechanical, acoustic and thermal insulation ([1] and authors therein). However, some of these properties become disadvantageous when plastic products become waste. According to Plastics Europe (2022) [2], global plastics production reached 390.7 million tonnes in 2021 (57.2 million tonnes in Europe). A considerable fraction of plastic waste ends up in various environmental compartments due to littering or mismanagement. The marine environment is considered the final sink for most of the waste dispersed on land. Since 2010, between 4.8 and 12.7 million tonnes of plastics produced by 192 coastal countries (93% of global population) have been discarded in the ocean [3]. Once there, plastics accumulate and can even remain for centuries, depending on the specific plastic's properties and composition as well as the environmental conditions [4], which negatively impacts marine ecosystems, biodiversity and, potentially, human health. According to the DeFishGear Project's report [5], which investigated the different types of waste present in the Adriatic and Ionian Seas, plastic bags (26.5%) were the most abundant floating wastes.

Many worldwide efforts have been put into place to regulate the growing plastic pollution crisis. In Europe, the 2018 European Strategy for Plastics in a Circular Economy [6] has provided a roadmap for tackling plastic pollution and leading the transition toward a sustainable and circular economy for plastic polymers. In this strategy, as in other international initiatives, the design, production and use of polymeric materials that ensure the performance of classic plastic types while having more degradation properties, is proposed as part of a potential solution to plastic pollution [7]. At present, different types of plastic materials are on the market whose degradation properties are based on different processes, primarily microbial-, oxidative-, hydro- and photodegradation ([8] and authors cited therein). Biodegradable and compostable plastics consumption was estimated at 335,000 tonnes in 2017, an amount that is expected to grow thanks to the commitment of national and international policies for a transition towards a circular, bio-economy [9,10].

Compostable plastics, which are generally marketed through certification or labeling (e.g., TŰV Austria, OK compost), are largely used in several commercial sectors, primarily in foodware, dishes, cutlery and bags. The spectrum of polymers includes aliphatic polyesters and copolyesters such as polylactic acid (PLA), polyhydroxyalkanoates (PHAs), polybutylene succinate adipate (PBSA), polycaprolactone (PCL), polyglycolic acid (PGA), aliphatic-aromatic polyesters such as polybutylene adipate terephthalate (PBAT) and carbohydrates (e.g., cellulose, cellulose acetate, starch and starch blends). As for almost all plastics compositions, these materials can be generated from fossil carbon sources (*compostable petroleum-based plastics*) or biological resources (*compostable bio-based plastics*). The origin of carbon does not influence the compostability of the material which, conversely, depends on the chemical structure of the polymers and, possibly, on the presence of specific additives (e.g., pro-oxidant additives [9,11]).

At present, there are several commercial *bio-based* plastics on the market that are certified as compostable in industrial facilities (e.g., PLA, PHAs, some of the starch blends, cellulose-based plastics and lignin-based polymer composites) [10]. Despite posing some risks, the use of compostable *bio-based* plastics in different sectors is strongly encouraged due to their potential for a double win, both in terms of sustainable end-of-life and a shift to renewable carbon feedstock and low greenhouse gas emissions [6]. However, items made of compostable *bio-based* plastics are often accidentally or intentionally found dispersed in terrestrial and aquatic environments as a result of improper disposal due to the widespread false belief that compostable plastics are completely degradable under environmental (uncontrolled) conditions.

While many manufacturers claim the degradability of *bio-based* materials under marine conditions based on the available degradation standard methods (e.g., ASTM D6691-17, ASTM D7081-05; "OK biodegradation MARINE" label based on the criteria of ASTM D7081), the scientific community is concerned that these approaches are not effectively assessing biodegradation under realistic scenarios, as they do not account for the heterogeneous and variable properties of marine environments potentially impacting degradation processes [9,12,13].

In this paper, we aim to explore the degradation process of compostable starch-based (i.e., *bio-based*) plastic shopping bags in the marine environment by using Fourier-transform infrared spectroscopy (FTIR) and isotope ratio mass spectrometry (IRMS). The methods commonly used for assessing degradability for certification purposes rely on measurements of carbon to CO_2 conversion (or to CH_4 under anaerobic conditions), biological demand for oxygen and secondary parameters (e.g., the evolution of weight loss, the decrease of molecular weight and the deterioration of mechanical properties) under different degradation environments (e.g., industrial compost, home compost, soil, sea, water and aerobic/anaerobic conditions) [9].

FTIR is an established methodology used to identify the polymeric composition of plastic material. The obtained infrared spectrum allows for the determination of the chemical structure of the mixture focusing on the observation of specific functional groups. IRMS has only recently begun to be explored for applications in plastics research and nowadays is an established methodology used to identify the polymeric composition of plastic materials [1]. It measures the relative abundances of stable isotopes, which depend on natural isotopic fractionation (the partitioning of the heavier and lighter isotopes due to different biotic and abiotic processes). The approach used in this study provides an easy and time/cost-saving technique that does not require laborious preparative steps and thus is more easily applicable in routine laboratory activities (compared to other techniques like *GC-MS and LC-MS*). The coupled application of these two analytical methodologies is intended to gain complementary evidence regarding: (i) the time-course degradability of different polymers usually present in the blend (i.e., starch and polyester) by measuring changes in the FTIR spectral signatures, and (ii) potential degradation mechanisms by analyzing changes in carbon isotopes rates (i.e., ^{13}C vs. ^{12}C, via IRMS). To our knowledge, there is no literature addressing the biodegradability of plastic polymers in marine environments when a coupling of these two analytical methods is used. The outputs of this study will contribute to a better understanding of degradation processes and the end-of-life of compostable (and, more generally, degradable) plastics in marine environments and thus support advancements in the development of more sustainable-by-design plastic materials.

2. Materials and Methods

2.1. Samples

In this study, two types of shopping bags were compared in terms of degradation processes: ultra-lightweight bags with a wall thickness below 15 microns (ULB) and lightweight bags (LSB) produced in compliance with the EN 13432 standard method with a wall thickness below 50 microns, both licensed by Mater-Bi® (Novamont, Novara, Italy). Shopping bags made of polyethylene (PE) were used as control samples. The analyzed bags were collected from large-scale retail distribution supermarkets in Italy.

2.2. Shopping Bags Degradation Experiment

Shopping bags were exposed to in situ marine conditions by placing them at a 1 m depth in the tourist marina of Chioggia (Venice, Italy) (45.223186 N, 12.281897 E), in order to simulate a dispersed bag floating in the water column. The experiment was carried out in the summer from 21 June to 1 September 2019 for a total of 73 days of exposure. Average physicochemical water conditions of the period were: temperature 26.8 ± 1.1 °C; salinity: 33.71 ± 1.03 PSU; pH: 8.17 ± 0.05; oxygen: 5.54 ± 0.67 mg/l (data available from the Hydrobiological Station of Chioggia, University of Padua (45.22340 N, 12.28434 E; https://chioggia.biologia.unipd.it/banche-dati/parametri-laguna/parametri-2019/, accessed on 30 September 2019). The bags were fixed on the surface of polyethylene supports previously filled with seawater in order to regulate their buoyancy, and the supports were immersed in the water column along ropes tied to the piers of the marina. Six samplings were carried out at regular intervals (T1: 4 July 2019; T2: 16 July 2019; T3: 31 July 2019; T4: 8 August 2019; T5: 21 August 2019; T6: 1 September 2019) during the exposure period of about two months. During each sampling, fragments of the different bags were collected, fouling organisms were gently removed by hand and samples were rinsed with deionized water. Samples were successively dried at room temperature and then subjected to isotopic (IRMS) and spectroscopic (FTIR) analyzes.

2.3. EA-IRMS

Isotopic analysis was performed by Delta V Advantage, an isotope ratio mass spectrometer, coupled with Flash 2000, an elemental analyzer (both Thermo Fisher Scientific, Bremen, Germany), operating in dynamic flash combustion mode.

Small pieces (<4 mm largest dimension) were cut from each sample and introduced into tin capsules. Capsules were gently crimped closed and introduced into the combustion reactor by means of MAS 200 R autosampler (Thermo Fisher Scientific, Bremen, Germany), and the reaction was supported by an oxygen pulse of 3 s. The furnace temperature was 900 °C. A helium flow of 100 mL/min carried the gases formed during combustion through the chromium oxide catalyst, the reduced copper and the silver cobaltous-cobaltic oxide layer. After water removal on a magnesium perchlorate trap, gases reached the GC separation column through the ConFlo IV interface and then were subjected to the isotopic composition determination on IRMS. The isotopic composition was equal to the ratio between the abundance of the heavy isotopic form compared to the light one ($^{13}C/^{12}C$) and was expressed as a deviation in parts per thousand (‰), from an internationally identified standard benchmark material (Vienna Pee Dee Belemnite—VPDB). Carbon isotopic ratios were expressed according to the Equation (1):

$$\delta^{13}C\ (‰) = [(^{13}C/^{12}C_{sample})/(^{13}C/^{12}C_{VPDB}) - 1] \times 1000 \quad (1)$$

where $^{13}C/^{12}C$ is the ratio between abundances of the masses of the heavy isotope and of the light one, $^{13}C/^{12}C_{sample}$ is the ratio measured in the sample and $^{13}C/^{12}C_{VPDB}$ is the ratio measured for the international standard. A positive value of $\delta^{13}C$ showed that the heavy isotope had been enriched in the sample compared to the standard, while a negative value showed that, in the sample, the heavy isotope had been depleted.

This number was then multiplied by 1000 so that $\delta^{13}C$ was expressed in units of parts per thousand (‰).

Reference materials sucrose IAEA CH6 (−10.449 ± 0.033‰, International Atomic Energy Agency, IAEA, Vienna, Austria), caffeine IAEA-600 (−27.771 ± 0.043‰, International Atomic Energy Agency, IAEA, Vienna, Austria) oxalic acid (−14.48 ± 0.21‰, International Atomic Energy Agency, IAEA, Vienna, Austria) and L-glutamic acid (−26.389 ± 0.042‰, National Institute of Standards and Technology, NIST, Gaithersburg, MD, USA) were used to calibrate the instrument.

The $\delta^{13}C$ values for the standards were linearly plotted against their expected values (certified) with a regression coefficient (R^2) ranging from 0.95 to 0.99 in order to ensure the complete combustion of the samples. The linear regression curve was then applied to correct the $\delta^{13}C$ results for the samples [14].

The measurement trueness was evaluated through the analysis of certified material CH7 polyethylene IAEA (−32.04 ± 0.04‰, International Atomic Energy Agency, IAEA, Vienna, Austria). The analytical precision of carbon isotope measurements, calculated as the mean of standard deviations (see Table 1) was 0.2‰.

2.4. FTIR

IR absorption spectra were acquired on a Nexus FTIR spectrometer (Thermo Nicolet, Madison, WI, USA) in attenuated total reflection (ATR) mode with an ATR ITX accessory (Thermo Nicolet, Madison, WI, USA) equipped with a diamond ATR crystal. A MIR Globar source was used and the detector was of the DTGS type. The spectral region spanned from 4000 to 650 cm^{-1}, with a resolution of 4 cm^{-1}. In the acquisition of the IR spectrum of each sample, 128 scans were collected. Nicolet Omnic software (version 7.2) was used for the treatment of spectra. No ATR correction was applied to IR data.

Integration of the peaks was performed displaying the spectra in the absorption scale. The same integration limits were chosen for all the samples. These limits were chosen to encompass about 95% of the total area, excluding the band edges [17,18].

In order to quantify and interpret the chemical modifications which generated the spectral evolution, three areas in the infrared spectra were selected: the region from 3100 to 3600 cm^{-1} related to the hydroxyl groups, the interval 1690–1810 cm^{-1} due to the carbonyl group in carboxylic acids and esters and, finally, the range between 1020 and 1070 cm^{-1} where the absorptions were due to C-O-C bonds from alcohols and esters [19,20]. In order to normalize and obtain data independent of the sample size, the integrated intensities listed above were divided by the area of the peak at 730 cm^{-1}, which was assigned to the backbone of the polyester chains (C-H and C=O out of plane bending). The functional groups were quantified by calculating the ratio between the area of the peak of the relevant functional group and the area of the reference peak at 730 cm^{-1}. Therefore, the hydroxyl group index was obtained by the ratio between the absorptions around 3300 cm^{-1} and 730 cm^{-1} and the carbonyl group index by the ratio of the peaks at about 1711 cm^{-1} and 730 cm^{-1} [21].

Table 1. $\delta^{13}C$ values for matrices analyzed: mean, standard deviation (SD) and number of replicates (n); * certified value −32.151 ± 0.050‰; ** only range reported.

Sample Type	Mean $\delta^{13}C$ (‰)	SD (‰)	n	Reference
Shopping bags				
Ultra-lightweight food bag—ULB T0	−21.14	0.13	5	This study
Lightweight shopping bag—LSB T0	−22.87	0.09	5	This study
Polyethylene bag—PE T0	−30.41	0.51	5	This study
Natural samples				
Corn starch	−12.23	0.10	3	This study
Corn	−11.38	0.55	3	[15]
Corn	−10.7 to −10.8 **		3	[16]
Potato starch	−25.52	0.00	2	This study
Potato	−27.80	0.01	2	[15]
Potato	−27.2 to −25.8 **		3	[16]
Sunflower oil	−29.04	0.10	2	This study
Rapeseed oil	−30.14	0.08	2	This study
Polymer samples				
Polylactic acid (PLA)	−9.48	0.40	2	This study
Polylactic acid (PLA)	−13.87	2.18	6	[15]
Polyhydroxybutyrate (PHB)	−11.83	0.20	3	This study
Polycaprolactone 10000 (PCL)	−27.15	0.40	3	This study
Polycaprolactone 80000 (PCL)	−27.33	0.20	3	This study
Polybutylene succinate (PBS)	−27.47	0.20	3	This study
Polyethylene terephtahalate (PET)	−27.84	1.71	7	[15]
IAEA CH-7 polyethylene standard (PE) *	−32.04	0.04	3	This study

2.5. Statistical Analysis

One-way ANOVA was used to determine whether there was a statistically significant difference among the isotopic carbon data of the three polymers (PE, ULB and LSB) of the exposed shopping bags. The method used to discriminate the means was Fisher's Least Significant Difference LSD, performed by Statgraphics Plus software (version 5.0). Statistically significant differences between the isotopic results obtained from the different exposition times of each bag's typology were also tested.

3. Results

3.1. IRMS Analysis

Data concerning the $\delta^{13}C$ values of the shopping bags analyzed in this study are reported in Table 1. Mater-Bi® bags were found to be made principally with corn starch (the only declared component) and PCL [22,23] or partially *bio-based* copolyester prepared from vegetable oils [24]. Instead, Eich et al. (2015) [25] mentioned that the Mater-Bi® bag analyzed in their study consisted of a starch-based biopolymer/polyethylene terephthalate (PET) blend. The $\delta^{13}C$ values of −21.14 ± 0.13‰ and −22.87 ± 0.09‰ for ULB and LSB, respectively, suggest a mixed composition of the used blends. In particular, the contribution of starch and polyester components to the bag's composition was estimated by applying a two end-member mixing model. Considering the literature, references [22–24,26,27] and $\delta^{13}C$ values determined for different polyesters and natural raw materials, corn starch and biodegradable polyesters PCL and PBS were chosen as end-members. Potential raw materials, such as polymers, potato and corn starches (C3 and C4 plants, respectively) and vegetable oils were considered in this study as possible end-members for isotope mixing analysis. A similar approach was adopted for several environmental studies ([28] and authors cited therein). The following equations were used for mixing calculations (Equations (2) and (3)):

$$F_{starch} + F_{polyester} = 1 \tag{2}$$

$$\delta^{13}C_{bag} = F_{starch} \times \delta^{13}C_{starch} + F_{polyester} \times \delta^{13}C_{polyester} \tag{3}$$

where F is the fraction percentage, $\delta^{13}C_{starch}$ and $\delta^{13}C_{polyester}$ are the two end-members (−11.45‰ and −27.32‰, respectively) calculated as the mean of values reported in the literature [15,16] and those determined in this study.

The results of mixing calculations showed a higher percentage of polyesters in LSB with respect to ULB (72% and 61%, respectively).

The changes in isotopic composition over time (immersion in seawater) of the shopping bags (ULB and LSB) showed that over the 73-day exposure period and under variable conditions of temperature and salinity (i.e., 24.24–28.76 °C and 31.19–36.28 PSU, respectively), the $\delta^{13}C$ values of both Mater-Bi® bags showed a gradual decrease towards more negative values (Figure 1). The recorded $\delta^{13}C$ variation ($\Delta\delta^{13}C$) was 2.16 and 2.84‰ for ULB and LSB, respectively. During the experiment, all ULB and LSB samples showed structural changes with the formation of numerous heterogeneous pinholes, cracks and grooves. The evidence that ULB fragments were not found at T6 is presumably linked to the mechanical action of waves, rather than to chemical degradation [13,29]. The PE control bags did not show a marked decrease of $\delta^{13}C$ and their observed $\delta^{13}C$ variation ($\Delta\delta^{13}C = 0.41$‰) was close to the uncertainty of the analytical method, which is different from what was observed for compostable bags.

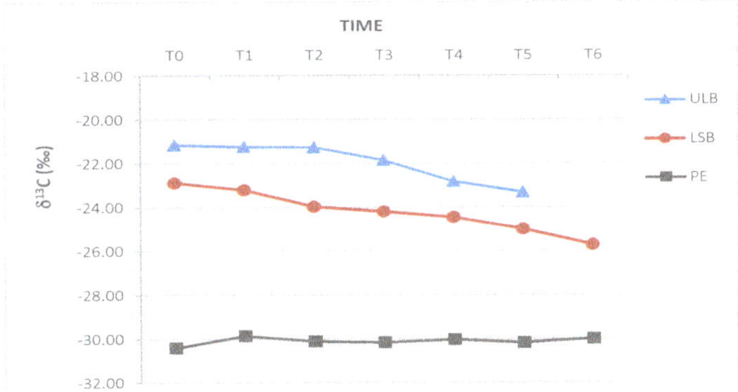

Figure 1. Changes in $\delta^{13}C$ values in polyethylene bag (PE), lightweight shopping bag (LSB) and ultra lightweight bag (ULB) during exposure time points.

By applying the mixing approach to the isotopic values of the samples (T0–T6), where T0 indicated values relative to the starting composition of the shopping bags, we observed that the percentage of starch fraction decreased at a similar rate in both types of shopping bags, independently of its initial proportion. Indeed, between T0 and T5 (T6 was analyzed only for LSB) 14% and 13% of starch were lost, respectively for ULB and LSB (Figure 2). This implied a progressive enrichment in polyesters in both types of bags, even if ULB showed an initial difficulty in degrading.

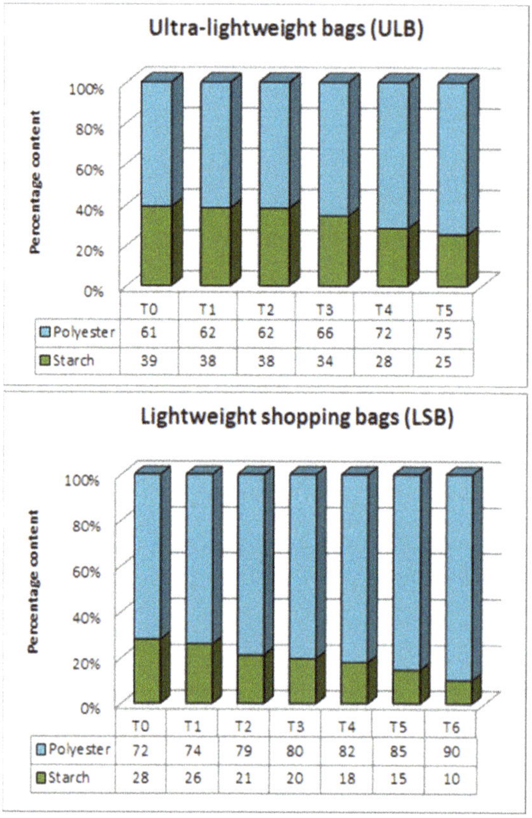

Figure 2. Changes in composition of ultra-lightweight (ULB) and lightweight (LSB) shopping bag samples during exposure to coastal marine environments.

The degradation process of starch follows zero-order kinetics (Figure 3). The uncertainty associated with different immersion times was found to range between 2% and 5% and the degradation process curve can be considered statically robust.

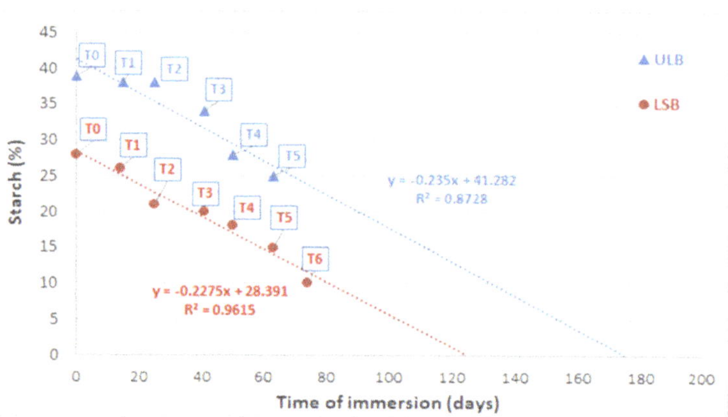

Figure 3. Estimation of starch component degradation time in ultra lightweight bag (ULB) and lightweight shopping bag (LSB).

Considering the intercept in the X-axis, we could predict the time necessary for the biodegradation of starch components considering the environmental conditions present during the experiment, estimating it in about 180 and 124 days for ULB and LSB, respectively.

3.2. FTIR Results

Integration of the peaks was performed displaying the spectra in the absorption scale. The same integration limits for all the samples were chosen. These limits were chosen to encompass about 95% of the total area, excluding the band edges [17,18]. Even though the formulation of the bioplastic used for the tested bag is proprietary and unknown, the spectra were consistent with those expected by a mixture of starch and a polyester (PEs) (Figure 4).

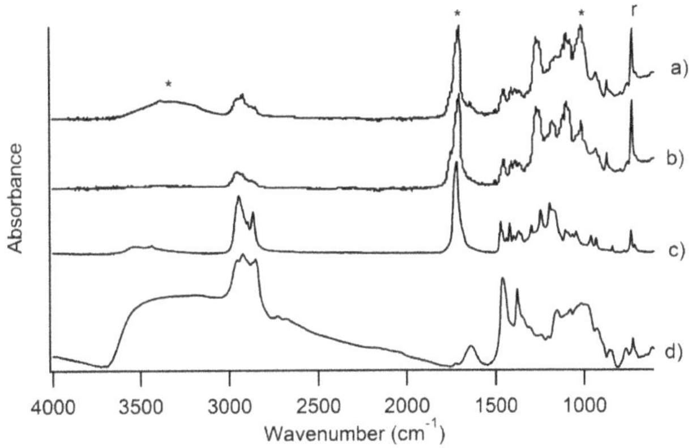

Figure 4. FTIR spectra of (**a**) ULB, (**b**) LSB, (**c**) PEs, (**d**) starch. The peaks which were integrated are denoted by * in the ULB spectrum; the peak taken as reference is indicated by r in the ULB spectrum.

Table 2 reports the absorption bands which characterize the functional groups for PEs, starch, ULB and LSB bags, respectively. The spectral results for ULB show an absorption band centered in 3391 cm^{-1} due to the O-H stretching and another one centered in 2921 cm^{-1} related to the C-H stretching; weak chemical bonds are also present in the spectral region between 1455 and 1388 cm^{-1} assigned to C-H bending. A very strong absorption band is centered at 1712 cm^{-1} due to C=O stretching of the ester group in association with a strong band at 1268 cm^{-1} due to C-O stretching of the O=C-O group. Some bands are present at 1119, 1102 cm^{-1}, assigned to aliphatic ethers, and at 1080 and 1018 cm^{-1}, due to the presence of primary and secondary alcohols, in association with a weak band at 1646 cm^{-1} due to C-O bending associated with the OH group. The spectral results for LSB are very similar to those of ULB, except for the intensity of the signals assigned to the primary and secondary alcohols that in the ULB spectrum are negligible and the CH_2 chains region between 1455 and 1388 cm^{-1} stronger than in ULB.

Table 2. Absorption bands of functional groups stretching for PEs, starch, ULB and LSB bags.

Bands	Absorption Bands (cm^{-1})	PEs	Starch	ULB	LSB
O-H groups (hydroxyl) stretching	3600–3300	✓	✓	✓	
C-H stretching	2930–2900	✓	✓	✓	✓
C=O stretching (carboxylic and ester groups)	1730–1705	✓		✓	✓
C-O bending associated with OH group	1650–1637		✓		✓
C-H bending	1460–1375	✓	✓	✓	✓
C-O stretching	1270–1050	✓	✓	✓	✓
C-O-C stretching (aliphatic ethers)	1180–1150	✓	✓	✓	✓
C-H and C=O bending (out of plane)	970–750	✓		✓	✓

PEs FTIR spectrum is characterized by C=O stretching centered in 1710 cm^{-1} of the carbonyl band of the ester group and C–O stretching in 1240 cm^{-1} and 1090 cm^{-1}. The bands at 2950, 2910 cm^{-1} are assigned to C-H symmetrical stretching in CH_2 chains and 1460–1375 cm^{-1} are assigned to C-H bending. C-H and C=O bending (out of plane) around 700 cm^{-1} [30–35] is typically present in polyesters. The broad band at around 3300 cm^{-1} for starch is due to the stretching mode of the O-H groups. The adsorption band at 1650 cm^{-1} is attributed to C-O bending associated with the OH group together with the band at 1150 cm^{-1} assigned to C-O stretching. The bands at 2955–2920 cm^{-1} and 1460–1375 cm^{-1} are assigned to C-H stretching and C-H bending, respectively. In addition, the characteristic C-O-C ring vibration on starch leads to an absorbance peak at around 1180–1150 cm^{-1}.

In summary, both ULB and LSB seem to be composed of polyester and starch, as already found by the isotopic analysis approach. The diverse relative proportion of the two components in ULB and LSB explain the spectral differences found among them, confirming the higher percentage of starch in ULB than that in LSB.

Figure 5 shows that the ratio I_{3300}/I_{730}, related to the quantity of hydroxyl groups, increases as a function of immersion time. This can be ascribed to the hydrolytic degradation of the constituting polymer. It is well known that the decrease in molecular weight brings about an increase in the number of terminal groups and, therefore, a strengthening of the OH signal.

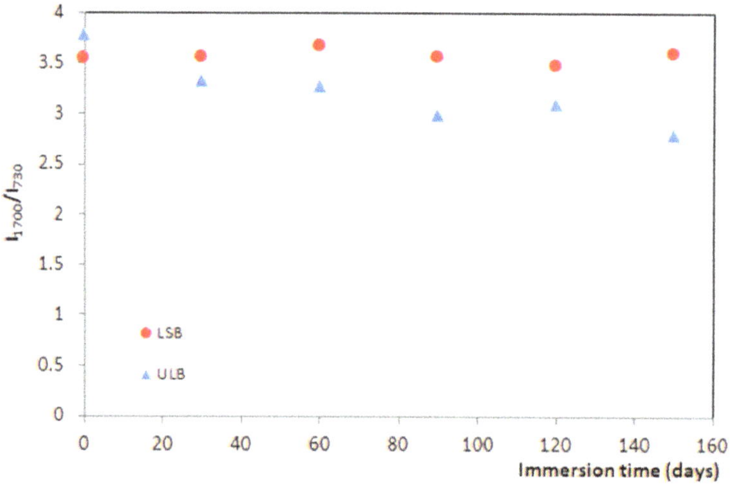

Figure 5. Ratio I_{3300}/I_{730}, related to the quantity of hydroxyl groups as a function of immersion time for ULB (blue triangle) and LSB (red circle).

Interesting to note is that no trend was observed for LSB when the C=O functional groups (I_{1700}/I_{730}) were monitored (Figure 6), while a mildly decreasing trend was observed for ULB. This indicates that the degradation process does not influence such functional groups or that it influences them very slowly. If hydrolysis happens at the expense of starch (Figure 4), the reduction in chain size brings about the formation of smaller polysaccharides, with no formation of significant further carbonyl groups, but with an increase in hydroxyl terminal groups. Therefore, in this scenario, the IR signal at 1700 cm^{-1} remains unaltered. However, when a polyester group is hydrolyzed, the C=O group is retained, but a shift should be observed from the characteristic frequencies of the esters and those of the carboxylic acid. However, the shape of the peak at about 1700 cm^{-1} does not significantly change, reflecting the fact that the polyester moiety is not involved in the degradation process. Therefore, IR data point towards a preferred degradation of the starch component of the formulation, with the polyester part remaining unaltered.

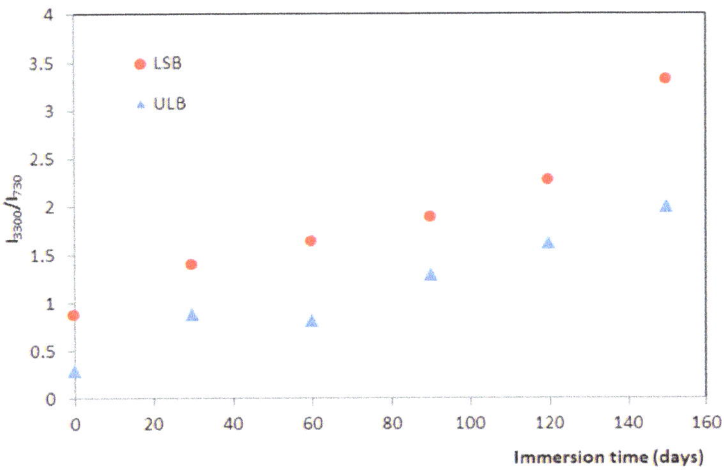

Figure 6. C=O functional groups (I_{1700}/I_{730}) as function of immersed time respectively for ULB (blue triangle) and LSB (red circle).

4. Discussion

The production of bioplastics has increased in recent years and these polymers continue to enter the environment. Understanding the qualitative and quantitative impact of plastic polymers is fundamental for addressing their potential environmental pollution and their toxicological impact [36]. In this regard, the identification and quantification of bioplastic degradation products (solid microparticles and soluble compounds) and residual components, by using different analytical techniques, is a very complex challenge.

Qualitative analysis of bioplastics has been performed using spectroscopic methods (i.e., fluorescence, nuclear magnetic resonance (NMR), Fourier-transform infrared spectroscopy (FTIR) and UV-VIS spectrophotometric techniques [37,38]), while the quantitative analysis of solid microdebris of bioplastics uses FTIR micro imaging, gas chromatography with mass spectrometry and thermal analysis [39–41].

The potential utility of IRMS as a complementary analytical method with respect to "traditional" analytical techniques for plastic analyses has been reported in our previous work [1]. In this preliminary study, carbon stable isotope analysis ($\delta^{13}C$) was applied. This method was used to discriminate different plastic polymers (petroleum- and plant-derived) in plastic and bioplastic materials (among which the shopping bags of Mater-Bi®, Novamont), as well as for the quali-quantitative evaluation of polymers degradability in marine environments. In the present study, the same approach, i.e., the assessment of isotopic values of several polymers and some possible natural raw constituents (corn starch, potato starch and vegetable oils), was applied to identify the carbon source used to manufacture shopping bags. Among the natural sources for biopolymers production, there are C3 plants and C4 plants, which differ in photosynthetic pathways and, consequently, their carbon fingerprint. It has been demonstrated that C4 plants such as sugar cane, millet, corn and tropical grasses, using the Hatch–Slack photosynthetic cycle, have $\delta^{13}C$ values ranging from −8 to −20‰, less negative respect to C3 plants (mostly dicotyledons such as wheat, rice, rye and cotton), that employ the Calvin–Benson cycle and have $\delta^{13}C$ values ranging from −22 to −35‰ [42]. Plant-derived plastic polymers used for food packaging, bags and bottles for drinking water showed a significant difference in isotopic values with respect to petroleum-derived plastic products ([15] and literature therein). Petroleum-derived packaging materials for food, such as shopping bags for fruits and vegetables (HDPE), were characterized by $\delta^{13}C$ mean values of -33.97 ± 1.15‰, whereas plant-derived supermarket shoppers ("BIO" bags) recorded $\delta^{13}C$ mean values of -25.30 ± 0.70‰, near values characteristic for C3 plants as a starch source. The different polymers analyzed, such as

PCL or PBS (Table 1), showed depleted values (from −27.47 to −27.15‰) with respect to C4-derived polymers (corn starch). PCL as well as PBS are often used in biodegradable polymer blends. PCL, which is a semicrystalline linear polyester [43] produced by polymerization of ε-caprolactone commonly derived from fossil sources, received great attention in the 1980s when Novamont created a PCL/starch composite under the trademark Mater-Bi® [44]. PBS, poly(butylene succinate-co-butylene adipate) was produced in 1990 through a polycondensation reaction of glycols with aliphatic dicarboxylic acids and their derivatives [45] and its processability and mechanical properties made it a candidate to replace other polymers such as polyethylene and polypropylene [46]. On the other hand, the less negative $\delta^{13}C$ values were evaluated for PLA and PHB, suggesting that the food bags (ULB and LSB) analyzed in this study (−21.14 ± 0.13‰ and −22.87 ± 0.09‰, for ULB and LSB, respectively), are not constituted of these polymers, but rather of polyesters such as PCL (repeatedly mentioned in the literature as a Mater-Bi® component) or PBS and corn starch (the only component declared by the producer). Our $\delta^{13}C$ results suggested a mixed composition of blends used in the shopping bags, as indicated by the results of the mixing model. This showed a higher percentage of polyesters in LSB with respect to ULB (72% and 61%, respectively), suggesting different purposes of the shopping bags' use. In fact, the LSB sample represents a bag type which requires higher tensile strength and flexibility [47]. However, the more negative $\delta^{13}C$ values reported for PE bags could suggest that bio-bags employed in the experiment do not contain PE, or at least not in large detectable quantities. It must, however, be taken into account that, according to the data obtained, vegetable oils of such common crops like sunflower and rapeseed would give co-polyesters an isotopic footprint between that of PCL (as well as PBS) and PE.

In summary, both ULB and LSB seem to be composed of polyester and starch, as also confirmed by FTIR analysis. In fact, the diverse relative proportion of the two components in ULB and LSB explains the spectral differences found among them, confirming the higher percentage of starch in ULB than that in LSB.

Regarding the changes in composition of the shopping bags (ULB and LSB) over the 73-day period of immersion in marine water, data showed a depletion of $\delta^{13}C$ values for ULB and LSB ($\Delta\delta^{13}C$ of −2.16 and −2.84‰, respectively), which was not observed for the PE control bags. The isotopic shift observed could be reasonably related to the degradative actions that occur in the marine environment (e.g., chemical and/or biological processes) such as photodegradation, thermo-oxidation, hydrolysi and biodegradation. These processes indeed could be responsible for the depolymerization and the subsequent assimilation and metabolism of the monomers by microorganisms ([15] and literature therein).

The lack of carbon depletion in the PE control bags suggested a very low PE degradation (if any) with respect to Mater-Bi® bags, as already observed by some authors [48,49]. In the work by Rutkowska et al. [48], PE bags exposed for 20 months at a depth of 2 m in the Baltic Sea showed no biodegradation and low changes in tensile elongation at break. Moreover, in the present study, the isotopic data of Mater-Bi® bags showed a faster loss of corn starch which has been suggested to be a more bioavailable and hydrophilic component compared to polyesters. Similar results were reported elsewhere [50].

Results obtained by FTIR spectroscopy showed that both ULB and LSB seem to be composed of polyester and starch confirming the outputs of the isotopic analysis approach. In particular, the absence of aromatic overtone signals in the region between 2000 and 1650 cm^{-1} could suggest the absence of co-polyesters containing aromatic monomers like terephthalic acid, in favor of the presence of PCL.

In fact, during the shopping bags degradation experiment, FTIR spectra showed an increase in hydroxyl groups (ratio I_{3300}/I_{730}), ascribed to the hydrolytic degradation of the constituting polymers. The decrease in molecular weight with an increase in the number of terminal groups, which contributes to the strengthening of the OH signals, is a well-known phenomenon. Hydrolysable ester bonds in a polymer backbone make polymers more susceptible to microbial degradation [16,45,51]. In particular, inorganic salts and bacteria have been reported to play a fundamental role in the degradation of PCL either through slow non-enzymatic bulk hydrolysis in aseptic water bodies or by rapid enzymatic degradation in bacteria-containing water bodies [52]. These authors hypothesized that the enzymatic degradation occurred, in particular at increasing degradation time, at the sample surface, while the inner part remained unaffected. On the other hand, if hydrolysis happens at the expense of starch, the reduction in chain size brings about the formation of smaller polysaccharides, with no formation of significant further carbonyl groups (Figure 6), but with an increase in hydroxyl terminal groups (Figure 5). Therefore, in this scenario, the IR signal at 1700 cm^{-1} remains unaltered. However, when a polyester group is hydrolyzed, the C=O group is retained, but a shift should be observed from the characteristic frequencies of the esters and those of the carboxylic acid. In this study, the shape of the peak at about 1700 cm^{-1} does not significantly change, reflecting the fact that the polyester moiety is not involved in the degradation process. Therefore, IR data point towards a preferred degradation of the starch component of the blends analyzed, with the polyester part remaining unaltered.

The preferential microbial degradation of starch in blends exposed to the marine environment reported in the literature [25,52,53], clearly depended on microbial abundance and composition. According to the trend we have observed, the preferential removal through degradation of isotopically heavier corn starch (medium value of −11.45‰) from the analyzed compostable blends lead to polyester enrichment in the remaining matrix. This results in the decrease of the $\delta^{13}C$ values over time, reaching lighter values characteristic of polyesters such as PCL. The degradation time of starch estimated in this study (180 and 124 days for ULB and LSB, respectively) during the present degradation experiment, is in accordance with observations of Niaounakis et al. (2019) [54], who reported that the hydrolysis of the starch component and loss of plasticizer after 8 months of seawater exposure. These authors also reported a severe loss of tensile properties of Mater-Bi® film.

However, it is not possible to accurately estimate how long it would take the polyester fraction to decompose. In this regard, Eich et al. (2015) [25] reported, using Scanning Electron Microscopy analysis, that the holes that appeared after 15 and 33 days of exposure to the marine environment in the plastic's surface might have developed due to dissolving or mineralization of the starch fraction. In this regard, the experiment conducted on PCL degradation under different environmental and laboratory conditions [52] showed that, after a one-year immersion in seawater, a weight loss of about 30% was observed, principally due to the superficial microbial enzymatic degradation, but no significant molecular weight changes were recorded, indicating that, under these conditions, the internal non-enzymatic hydrolysis is negligible.

Avella et al. (2000) [55] reported that in composting conditions with controlled temperature and moisture, pure PCL degrades in a few tens of days, while its blend with starch at 50% disappears in about 3 weeks. However, under marine conditions, PCL items exposed to seawater were still recoverable after 12 months, even if numerous cracks in the surface and the complete loss of tensile strength were observed [56].

All of these findings led us to hypothesize that biodegradable polymers can contribute to increasing the pool of microplastics in the marine environment. In fact, even though PCL-degrading microorganisms were found in a wide range of marine habitats, like bacteria belonging to the genus *Pseudomonas* [57], some PCL fragments were identified as micro- and meso-plastics polluting the Mediterranean Sea ([58] and literature therein).

5. Conclusions

Studies addressing the degradability of polymers in marine environments are very recent and focus mainly on a short-term scale, whereas long-term biodegradation dynamics and the effects on the marine biota are still missing. Furthermore, standard analytical methods to trace and quantify the biodegradation pathways are scarce and need improvements, considering that most studies are at the laboratory scale and not representative of different environmental conditions and habitats. Hence, the lack of knowledge about long-term degradation and the occurrence of residual biopolymers in the marine environment can generate speculations about their sustainability and contribution to marine plastic pollution.

This study revealed that two types of Mater-Bi® materials seem to be composed of polyester and starch, and that, once in the marine environment, the starch fraction is subjected to fast degradation (hydrolysis) estimated at about 124–180 days.

The combination of the IRMS technique with FTIR has been revealed as a promising and time/money-saving tool to more efficiently address the biodegradation of polymers in the marine environment and infer their degradation times and kinetics.

Author Contributions: Conceptualization: D.B. and F.R.; methodology: F.R., S.N., C.G., M.F., N.C. and A.D.L.; data validation: D.B. and L.B.; formal analysis: F.R., S.N., C.G., M.F., N.C. and V.C.; statistical analyses: N.C.; data curation: all authors; writing—original draft preparation: F.R. and D.B.; writing—review and editing: all authors. All authors have read and agreed to the published version of the manuscript.

Funding: This research received no external funding.

Institutional Review Board Statement: Not applicable.

Informed Consent Statement: Not applicable.

Data Availability Statement: Data can be made available upon request to the corresponding author.

Acknowledgments: The authors thank the Nautical Club of Chioggia for making the space available for the environmental exposure experiment.

Conflicts of Interest: The authors declare no conflict of interest.

References

1. Berto, D.; Rampazzo, F.; Gion, C.; Noventa, S.; Formalewicz, M.; Ronchi, F.; Traldi, U.; Giorgi, G. Elemental Analyzer/Isotope Ratio Mass Spectrometry (EA/IRMS) as a Tool to Characterize Plastic Polymers in a Marine Environment. In *Plastics in the Environment*; Gomiero, A., Ed.; IntechOpen: Rijeka, Croatia, 2019.
2. Plastics Europe Plastics—The Facts 2022. Available online: https://plasticseurope.org/knowledge-hub/plastics-the-facts-2022-2/ (accessed on 19 May 2023).
3. Jambeck, J.R.; Geyer, R.; Wilcox, C.; Siegler, T.R.; Perryman, M.; Andrady, A.; Narayan, R.; Law, K.L. Plastic Waste Inputs from Land into the Ocean. *Science* **2015**, *347*, 768–771. [CrossRef]
4. Kershawm, P.; Katsuhiko, S.; Lee, S.; Samseth, J.; Woodring, D. Plastic Debris in the Ocean. In *UNEP YEAR BOOK 2011*; UNEP/Earthprint: Nairobi, Kenya, 2011. Available online: https://citeseerx.ist.psu.edu/viewdoc/download;jsessionid=03F07686A4BFC15E8676A0D9A580E3ED?doi=10.1.1.691.5994&rep=rep1&type=pdf (accessed on 16 September 2021).
5. Vlachogianni, T.; Anastasopoulou, A.; Fortibuoni, T.; Ronchi, F.; Zeri, C.; Marine Litter Assessment in the Adriatic and Ionian Seas. In IPA-Adriatic DeFishGear Project, MIO-ECSDE, HCMR and ISPRA; 2017; Volume 168. Available online: https://mio-ecsde.org/wp-content/uploads/2017/02/Final-MLA-pages_final.pdf (accessed on 16 September 2021).
6. European Commission. A European Strategy for Plastics in a Circular Economy. In *Communication from the Commission to the European Parliament, the Council, the European Economic and Social Committee and the Committee of the Regions*; European Commission: Brussels, Belgium, 2018.
7. Science Advice for Policy by European Academies (SAPEA). *Biodegradability of Plastics in the Open Environment*; SAPEA: Berlin, Germany, 2020. [CrossRef]
8. Lambert, S.; Wagner, M. Environmental Performance of Bio-Based and Biodegradable Plastics: The Road Ahead. *Chem. Soc. Rev.* **2017**, *46*, 6855–6871. [CrossRef] [PubMed]
9. Ghosh, K.; Jones, B.H. Roadmap to Biodegradable Plastics—Current State and Research Needs. *ACS Sustain. Chem. Eng.* **2021**, *9*, 6170–6187. [CrossRef]
10. Brizga, J.; Hubacek, K.; Feng, K. The Unintended Side Effects of Bioplastics: Carbon, Land, and Water Footprints. *One Earth* **2020**, *3*, 45–53. [CrossRef]
11. Mekonnen, T.; Mussone, P.; Khalil, H.; Bressler, D. Progress in Bio-Based Plastics and Plasticizing Modifications. *J. Mater. Chem. A* **2013**, *1*, 13379–13398. [CrossRef]
12. Imam, S.H.; Gordon, S.H.; Shogren, R.L.; Tosteson, T.R.; Govind, N.S.; Greene, R. V Degradation of Starch–Poly(β-Hydroxybutyrate-Co-β-Hydroxyvalerate) Bioplastic in Tropical Coastal Waters. *Appl. Environ. Microbiol.* **1999**, *65*, 431–437. [CrossRef]
13. Accinelli, C.; Saccà, M.L.; Mencarelli, M.; Vicari, A. Deterioration of Bioplastic Carrier Bags in the Environment and Assessment of a New Recycling Alternative. *Chemosphere* **2012**, *89*, 136–143. [CrossRef]

14. Machavaram, M.; Mills, M. *NRMRL LRPCD SOP for 13C and/or 15N Isotope Ratio Analyses of Environmental Samples (Solid and Non-Aqueous Liquid) by Elemental Analyzer—Isotope Ratio Mass-Spectrometer (EA-IRMS)*; EPA: Washington, DC, USA, 2017.
15. Berto, D.; Rampazzo, F.; Gion, C.; Noventa, S.; Ronchi, F.; Traldi, U.; Giorgi, G.; Cicero, A.M.; Giovanardi, O. Preliminary Study to Characterize Plastic Polymers Using Elemental Analyser/Isotope Ratio Mass Spectrometry (EA/IRMS). *Chemosphere* **2017**, *176*, 47–56. [CrossRef]
16. Suzuki, Y.; Akamatsu, F.; Nakashita, R.; Korenaga, T. A Novel Method to Discriminate between Plant- and Petroleum-Derived Plastics by Stable Carbon Isotope Analysis. *Chem. Lett.* **2010**, *39*, 998–999. [CrossRef]
17. Tungol, M.W.; Bartick, E.G.; Montaser, A. Forensic Analysis of Acrylic Copolymer Fibers by Infrared Microscopy. *Appl. Spectrosc.* **1993**, *47*, 1655–1658. [CrossRef]
18. Causin, V.; Marega, C.; Guzzini, G.; Marigo, A. Forensic Analysis of Poly(Ethylene Terephthalate) Fibers by Infrared Spectroscopy. *Appl. Spectrosc.* **2004**, *58*, 1272–1276. [CrossRef] [PubMed]
19. El-Ghaffar, M.A.A.; Youssef, E.A.M.; Darwish, W.M.; Helaly, F.M. A Novel Series of Corrosion Inhibitive Polymers for Steel Protection. *J. Elastomers Plast.* **1998**, *30*, 68–94. [CrossRef]
20. Rajakumar, K.; Sarasvathy, V.; Thamarai Chelvan, A.; Chitra, R.; Vijayakumar, C.T. Natural Weathering Studies of Polypropylene. *J. Polym. Environ.* **2009**, *17*, 191–202. [CrossRef]
21. Brandon, J.; Goldstein, M.; Ohman, M.D. Long-Term Aging and Degradation of Microplastic Particles: Comparing in Situ Oceanic and Experimental Weathering Patterns. *Mar. Pollut. Bull.* **2016**, *110*, 299–308. [CrossRef]
22. Alvarez, V.A.; Ruseckaite, R.A.; Vázquez, A. Degradation of Sisal Fibre/Mater Bi-Y Biocomposites Buried in Soil. *Polym. Degrad. Stab.* **2006**, *91*, 3156–3162. [CrossRef]
23. Bastioli, C. Properties and Applications of Mater-Bi Starch-Based Materials. *Polym. Degrad. Stab.* **1998**, *59*, 263–272. [CrossRef]
24. Tosin, M.; Weber, M.; Siotto, M.; Lott, C.; Degli-Innocenti, F. Laboratory Test Methods to Determine the Degradation of Plastics in Marine Environmental Conditions. *Front. Microbiol.* **2012**, *3*, 225. [CrossRef]
25. Eich, A.; Mildenberger, T.; Laforsch, C.; Weber, M. Biofilm and Diatom Succession on Polyethylene (PE) and Biodegradable Plastic Bags in Two Marine Habitats: Early Signs of Degradation in the Pelagic and Benthic Zone? *PLoS ONE* **2015**, *10*, e0137201. [CrossRef]
26. Bastioli, C.; Innocenti, F.D.; Guanella, I.; Romano, G. Compostable Films of Mater-Bi Z Grades. *J. Macromol. Sci. Part A* **1995**, *32*, 839–842. [CrossRef]
27. Barbale, M.; Chinaglia, S.; Gazzilli, A.; Pischedda, A.; Pognani, M.; Tosin, M.; Degli-Innocenti, F. Hazard Profiling of Compostable Shopping Bags. Towards an Ecological Risk Assessment of Littering. *Polym. Degrad. Stab.* **2021**, *188*, 109592. [CrossRef]
28. Giani, M.; Berto, D.; Rampazzo, F.; Savelli, F.; Alvisi, F.; Giordano, P.; Ravaioli, M.; Frascari, F. Origin of Sedimentary Organic Matter in the North-Western Adriatic Sea. *Estuar. Coast. Shelf Sci.* **2009**, *84*, 573–583. [CrossRef]
29. O'Brine, T.; Thompson, R.C. Degradation of Plastic Carrier Bags in the Marine Environment. *Mar. Pollut. Bull.* **2010**, *60*, 2279–2283. [CrossRef] [PubMed]
30. Chércoles Asensio, R.; San Andrés Moya, M.; de la Roja, J.M.; Gómez, M. Analytical Characterization of Polymers Used in Conservation and Restoration by ATR-FTIR Spectroscopy. *Anal. Bioanal. Chem.* **2009**, *395*, 2081–2096. [CrossRef] [PubMed]
31. Mecozzi, M.; Pietroletti, M.; Monakhova, Y.B. FTIR Spectroscopy Supported by Statistical Techniques for the Structural Characterization of Plastic Debris in the Marine Environment: Application to Monitoring Studies. *Mar. Pollut. Bull.* **2016**, *106*, 155–161. [CrossRef]
32. Jung, M.R.; Horgen, F.D.; Orski, S.V.; Rodriguez, C.V.; Beers, K.L.; Balazs, G.H.; Jones, T.T.; Work, T.M.; Brignac, K.C.; Royer, J.-C.; et al. Validation of ATR FT-IR to Identify Polymers of Plastic Marine Debris, Including Those Ingested by Marine Organisms. *Mar. Pollut. Bull.* **2018**, *127*, 704–716. [CrossRef]
33. Veerasingam, S.; Ranjani, M.; Venkatachalapathy, R.; Bagaev, A.; Mukhanov, V.; Litvinyuk, D.; Mugilarasan, M.; Gurumoorthi, K.; Guganathan, L.; Aboobacker, V.M.; et al. Contributions of Fourier Transform Infrared Spectroscopy in Microplastic Pollution Research: A Review. *Crit. Rev. Environ. Sci. Technol.* **2021**, *51*, 2681–2743. [CrossRef]
34. Verleye, G.A.L.; Roeges, N.P.G.; De Moor, M.O. *Easy Identification of Plastics and Rubbers*; Rapra Technology Ltd.: Shawbury, UK, 2001; ISBN 978-1859572689.
35. Noda, I.; Dowrey, A.E.; Haynes, J.L.; Marcott, C. Group Frequency Assignments for Major Infrared Bands Observed in Common Synthetic Polymers. In *Physical Properties of Polymers Handbook*; Mark, J.E., Ed.; Springer: New York, NY, USA, 2007; pp. 395–406, ISBN 978-0-387-69002-5.
36. González-Pleiter, M.; Tamayo-Belda, M.; Pulido-Reyes, G.; Amariei, G.; Leganés, F.; Rosal, R.; Fernández-Piñas, F. Secondary Nanoplastics Released from a Biodegradable Microplastic Severely Impact Freshwater Environments. *Environ. Sci. Nano* **2019**, *6*, 1382–1392. [CrossRef]
37. Kumar, M.; Singhal, A.; Verma, P.K.; Thakur, I.S. Production and Characterization of Polyhydroxyalkanoate from Lignin Derivatives by Pandoraea Sp. ISTKB. *ACS Omega* **2017**, *2*, 9156–9163. [CrossRef]
38. Akdoğan, M.; Çelik, E. Purification and Characterization of Polyhydroxyalkanoate (PHA) from a Bacillus Megaterium Strain Using Various Dehydration Techniques. *J. Chem. Technol. Biotechnol.* **2018**, *93*, 2292–2298. [CrossRef]
39. Sikorska, W.; Zięba, M.; Musioł, M.; Kowalczuk, M.; Janeczek, H.; Chaber, P.; Masiuchok, O.; Demchenko, V.; Talanyuk, V.; Iurzhenko, M.; et al. Forensic Engineering of Advanced Polymeric Materials—Part VII: Degradation of Biopolymer Welded Joints. *Polymers* **2020**, *12*, 1167. [CrossRef]

40. Ye, Y.; Yu, K.; Zhao, Y. The Development and Application of Advanced Analytical Methods in Microplastics Contamination Detection: A Critical Review. *Sci. Total Environ.* **2022**, *818*, 151851. [CrossRef] [PubMed]
41. Yusuf, A.; Sodiq, A.; Giwa, A.; Eke, J.; Pikuda, O.; Eniola, J.O.; Ajiwokewu, B.; Sambudi, N.S.; Bilad, M.R. Updated Review on Microplastics in Water, Their Occurrence, Detection, Measurement, Environmental Pollution, and the Need for Regulatory Standards. *Environ. Pollut.* **2022**, *292*, 118421. [CrossRef] [PubMed]
42. Muccio, Z.; Jackson, G.P. Isotope Ratio Mass Spectrometry. *Analyst* **2009**, *134*, 213–222. [CrossRef] [PubMed]
43. Lovera, D.; Márquez, L.; Balsamo, V.; Taddei, A.; Castelli, C.; Müller, A.J. Crystallization, Morphology, and Enzymatic Degradation of Polyhydroxybutyrate/Polycaprolactone (PHB/PCL) Blends. *Macromol. Chem. Phys.* **2007**, *208*, 924–937. [CrossRef]
44. Ali Akbari Ghavimi, S.; Ebrahimzadeh, M.H.; Solati-Hashjin, M.; Abu Osman, N.A. Polycaprolactone/Starch Composite: Fabrication, Structure, Properties, and Applications. *J. Biomed. Mater. Res. Part A* **2015**, *103*, 2482–2498. [CrossRef]
45. Tserki, V.; Matzinos, P.; Pavlidou, E.; Vachliotis, D.; Panayiotou, C. Biodegradable Aliphatic Polyesters. Part I. Properties and Biodegradation of Poly(Butylene Succinate-Co-Butylene Adipate). *Polym. Degrad. Stab.* **2006**, *91*, 367–376. [CrossRef]
46. Fujimaki, T. Processability and Properties of Aliphatic Polyesters, 'BIONOLLE', Synthesized by Polycondensation Reaction. *Polym. Degrad. Stab.* **1998**, *59*, 209–214. [CrossRef]
47. Correa, A.C.; Carmona, V.B.; Simão, J.A.; Capparelli Mattoso, L.H.; Marconcini, J.M. Biodegradable Blends of Urea Plasticized Thermoplastic Starch (UTPS) and Poly(ε-Caprolactone) (PCL): Morphological, Rheological, Thermal and Mechanical Properties. *Carbohydr. Polym.* **2017**, *167*, 177–184. [CrossRef]
48. Rutkowska, M.; Heimowska, A.; Krasowska, K.; Janik, H. Biodegradability of Polyethylene Starch Blends in Sea Water. *Polish J. Environ. Stud.* **2002**, *11*, 267–272.
49. Pegram, J.E.; Andrady, A.L. Outdoor Weathering of Selected Polymeric Materials under Marine Exposure Conditions. *Polym. Degrad. Stab.* **1989**, *26*, 333–345. [CrossRef]
50. di Franco, C.R.; Cyras, V.P.; Busalmen, J.P.; Ruseckaite, R.A.; Vázquez, A. Degradation of Polycaprolactone/Starch Blends and Composites with Sisal Fibre. *Polym. Degrad. Stab.* **2004**, *86*, 95–103. [CrossRef]
51. Lucas, N.; Bienaime, C.; Belloy, C.; Queneudec, M.; Silvestre, F.; Nava-Saucedo, J.-E. Polymer Biodegradation: Mechanisms and Estimation Techniques—A Review. *Chemosphere* **2008**, *73*, 429–442. [CrossRef] [PubMed]
52. Lu, B.; Wang, G.-X.; Huang, D.; Ren, Z.-L.; Wang, X.-W.; Wang, P.-L.; Zhen, Z.-C.; Zhang, W.; Ji, J.-H. Comparison of PCL Degradation in Different Aquatic Environments: Effects of Bacteria and Inorganic Salts. *Polym. Degrad. Stab.* **2018**, *150*, 133–139. [CrossRef]
53. Chen, X.; Wang, L.; Shi, J.; Shi, H.; Liu, Y. Environmental Degradation of Starch/Poly(Lactic Acid) Composite in Seawater. *Polym. Polym. Compos.* **2011**, *19*, 559–566. [CrossRef]
54. Niaounakis, M.; Kontou, E.; Pispas, S.; Kafetzi, M.; Giaouzi, D. Aging of Packaging Films in the Marine Environment. *Polym. Eng. Sci.* **2019**, *59*, E432–E441. [CrossRef]
55. Avella, M.; Errico, M.E.; Laurienzo, P.; Martuscelli, E.; Raimo, M.; Rimedio, R. Preparation and Characterisation of Compatibilised Polycaprolactone/Starch Composites. *Polymer* **2000**, *41*, 3875–3881. [CrossRef]
56. Suzuki, M.; Tachibana, Y.; Kasuya, K. Biodegradability of Poly(3-Hydroxyalkanoate) and Poly(ε-Caprolactone) via Biological Carbon Cycles in Marine Environments. *Polym. J.* **2021**, *53*, 47–66. [CrossRef]
57. Molitor, R.; Bollinger, A.; Kubicki, S.; Loeschcke, A.; Jaeger, K.-E.; Thies, S. Agar Plate-Based Screening Methods for the Identification of Polyester Hydrolysis by Pseudomonas Species. *Microb. Biotechnol.* **2020**, *13*, 274–284. [CrossRef]
58. Manfra, L.; Marengo, V.; Libralato, G.; Costantini, M.; De Falco, F.; Cocca, M. Biodegradable Polymers: A Real Opportunity to Solve Marine Plastic Pollution? *J. Hazard. Mater.* **2021**, *416*, 125763. [CrossRef]

applied sciences

MDPI

Article

Shell Organic Matrix (*Conchix*) of the Mediterranean Mussel *Mytilus galloprovincialis* L. as the Medium for Assessment of Trace Metals in the Boka Kotorska Bay

Rajko Martinović [1,*], Danijela Joksimović [1], Ana Perošević-Bajčeta [1], Ivana Čabarkapa [2] and Hermann Ehrlich [3]

1 Institute of Marine Biology, University of Montenegro, Put I Bokeljske Brigade 68, 85330 Kotor, Montenegro
2 Institute of Food Technology, University of Novi Sad, 21000 Novi Sad, Serbia
3 Center for Advanced Technologies, Adam Mickiewicz University, Uniwersytetu Poznańskiego 10, 61-614 Poznan, Poland
* Correspondence: rajko.mar@ucg.ac.me

Abstract: The content of trace metals, namely Zn, Cu, Fe, Mn, Pb, Cd, and Hg, in four types of media, i.e., soft tissues, shells, and the products of shell demineralization (organic matrix—*conchix* and extract) of the Mediterranean mussel *Mytilus galloprovicialis* L., at three sites in the Boka Kotorska Bay of the Adriatic Sea were determined. The main aim was to investigate the accumulation patterns of trace metals in *conchix* and their possible relationship with other tested media. *Conchix* weight within a group of mussels from Sv. Nedjelja was significantly higher in comparison with the IMB, while *conchix* % in the shell showed a negative correlation with dry shell weight. The highest metal pollution index (MPI) values found in the soft tissues of mussels from Sv. Nedjelja, Cogi, and the IMB were 2.319, 2.711, and 2.929 μg g^{-1}, respectively. PCA analysis showed similarities in trace metal accumulation in all media except *conchix*. According to CCA analysis, *conchixes* were grouped around Cu, Fe, and Hg, while Cd and Zn were in correlation with the soft tissues. Moreover, the shells were in correlation with Mn. Simple isolation with high yield, close contact to the environment in comparison with calcified shell layers, and susceptibility to possible pollution sources due to the accumulation of specific metals are the main reasons to consider *conchix* of *M. galloprovincialis* as a medium with potential in trace metal assessments of marine ecosystems.

Keywords: organic matrix; periostracum; trace metals; shell; Adriatic Sea

Citation: Martinović, R.; Joksimović, D.; Perošević-Bajčeta, A.; Čabarkapa, I.; Ehrlich, H. Shell Organic Matrix (*Conchix*) of the Mediterranean Mussel *Mytilus galloprovincialis* L. as the Medium for Assessment of Trace Metals in the Boka Kotorska Bay. *Appl. Sci.* **2023**, *13*, 7582. https://doi.org/10.3390/app13137582

Academic Editor: Simone Morais

Received: 28 April 2023
Revised: 14 June 2023
Accepted: 19 June 2023
Published: 27 June 2023

1. Introduction

Bivalves are globally distributed along the coasts of many countries inhabited by more than the half of world's population. Several features, such as sessile nature, filter-feeding, and bioaccumulation of trace metals in much higher concentrations in comparison with the ambient environment make bivalves suitable bioindicators of trace metal contamination [1]. During the last decade, the Mediterranean mussel *Mytilus galloprovincialis* (L.) has been used in many trace metal studies [2–5]. Moreover, *M. galloprovincialis* L. was defined as a common indicator for trace/heavy metal pollution according to the United Nations' environmental program for the Mediterranean region [6].

The adverse effects of trace metals on marine organisms and indirectly on humans via the consumption of seafood are well known [7–9]. The anthropogenic impacts on many coastal waters are becoming increasingly pronounced due to large amounts of effluents enriched with toxic trace metals [10]. Physicochemical and biogeochemical processes may produce the heterogeneous distribution of trace metals in different environmental matrices of the Adriatic estuaries [11]. The Boka Kotorska Bay, southern Adriatic Sea, records constant growth of maritime transport with higher numbers of cruise ship visits per year [12]. These large ships contribute to increased sediment re-suspension, which can lead to the transition of trace metals from sediments to the water column. So far, the investigation of trace metal content in the Boka Kotorska Bay in seawater, sediments, sea grasses, and bivalves [13–15] has been conducted.

Bivalves' soft tissues, rather than shells, were more popular among researchers in metal bioaccumulation studies [16]. Despite less involvement, trace metal assessment in bivalves' shells was widely accepted and applied in aquatic ecosystem monitoring [17–19]. However, short-term laboratory treatments indicated that bivalve's shells could be less suitable for metal detection due to a slower accumulation rate in comparison with the soft tissues [20]. Although bivalves were recognized as good bioindicators, a more accurate methodology of using bivalves in trace metal biomonitoring should be developed [21]. To improve the methodology, trace metal assessment in different mediums/parts of bivalves, namely the calcite and aragonite shell layers separately, both calcareous shell layers without periostracum, and separated periostracum and byssus threads [22–26], was carried out.

The periostracum is a thin yellow to brownish organic layer at the shell surface that envelops calcified parts, protects bivalves against dissolution in acidic water, prevents fouling by microorganisms, and tends to accumulate trace metals from the environment [27]. The surfaces of periostracum and deposited crystals are covered by a proteinaceous organic matrix [28]. Recently, the isolation procedure of organic matrices (*conchixes*) from the Mediterranean mussel has been published [29]. During the process, an intact shell organic matrix, including the periostracum, was obtained. Considering the limitations of the soft tissues and shells to a certain degree, our idea was to examine *conchix* as a relatively novel medium to improve the methodology for trace metal assessment.

Keeping in mind the above, the main aim of this study was to investigate the accumulation patterns of certain trace metals in the products of shell demineralization (*conchix* and shell extract), soft tissues, and the shells of the Mediterranean mussel at three sites within the Boka Kotorska Bay, Montenegro for the marine ecosystem assessment.

2. Materials and Methods

2.1. Description of Sampling Sites

We collected the Mediterranean mussels *M. galloprovincialis* on the 2nd of November 2019 from the three mussel farms: Sv. Nedjelja (42°27′29.60″ N 18°40′20.90″ E), Cogi (42°29′8.20″ N 18°44′46.06″ E), and the Institute of Marine Biology Kotor (IMB) within the Boka Kotorska Bay, Adriatic Sea, Montenegro (Figure 1). The first site, Sv. Nedjelja, is located in the Tivat Bay, at the entrance to Kotor Bay with a sandy bottom and a higher level of seawater exchange during the year. The second site, Cogi, is a fish and mussel farm near Orahovac, in the smaller Kotor Bay, while the IMB is placed near the port of Kotor. The Cogi and IMB sites were muddy sediments, had a lower level of seawater exchange due to higher inner positioning within the bay, and had higher seasonal salinity oscillations caused by freshwater inflow from the land and underwater springs. All three sampling sites had the same level and sources of anthropogenic impact in the Bay area, such as land-based sources, air-based sources, and maritime transportation [30].

2.2. Sampling and Preparation

Mussels of a similar shell length (50–70 mm), 50 per site, were collected, placed into polyethylene bags, kept in a cooler box with ice, and transported to the laboratory. After cleaning and rinsing with Milli-Q water (Millipore SAS, Molsheim, France) the soft tissues were carefully separated from the shell using a ceramic knife while the byssus and the ligament were removed. The shells and the soft tissues (Figure 2b) of the individual mussels were analyzed. The shells were divided into two halves; one shell valve was used for the determination of trace metals, while the other shell valve (Figure 2a) was subjected to the demineralization process. The soft tissues were frozen at −18 °C, freeze-dried at −40 °C for 48 h (Alpha 2–4 LD plus, CHRIST, Hagen, Germany), reduced to powder, and homogenized.

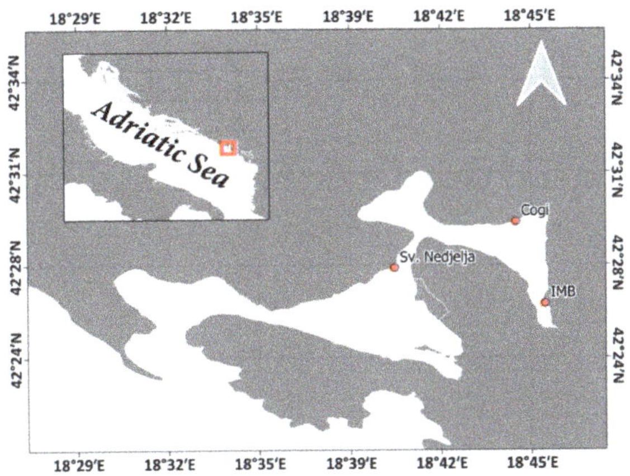

Figure 1. Sampling sites in the Boka Kotorska Bay, Montenegro. Scale: the sides of the large square grid represent a length c. 10 km.

Figure 2. The samples of the Mediterranean mussel *M. galloprovincialis* L. used for trace metal analyses: (**a**) shell valve; (**b**) soft tissue, i.e., the products of shell demineralization; (**c**) *conchix*; and (**d**) shell extract. Scale bar = 1 cm.

2.3. Isolation of Conchix

Decalcification of three groups of the Mediterranean mussel was performed using Ethylenediaminetetraacetic (EDTA)-disodium-salt-based solution. Shell valves used for the isolation of *conchixes* were placed in a glass with a previously prepared EDTA-based solution (pH 7.25) and stored in an incubator at 39 °C. At every 24 h, shells were rinsed with dH_2O, and the solution was replaced. After the replacement, the used solution was deposited in the tank and dried for 3–4 weeks to obtain shell extract (Figure 2d). Isolation was completed after five days, and the obtained *conchixes* (Figure 2c) were carefully rinsed with dH_2O and dried in the air. The complete procedure of solution preparation used for demineralization and *conchix* isolation was thoroughly described in Ehrlich et al. [29].

2.4. Size and Weight Measurements

Allometric parameters, such as size and weight of the whole shell, shell valves, wet and dry tissues, and dried *conchixes* of the Mediterranean mussel, were measured. The length and width were measured using an electronic digital caliper (Maurer, Padova, Italy), while an electronic balance (Kern & Sohn GmbH, Balingen, Germany) was used for the weight measurement. According to values of *conchix* weight and dry shell valve weight, *conchix* % in the shell valve was calculated using Microsoft Office Excel 2007.

2.5. Condition Index

The condition index (CI) was calculated according to [31] as the ratio between dry soft tissue weight (DSTW) and dry shell weight (DSW): CI = DSTW (g)/DSW (g).

2.6. Trace Metal Determination

The mussel samples were analyzed for Fe, Mn, Zn, and Cu contents according to the International Organization for Standardization method [32] by atomic absorption spectrometry using a flame atomic absorption spectrometer (Varian SPECTRA AA-10, Varian Techtron Pty Limited, Mulgrave, VC, Australia), equipped with a flame furnace and operated with an air acetylene flame after mineralization by dry ashing. Dry ashing was conducted according to Tomovic et al. [33]. Briefly, after drying, the sample was charred on a hot plate and then incinerated in a muffle furnace at 450 °C overnight. After ash was obtained, it was moistened, treated with 10 mL of HCl/diH2O (1:1, *v*/*v*), and evaporated to dryness. The ash was redissolved with 10 mL of HCl/diH_2O (1:9, *v*/*v*), transferred to a 50 mL volumetric flask, and diluted to volume with diH_2O. Heavy metal analyses of Pb and Cd were performed by atomic absorption spectrometry after dry ashing according to the EN method [34], while Hg was determined using a direct mercury analyzer, namely the AMA 254 Advanced mercury analyzer (Altec, Prague, Czech Republic), based on the principles of sample thermal decomposition. The obtained results of the investigated elements in mussels were expressed in µg/g of sample dry weight. Each obtained value is the mean of five determinations. Procedural blanks were analyzed once for every five samples. To avoid possible contamination, all employed glassware and equipment were acid-washed.

2.7. Metal Pollution Index

To compare the total content of trace metals at the 3 sampling sites in different mediums, the metal pollution index (MPI) was used. The MPI was obtained using the following equation: $MPI = (M1 \times M2 \times M3 \times \ldots \times Mn)^{1/n}$, where M1 is the concentration value of the first metal; M2 is the concentration value of the second metal; and Mn is the concentration of the metal n expressed in $\mu g\ g^{-1}$. The number of metals was (n) = 7. Any MPI > 1 indicates that the investigated site is polluted, whereas any MPI < 1 indicates no pollution [35–37].

2.8. Statistical Analyses

Statistical analysis of the results was carried out using Statistica 14 Software Inc., (Palo Alto, CA, USA) [38]. The Kolmogorov–Smirnov test for normality of distribution was used prior to statistical analysis. Since the data were not in line with the requirements for the application of parametric tests, differences between each group were tested using the Mann–Whitney U test. Correlation analyses were carried out using the Spearman correlation test with a significance level of $p < 0.05$. Principal component analysis (PCA) was used to assess differentiation among the sites and mediums analyzed based on the trace metal level. The sites were compared through canonical correlation analysis (CCA) to determine their relationships relative to the levels of trace metal accumulation in the following four media: soft tissues; shells; shell organic matrix, i.e., *conchix*; and shell extracts of the Mediterranean mussel.

3. Results

The *conchix* weight within the group of mussels from Sv. Nedjelja was significantly higher in comparison with that from the IMB (p = 0.043), while significant differences between other groups, analyzed by the Mann–Whitney U test, were not observed ($p > 0.05$; Figure 3a). The weight of the dry shell valve in mussels from Cogi was significantly lower in comparison to that from the other two sites ($p < 0.05$), while the differences between Sv. Nedjelja and Cogi were not significant ($p > 0.05$).

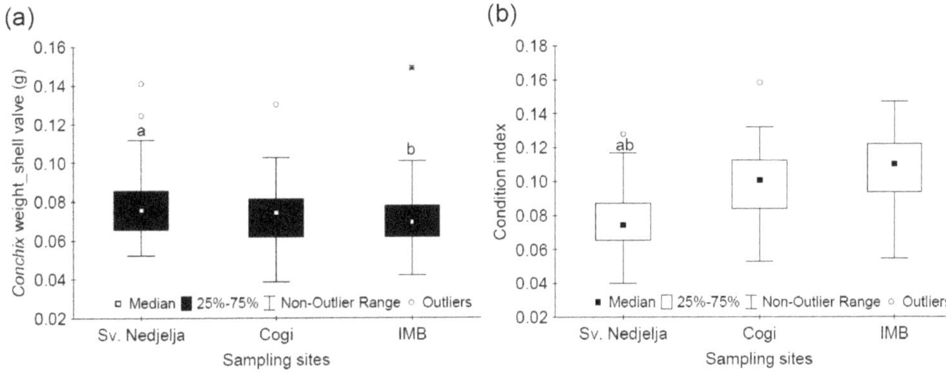

Figure 3. (**a**) *Conchix* weight in the shell valve; (**b**) condition index within the group of *M. galloprovincialis* L. specimens sampled from Sv. Nedjelja, Cogi, and the IMB; a and b—statistically significant difference ($p < 0.05$) in intercomparison; ab—statistically significant difference ($p < 0.05$) in comparison with the other sampling sites. *—extremes. Statistical significance was tested using the Mann–Whitney U test.

In general, a significant correlation ($p < 0.05$; $r = 0.562$) between the *conchix* weight and the weight of the dry shell valve within the mussels' samples from all studied sites was determined using the Spearman correlation test.

The values of the condition index (CI) in mussels from Sv. Nedjelja were significantly lower in comparison with Cogi and the IMB ($p < 0.05$), while there were no statistically significant differences between Cogi and the IMB ($p > 0.05$; Figure 3b).

A statistically significant correlation between the *conchix* weight and CI within the mussels' samples from all studied sites was not observed ($p = 0.298$; $r = -0.085$).

Conchix % in the shell valve within the group of mussels from Cogi was significantly higher in comparison with the other two sites ($p < 0.05$), while the differences between Sv. Nedjelja and Cogi were not statistically significant ($p > 0.05$; Figure 4a).

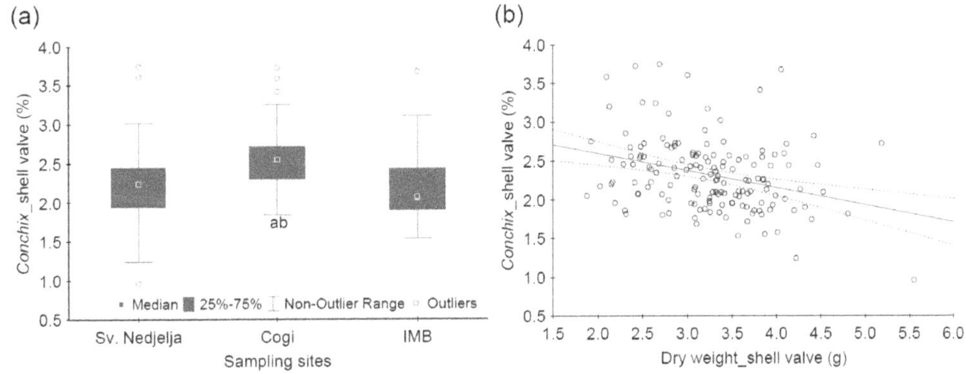

Figure 4. (**a**) *Conchix* % in the shell valve and (**b**) correlation of *conchix* % with a dry weight of the shell valve within the group of *M. galloprovincialis* L. specimens sampled from Sv. Nedjelja, Cogi, and the IMB. ab—statistically significant difference ($p < 0.05$) in comparison with the other sampling sites. Dashed lines represent confidence bands lines while the fitting line is in the middle. Statistical significance ($p < 0.05$) was tested using the Mann–Whitney U test, while the correlation was determined by the Spearman correlation test.

Data on *conchix* % in the shell valve from all studied groups showed a significant correlation with the CI ($p = 0.044$; $r = 0.163$) and a negative correlation ($p < 0.05$; $r = -0.316$) with the weight of the dry shell valve (Figure 4b).

Table 1 presents trace metal concentrations and the MPI in four different media of *M. galloprovincialis* L., namely shell (KS, KC, KI), soft tissue (TS, TC, TI), *conchix* (OS, OC, OI), and shell extract (ES, EC, EI) sampled from Sv. Nedjelja (S), Cogi (C), and the IMB (I). The highest contents of Zn, Mn, and Cd were found in the soft tissues, while the highest contents of Cu, Fe, and Hg were found in the *conchixes*. Lead contents were generally highest in soft tissues with a mean value of 0.92 μg g^{-1}, although the highest single concentration (1.22 μg g^{-1}) was found in *conchixes* from Sv. Nedjelja. The lowest concentrations were generally found in extracts, although Pb (<0.25), Cd (<0.03), and Hg (<0.001) contents were below the detection limits in both extracts and shells, except for Pb content in the shell sample from the IMB (0.31 μg g^{-1}). In addition, the lowest Fe contents were found in the shells as well. The highest MPI values were found in the soft tissues of mussels from Sv. Nedjelja, Cogi, and the IMB were 2.319, 2.711 and 2.929 μg g^{-1}, respectively (Table 1). The *conchixes* from Sv. Nedjelja had higher MPI values (2.158 μg g^{-1}) in comparison with Cogi and the IMB, which were 1.214 and 1.163 μg g^{-1}, respectively. MPI values of the shells and shell extracts at all three sites were lower than 1.

Table 1. Trace metal concentrations and metal pollution index (MPI) within the group of *M. galloprovincialis* L. specimens sampled from Sv. Nedjelja (S), Cogi (C), and the IMB (I) in the shell (KS, KC, KI), soft tissue (TS, TC, TI), *conchix* (OS, OC, OI), and shell extract (ES, EC, EI).

Trace Metals (μg g^{-1})	Medium/ Site											
	KS	KC	KI	TS	TC	TI	OS	OC	OI	ES	EC	EI
Zn	11.6	11.6	11.0	125	135	104	7.85	13.6	7.59	5.3	12.7	4.15
Cu	2.09	3.91	2.74	2.93	3.39	3.53	15.8	11.0	12.6	2.02	1.05	2.14
Fe	0.69	0.81	0.64	21.0	11.0	35.4	68.3	21.6	40.8	3.92	4.82	3.86
Mn	2.44	2.57	2.31	2.78	3.63	3.62	2.07	1.60	2.80	1.51	1.48	1.62
Pb	<0.25	<0.25	0.31	0.54	1.16	1.06	1.22	<0.25	<0.25	<0.25	<0.25	<0.25
Cd	<0.03	<0.03	<0.03	0.69	0.73	0.41	0.06	0.067	<0.03	<0.03	<0.03	<0.03
Hg	<0.001	<0.001	<0.001	0.077	0.065	0.095	0.168	0.100	0.145	<0.001	<0.001	<0.001
MPI	0.258	0.291	0.297	2.319	2.711	2.929	2.158	1.214	1.163	0.275	0.291	0.269

Comparing location differences across the same sample groups, the highest contents of Zn (135 μg g^{-1}), Pb (1.16 μg g^{-1}), and Cd (0.73 μg g^{-1}) were found in the soft tissues of mussels from Cogi, while Cu (3.53 μg g^{-1}), Fe (35.4 μg g^{-1}) and Hg (0.095 μg g^{-1}) contents were the highest in IMB soft tissues. Manganese concentrations were almost identical for Cogi (3.63 μg g^{-1}) and IMB samples (3.62 μg g^{-1}), and both were higher than that in the soft tissues of Sv. Nedjelja mussel samples. Conversely, *conchixes* from Sv. Nedjelja mussels had the highest concentrations of Cu (15.8 μg g^{-1}), Fe (68.3 μg g^{-1}), Pb (1.22 μg g^{-1}), and Hg (0.168 μg g^{-1}). Shells from Cogi had the highest Cu (11.6 μg g^{-1}), Fe (0.81 μg g^{-1}), and Mn (2.57 μg g^{-1}) concentrations. Furthermore, the extracts from Cogi had the highest Zn (12.7 μg g^{-1}) and Fe (4.82 μg g^{-1}) concentrations, while Cu (2.14 μg g^{-1}) and Mn (1.62 μg g^{-1}) contents were the highest in extracts from the IMB.

Figure 5 presents PCA and CCA plots of trace metal concentrations in four different media (shell, soft tissue, *conchix*, and shell extract) of *M. galloprovincialis* L. sampled from Sv. Nedjelja (S), Cogi (C), and the IMB (I). PCA analysis showed similarities in trace metal accumulation in all media except *conchix*. However, the *conchix* sample from Cogi (OC) was slightly similar to extracts from the IMB and Sv. Nedjelja mussels. According to CCA analysis, *conchixes* are grouped around Cu, Fe, and Hg, while Cd and Zn are in correlation with soft tissues. Moreover, the shells are in correlation with Mn, although the highest Mn concentrations were found in the soft tissues (Table 1).

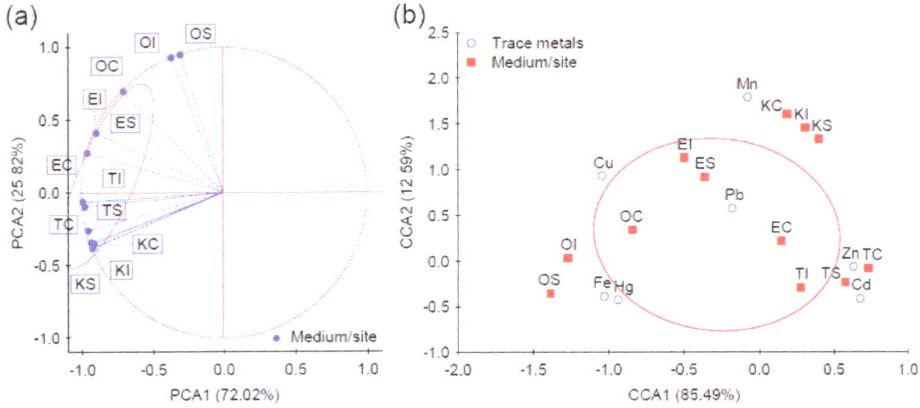

Figure 5. (**a**) The PCA and (**b**) CCA plots of trace metal concentrations within the group of *M. galloprovincialis* L. specimens sampled from Sv. Nedjelja (S), Cogi (C), and the IMB (I) in the shell (KS, KC, KI), soft tissue (TS, TC, TI), *conchix* (OS, OC, OI), and shell extract (ES, EC, EI).

4. Discussion

Despite a significant correlation between the *conchix* weight and dry shell weight, *conchix* % in the shell showed a negative correlation with dry shell weight. Thus, the values of *conchix* % in the shells of *M. galloprovincialis* L. were relatively stable due to their reduced contribution to the overall increase in shell weight during the mussels' growth in comparison with calcified shell layers. This could be explained by the much lower content of organic (*conchix*) than inorganic material in the shells from this study.

The concentrations of all metals analyzed were higher in the soft tissues and *conchixes* in comparison with the extracts and shells. Only Mn content in the shells from Sv. Nedjelja (2.44 $\mu g\ g^{-1}$) and Cogi (2.57 $\mu g\ g^{-1}$) were higher than in *conchixes* of mussels from the same locations (2.07 $\mu g\ g^{-1}$ and 1.6 $\mu g\ g^{-1}$, respectively). Mussels filter high amounts of water daily; therefore, they accumulate trace metals, which contribute to much higher content in the soft tissues than in ambient water [39]. However, despite different accumulation capacities, both soft tissues and shells can be used as bioindicators due to their ability to accumulate metals [40]. Trace metal concentration in the tissues is affected by the season, i.e., the reproductive cycle, as well as by environmental conditions, which can lead to higher concentrations of trace metals in winter and lower concentrations during the summer months [41]. It was explained by the dilution of metals during spring and summer due to the increased gonads' weight, which do not accumulate trace metals [42]. On the other hand, shells are considered more reliable indicators of contamination due to their lower trace metal variability in comparison with the soft tissues [43]. However, in our study, MPI values of organic mediums, such as soft tissues and *conchixes*, were >1, which indicated pollution, while the shells and shell extracts were unable to detect pollution (MPI < 1). The MPI values for the soft tissues at the investigated sites were in line with the data on the CI. A study by Andral et al. [44] showed that trace metals were inversely proportional to the CI, which was not the case in our study. Higher contents of most elements in mussels' soft tissues could be expected due to the sampling period in late autumn long after the spawning period.

The highest concentrations of Cu, Fe, and Hg were found in *conchixes*. The essential elements, including Fe and Cu, were more actively accumulated during the early ontogeny of the bivalve shells [45], indicating higher concentrations of these elements in the environment in the period of mussels' early life stages. Another study showed Fe distribution through the shell layers of *Arctica islandica*, with accumulation mainly in the periostracum [46]. Higher content of Fe in periostracum could contribute against dissolution by an acidic environment [47]. Moreover, higher Cu and Fe contents in *conchixes* compared to tissues, as well as slightly higher Hg contents, could be attributed to the redistribution of these elements from soft tissues during detoxification processes [27]. Hg content in *conchix* might be increased due to Hg variations during mussels' growth. The same processes can also explain site-related differences, i.e., highest Cu, Fe, Pb, and Hg concentrations in *conchixes* from Sv. Nedjelja samples, while soft tissues from the same mussels had the lowest concentrations of Cu, Mn, and Pb. The contents of the rest of the analyzed elements were also higher in the soft tissues from one of the other sites, Cogi or the IMB. Moreover, none of the analyzed metals in shells and extracts were found in the highest contents in Sv. Nedjelja samples, with the exception of Zn in shells (11.6 $\mu g\ g^{-1}$ in both Sv. Nedjelja and Cogi samples).

PCA and CCA analysis showed previously explained differences between *conchixes* and other media in trace metal contents. Clear separation of accumulation patterns in *conchixes* in comparison with all other media at PCA plot could be explained by the higher affinity and specificity of certain metals, such as Cu, Fe, and Hg, to accumulate in the *conchix*. Furthermore, similarities in trace metal accumulation of shells and shell extracts could be expected since both contain inorganic components. However, trace metal accumulation in the shells and shell extracts of *M. galloprovincialis* was much lower in comparison with organic mediums, indicating a lower potential of inorganic mediums for the detection of trace metals in this study. In addition to Cu, Fe, and Hg correlations with *conchixes*, the correlation of Zn and Cd with the soft tissues can be observed by CCA plots. Higher concentrations of Cu in the periostracum and Zn in the soft tissues in many populations of green-lipped mussel were found [25], which is in compliance with Cu and Zn distribution in the tissues of the Mediterranean mussel in our study. Furthermore, CCA indicated Mn correlation with the shells. Different studies found that Mn can be incorporated into aragonite bivalve shells, substituting Ca in the crystal lattice of $CaCO_3$ [48,49].

In accordance with the above, the *conchix* of the Mediterranean mussel as an organic material in close contact with the environment, in comparison with calcified shell layers and relatively independent from the shell weight changes, could detect differences in the accumulation of certain metals in shorter periods than the whole shell. Moreover, it was found that metals accumulated in the periostracum were all assimilated from the soft tissues, and it is less likely to be affected by the bivalve's reproductive cycle [50]. This could be the advantage of *conchix* over the soft tissues in trace metal assessment. Great importance could be given to the method of *conchix* isolation suggested by our team [29] since it provides a high yield of shell organic matrix and could contribute to higher detection levels in trace metal assessments. Particularly, other authors indicated difficulties in periostracum separation due to peeling, which led to incomplete isolation [51]. However, more studies on trace metal accumulation patterns in the *conchix* of *M. galloprovincialis* in highly polluted environments and in different seasons need to be conducted.

Author Contributions: Conceptualization, R.M. and H.E.; methodology, I.Č. and D.J.; investigation, R.M.; resources, D.J.; writing—original draft preparation, R.M.; writing—review and editing, A.P.-B.; visualization, A.P.-B.; supervision, H.E.; funding acquisition, D.J. All authors have read and agreed to the published version of the manuscript.

Funding: This research was funded by Ministry of Education, Science, Culture and Sports of Montenegro in the frame of ShellMED project (grant no. 01-789). This research was partially funded also by the National Science Centre within the framework of the project OPUS 19 grant of National Science Centre, Poland (2020/37/B/ST5/01909) and project MAESTRO 12 (2020/38/A/ST5/00151).

Data Availability Statement: Not applicable.

Acknowledgments: The authors are grateful to Zoran Gačić and Zdravko Ikica who provided improvement to our statistical analyses and technical support.

Conflicts of Interest: The authors declare no conflict of interest. The funders had no role in the design of the study; in the collection, analyses, or interpretation of data; in the writing of the manuscript; or in the decision to publish the results.

References

1. Krishnakumar, P.K.; Qurban, M.A.; Sasikumar, G. Biomonitoring of trace metals in the coastal waters using bivalve molluscs. In *Trace Elements—Human Health and Environment*; IntechOpen: London, UK, 2018; Volume 153. [CrossRef]
2. Spada, L.; Annicchiarico, C.; Cardellicchio, N.; Giandomenico, S.; Di Leo, A. Heavy Metals Monitoring in mussels *Mytilus halloprovincialis* from the Apulian Coasts (Southern Italy). *Mediterr. Mar. Sci.* **2013**, *14*, 99. [CrossRef]
3. Ujević, I.; Vuletić, N.; Lušić, J.; Nazlić, N.; Kušpilić, G. Bioaccumulation of Trace Metals in Mussel (*Mytilus galloprovincialis*) from Mali Ston Bay during DSP Toxicity Episodes. *Molecules* **2015**, *20*, 13031–13040. [CrossRef] [PubMed]
4. Yigit, M.; Celikkol, B.; Yilmaz, S.; Bulut, M.; Ozalp, B.; Dwyer, R.L.; Maita, M.; Kizilkaya, B.; Yigit, Ü.; Ergün, S.; et al. Bioaccumulation of Trace Metals in Mediterranean Mussels (*Mytilus galloprovincialis*) from a Fish Farm with Copper-Alloy Mesh Pens and Potential Risk Assessment. *Hum. Ecol. Risk Assess. Int. J.* **2017**, *24*, 465–481. [CrossRef]
5. Kouali, H.; Chaouti, A.; Achtak, H.; Elkalay, K.; Dahbi, A. Trace Metal Contents in the Mussel *Mytilus galloprovincialis* from Atlantic Coastal Areas in Northwestern Morocco: Levels of Contamination and Assessment of Potential Risks to Human Health. *Mar. Pollut. Bull.* **2022**, *179*, 113680. [CrossRef] [PubMed]
6. IMAP Common Indicator Guidance Fact Sheets. *Environment Programme Mediterranean Action Plan*; Integrated Monitoring and Assessment Programme: Athens, Greece, 2017.
7. Kadim, M.K.; Risjani, Y. Biomarker for Monitoring Heavy Metal Pollution in Aquatic Environment: An Overview toward Molecular Perspectives. *Emerg. Contam.* **2022**, *8*, 195–205. [CrossRef]
8. Kumar, A.; Kumar, V.; Pandita, S.; Singh, S.; Bhardwaj, R.; Varol, M.; Rodrigo-Comino, J. A Global Meta-Analysis of Toxic Metals in Continental Surface Water Bodies. *J. Environ. Chem. Eng.* **2023**, *11*, 109964. [CrossRef]
9. Kumar, A.; Cabral-Pinto, M.; Kumar, A.; Kumar, M.; Dinis, P.A. Estimation of Risk to the Eco-Environment and Human Health of Using Heavy Metals in the Uttarakhand Himalaya, India. *Appl. Sci.* **2020**, *10*, 7078. [CrossRef]
10. Kumar, A.; Song, H.-W.; Mishra, S.; Zhang, W.; Zhang, Y.-L.; Zhang, Q.-R.; Yu, Z.-G. Application of Microbial-Induced Carbonate Precipitation (MICP) Techniques to Remove Heavy Metal in the Natural Environment: A Critical Review. *Chemosphere* **2023**, *318*, 137894. [CrossRef]
11. Pavoni, E.; Crosera, M.; Petranich, E.; Faganeli, J.; Klun, K.; Oliveri, P.; Covelli, S.; Adami, G. Distribution, Mobility and Fate of Trace Elements in an Estuarine System Under Anthropogenic Pressure: The Case of the Karstic Timavo River (Northern Adriatic Sea, Italy). *Estuaries Coasts* **2021**, *44*, 1831–1847. [CrossRef]
12. Nikčević, J. Strengthening the Role of Local Government to Ensure Sustainable Development of the Cruise Sector: The Case of Kotor. *Mar. Policy* **2019**, *109*, 103693. [CrossRef]
13. Joksimović, D.; Perošević, A.; Castelli, A.; Pestorić, B.; Šuković, D.; Đurović, D. Assessment of Heavy Metal Pollution in Surface Sediments of the Montenegrin Coast: A 10-Year Review. *J. Soils Sediments* **2019**, *20*, 2598–2607. [CrossRef]
14. Joksimovic, D.; Stankovic, S. The Trace Metals Accumulation in Marine Organisms of the Southeastern Adriatic Coast, Montenegro. *J. Serb. Chem. Soc.* **2012**, *77*, 105–117. [CrossRef]
15. Perošević, A.; Pezo, L.; Joksimović, D.; Đurović, D.; Milašević, I.; Radomirović, M.; Stanković, S. The Impacts of Seawater Physicochemical Parameters and Sediment Metal Contents on Trace Metal Concentrations in Mussels—A Chemometric Approach. *Environ. Sci. Pollut. Res.* **2018**, *25*, 28248–28263. [CrossRef] [PubMed]
16. Zuykov, M.; Pelletier, E.; Harper, D.A.T. Bivalve Mollusks in Metal Pollution Studies: From Bioaccumulation to Biomonitoring. *Chemosphere* **2013**, *93*, 201–208. [CrossRef] [PubMed]
17. Protasowicki, M.; Dural, M.; Jaremek, J. Trace Metals in the Shells of Blue Mussels (*Mytilus edulis*) from the Poland Coast of Baltic Sea. *Environ. Monit. Assess.* **2007**, *141*, 329–337. [CrossRef] [PubMed]
18. El-Sorogy, A.S.; Youssef, M. Assessment of Heavy Metal Contamination in Intertidal Gastropod and Bivalve Shells from Central Arabian Gulf Coastline, Saudi Arabia. *J. Afr. Earth Sci.* **2015**, *111*, 41–53. [CrossRef]
19. Lyubas, A.A.; Tomilova, A.A.; Chupakov, A.V.; Vikhrev, I.V.; Travina, O.V.; Orlov, A.S.; Zubrii, N.A.; Kondakov, A.V.; Bolotov, I.N.; Pokrovsky, O.S. Iron, phosphorus and trace elements in mussels' shells, water, and bottom sediments from the Severnaya Dvina and the Onega river basins (Northwestern Russia). *Water* **2021**, *13*, 3227. [CrossRef]
20. Giusti, L.; Williamson, A.C.; Mistry, A. Biologically Available Trace Metals in *Mytilus edulis* from the Coast of Northeast England. *Environ. Int.* **1999**, *25*, 969–981. [CrossRef]
21. Yap, C.K.; Sharifinia, M.; Cheng, W.H.; Al-Shami, S.A.; Wong, K.W.; Al-Mutairi, K.A. A Commentary on the Use of Bivalve Mollusks in Monitoring Metal Pollution Levels. *Int. J. Environ. Res. Public Health* **2021**, *18*, 3386. [CrossRef]
22. Putten, E.V.; Dehairs, F.; Keppens, E.; Baeyens, W. High Resolution Distribution of Trace Elements in the Calcite Shell Layer of Modern *Mytilus edulis*: Environmental and Biological Controls. *Geochim. Cosmochim. Acta* **2000**, *64*, 997–1011. [CrossRef]
23. Puente, X.; Villares, R.; Carral, E.; Carballeira, A. Nacreous Shell of *Mytilus galloprovincialis* as a Biomonitor of Heavy Metal Pollution in Galiza (NW Spain). *Sci. Total Environ.* **1996**, *183*, 205–211. [CrossRef]
24. Bellotto, V.R.; Miekeley, N. Trace Metals in Mussel Shells and Corresponding Soft Tissue Samples: A Validation Experiment for the Use of *Perna perna* Shells in Pollution Monitoring. *Anal. Bioanal. Chem.* **2007**, *389*, 769–776. [CrossRef] [PubMed]
25. Yap, C.K. Distributions of Cu and Zn in the Shell Lipped Part Periostracum and Soft Tissues of *Perna viridis*: The potential of Periostracum as a Biomonitoring Material for Cu Contamination. *Pertanika J. Trop. Agric. Sci.* **2012**, *35*, 413–426.
26. Yap, C.K.; Al-Mutairi, K.A. Byssus of Green-Lipped Mussel *Perna viridis* as a Biomonitoring Biopolymer for Zinc Pollution in Coastal Waters. *Biology* **2023**, *12*, 523. [CrossRef] [PubMed]

27. Schöne, B.R.; Krause, R.A. Retrospective Environmental Biomonitoring—Mussel Watch Expanded. *Glob. Planet. Chang.* **2016**, *144*, 228–251. [CrossRef]
28. Wilbur, K.M.; Saleuddin, A.S.M. Shell formation. In *The Mollusca*; Academic Press: Cambridge, MA, USA, 1983; pp. 235–287.
29. Ehrlich, H.; Martinović, R.; Joksimović, D.; Petrenko, I.; Schiaparelli, S.; Wysokowski, M.; Tsurkan, D.; Stelling, A.L.; Springer, A.; Gelinsky, M.; et al. Conchixes: Organic Scaffolds Which Resemble the Size and Shapes of Mollusks Shells, Their Isolation and Potential Multifunctional Applications. *Appl. Phys. A* **2020**, *126*, 562. [CrossRef]
30. Joksimović, D.; Castelli, A.; Perošević, A.; Djurović, D.; Stanković, S. Determination of Trace Metals in *Mytilus galloprovincialis* along the Boka Kotorska Bay, Montenegrin Coast. *J. Trace Elem. Med. Biol.* **2018**, *50*, 601–608. [CrossRef]
31. Cravo, A.; Bebianno, M.J. Bioaccumulation of Metals in the Soft Tissue of *Patella aspera*: Application of Metal/Shell Weight Indices. *Estuar. Coast. Shelf Sci.* **2005**, *65*, 571–586. [CrossRef]
32. *ISO 6869:2000*; Animal Feeding Stuffs—Determination of the Contents of Calcium, Copper, Iron, Magnesium, Manganese, Potassium, Sodium and Zinc—Method Using Atomic Absorption Spectrometry. Institute for Standardization of Serbia: Belgrade, Serbia, 2000. Available online: https://iss.rs/en/project/show/iso:proj:33707 (accessed on 28 March 2023).
33. Tomović, V.M.; Petrović, L.S.; Tomović, M.S.; Kevrešan, Ž.S.; Džinić, N.R. Determination of Mineral Contents of Semimembranosus Muscle and Liver from Pure and Crossbred Pigs in Vojvodina (Northern Serbia). *Food Chem.* **2011**, *124*, 342–348. [CrossRef]
34. *ISO 14082:2008*; Determination of Trace Elements-Determination of Lead, Cadmium, Zinc, Copper, Iron and Chromium by Atomic Absorption Spectrometry (AAS) after Dry Ashing. Institute for Standardisation of Serbia: Belgrade, Serbia, 2008. Available online: https://iss.rs/en/project/show/iss:proj:20335 (accessed on 28 March 2023).
35. Usero, J.; Morillo, J.; Gracia, I. Heavy Metal Concentrations in Molluscs from the Atlantic Coast of Southern Spain. *Chemosphere* **2005**, *59*, 1175–1181. [CrossRef]
36. Rodríguez-Barroso, M.R.; Benhamou, Y.; El Moumni, B.; El Hatimi, I.; García-Morales, J.L. Evaluation of Metal Contamination in Sediments from North of Morocco: Geochemical and Statistical Approaches. *Environ. Monit. Assess.* **2008**, *159*, 169–181. [CrossRef]
37. Uluturhan, E. Heavy Metal Concentrations in Surface Sediments from Two Regions (Saros and Gökova Gulfs) of the Eastern Aegean Sea. *Environ. Monit. Assess.* **2009**, *165*, 675–684. [CrossRef]
38. TIBCO Software Inc. Data Science Workbench, Version 14. 2020. Available online: http://tibco.com (accessed on 4 February 2023).
39. Shirneshan, G.; Riyahi Bakhtiari, A. Accumulation and Distribution of Cd, Cu, Pb and Zn in the Soft Tissue and Shell of Oysters Collected from the Northern Coast of Qeshm Island, Persian Gulf, Iran. *Chem. Speciat. Bioavailab.* **2012**, *24*, 129–138. [CrossRef]
40. Forleo, T.; Zappi, A.; Melucci, D.; Ciriaci, M.; Griffoni, F.; Bacchiocchi, S.; Siracusa, M.; Tavoloni, T.; Piersanti, A. Inorganic Elements in *Mytilus galloprovincialis* Shells: Geographic Traceability by Multivariate Analysis of ICP-MS Data. *Molecules* **2021**, *26*, 2634. [CrossRef] [PubMed]
41. Azizi, G.; Layachi, M.; Akodad, M.; Baghour, M.; Ghalit, M.; Gharibi, E.; Ngadi, H.; Moumen, A. The Accumulation of Al, As, Li, Mg, Mn, S, Si, Ti, and V in the Mussel *Mytilus galloprovincialis* from the Moroccan Mediterranean Coastal Areas: Trends Pertaining to Seasons and Levels. *Ocean Sci. J.* **2020**, *55*, 405–418. [CrossRef]
42. Savari, A.; Lockwood, A.P.M.; Sheader, M. Effects of Season and Size (Age) on Heavy Metal Concentrations of the Common Cockle (*Cerastoderma edule* (L.)) from Southampton Water. *J. Molluscan Stud.* **1991**, *57*, 45–57. [CrossRef]
43. Yap, C.K.; Ismail, A.; Tan, S.G.; Abdul Rahim, I. Can the Shell of the Green-Lipped Mussel *Perna viridis* from the West Coast of Peninsular Malaysia Be a Potential Biomonitoring Material for Cd, Pb and Zn? *Estuar. Coast. Shelf Sci.* **2003**, *57*, 623–630. [CrossRef]
44. Andral, B.; Stanisiere, J.Y.; Sauzade, D.; Damier, E.; Thebault, H.; Galgani, F.; Boissery, P. Monitoring Chemical Contamination Levels in the Mediterranean Based on the Use of Mussel Caging. *Mar. Pollut. Bull.* **2004**, *49*, 704–712. [CrossRef] [PubMed]
45. Demina, L.L.; Galkin, S.V.; Dara, O.M. Trace Metal Bioaccumulation in the Shells of Mussels and Clams at Deep-Sea Hydrothermal Vent Fields. *Geochem. Int.* **2012**, *50*, 133–147. [CrossRef]
46. Dunca, E.; Mutvei, H.; Göransson, P.; Mörth, C.-M.; Schöne, B.R.; Whitehouse, M.J.; Elfman, M.; Baden, S.P. Using Ocean Quahog (*Arctica islandica*) Shells to Reconstruct Palaeoenvironment in Öresund, Kattegat and Skagerrak, Sweden. *Int. J. Earth Sci.* **2008**, *98*, 3–17. [CrossRef]
47. Swinehart, J.H.; Smith, K.W. Iron and Manganese Deposition in the Periostraca of Several Bivalve Molluscs. *Biol. Bull.* **1979**, *156*, 369–381. [CrossRef]
48. Soldati, A.L.; Jacob, D.E.; Glatzel, P.; Swarbrick, J.C.; Geck, J. Element Substitution by Living Organisms: The Case of Manganese in Mollusc Shell Aragonite. *Sci. Rep.* **2016**, *6*, 22514. [CrossRef] [PubMed]
49. Schöne, B.R.; Huang, X.; Zettler, M.L.; Zhao, L.; Mertz-Kraus, R.; Jochum, K.P.; Walliser, E.O. Mn/Ca in Shells of *Arctica islandica* (Baltic Sea)—A Potential Proxy for Ocean Hypoxia? *Estuar. Coast. Shelf Sci.* **2021**, *251*, 107257. [CrossRef]
50. Yap, C.K.; Tan, S.G. A study on the potential of the periostracum of *Perna viridis* as a biomonitoring material for Pb in tropical coastal waters based on correlation analysis. *Sains Malays.* **2011**, *40*, 809–819.
51. Dermott, R.M.; Lum, K.R. Metal Concentrations in the Annual Shell Layers of the Bivalve *Elliptio complanata*. *Environ. Pollut. Ser. B Chem. Phys.* **1986**, *12*, 131–143. [CrossRef]

applied sciences

MDPI

Article

Integrated Assessment of CO_2-Induced Acidification Lethal and Sub-Lethal Effects on Tropical Mussels *Perna perna*

Lorena da Silva Souza [1], Estefanía Bonnail [2], Luis Felipe de Almeida Duarte [3,4], Augusto Cesar [4], Inmaculada Riba [1,4] and Camilo Dias Seabra Pereira [3,4,*]

1 Faculty of Marine and Environmental Sciences, University of Cádiz, 11510 Cádiz, Spain
2 Centro de Investigaciones Costeras, Universidad de Atacama (CIC-UDA), Avenida Copayapu 485, Copiapó 1530000, Chile
3 Department of Ecotoxicology, Santa Cecília University (UNISANTA), Santos 11045-907, São Paulo, Brazil
4 Department of Marine Sciences, Federal University of São Paulo (UNIFESP), Santos 11030-100, São Paulo, Brazil
* Correspondence: camilo.seabra@unifesp.br or camilo@unisanta.br

Abstract: Leakages of CO_2 capture and storage systems from the seabed are able to cause significant adverse biological effects in marine species. Adult mussels were exposed to different CO_2 enrichment scenarios (pH from 8.3 to 6.0) for 96 h, and endpoints (lysosomal membrane deterioration, lipid peroxidation and primary damages in DNA) were assessed. Mortality and reduced health status can occur after short exposure of the tropical mussel *Perna perna* to pH levels lower than 7.5. Results pointed out cytogenotoxic effects in the hemolymph and gills after 48 and 96 h of exposure, respectively. These findings should be considered when environmental monitoring approaches are performed in tropical marine areas employing CCS strategies.

Keywords: CO_2 enrichment; mollusks; ocean acidification; biomarkers; carbon capture storage

Citation: Souza, L.d.S.; Bonnail, E.; de Almeida Duarte, L.F.; Cesar, A.; Riba, I.; Pereira, C.D.S. Integrated Assessment of CO_2-Induced Acidification Lethal and Sub-Lethal Effects on Tropical Mussels *Perna perna*. *Appl. Sci.* **2023**, *13*, 7199. https://doi.org/10.3390/app13127199

Academic Editor: Dimitris Mourtzis

Received: 25 January 2023
Revised: 2 June 2023
Accepted: 12 June 2023
Published: 16 June 2023

1. Introduction

Rising atmospheric carbon dioxide (CO_2) concentration is causing global warming and ocean acidification [1–3], which increasingly are recognized as important drivers of change in biological systems [4–7]. In aquatic environments, the CO_2 undergoes a series of chemical transformations, and elements recombine giving rise to new compounds [8]. First, the carbonic gas mixes with the water, producing aqueous CO_2 and carbonic acid (H_2CO_3). This compound, classified as a weak acid, quickly dissociates, producing bicarbonate (HCO_3^-) and protons (H^+), which react with other molecules present in the aquatic environment. Bicarbonate ions can also undergo chemical reactions, transforming into CO_3^{2-} and releasing hydrogen (H^+) in the water column, lowering the pH [9,10].

Since the industrial era, oceanic uptake of CO_2 has resulted in acidification of the ocean, and the pH of ocean surface water has decreased by 0.1, corresponding to a 26% increase in acidity [11]. In order to reduce atmospheric CO_2 levels, many mitigation strategies have been developed and proposed. One such strategy is large-scale carbon dioxide capture and storage strategies (CCSs) in geological formations. According to the International Energy Agency [12], it could contribute to a reduction of 19% in CO_2 emissions by 2050. This technology consists of trapping CO_2 from industrial and energy-related sources, transporting it to a storage site, injecting and storing it for a long time instead of releasing this gas into the atmosphere [13].

Transportation of CO_2 in high-pressure pipelines from source to storage location constitutes an important link in the CCS chain, especially when transporting large quantities of CO_2 over long distances [14]. The possibility of leakage from the sub-seabed reservoirs and transportation pipelines into the waters and atmosphere causes great concern, with the timeframe for safe CO_2 storage being relatively unknown [15]. Considering that oceans have the largest capacity for CO_2 storage, sub-seabed geological formations, such as depleted oil and gas reservoirs and saline aquifers, have been designated as potential storage locations for CO_2 sequestration. The complexity in foreseeing the location and magnitude of possible seepages makes difficult the evaluation of potential effects on aquatics ecosystems. Two main potential sources of CO_2 escape are transport facilities and storage areas [16]. The effects of CO_2 leakage will depend on the amount and/or rate of leakage, the transport, the dispersion processes and the chemical buffering capacity of the sedimentary or water system, contributing to the imbalance of seawater's chemistry. There are also natural sources of CO_2 enrichment, such as natural CO_2 vents [17,18], bacterial organic matter degradation, diagenesis process [19] and submarine eruptions such as in the Canary Islands (Spain), where pH values between 5.13 and 8.04 have been measured [20,21].

In this context, many studies have been performed in order to assess the impacts of changes in the marine carbonate system as well as pH reduction via CO_2 enrichment to organisms [22–26]. Several authors mimicked a CO_2 leakage in the lab, assessing different responses related to decreases in pH values in diversified organisms, for instance: growth rate of bacteria (*Roseobacter* sp. CECT 7117 and *Pseudomonas litoralis* CECT 7670) [27] and bacterial communities [28]; mortality rate applied to amphipods (*Ampelisca brevicornis* and *Hyale youngi*) [29]; impact on the early life stages of marine mussel *Perna perna* [30,31]; and effects on growth, cell viability and oxidative stress using three microalgae species (*Tetraselmis chuii, Phaeodactylum tricornutum* and *Nannochloropsis gaditana*) [32]. Histopathological effects and lysosomal membrane stability were also assessed in mollusks such as clams (*Ruditapes philippinarum*) and mussels (*Mytilus edulis*) [33,34] apart from a battery of biomarkers using the clam *Scrobicularia plana* [35].

Ocean acidification impacts other aspects of a marine organism's physiology, including acid–base balance, energy metabolism, redox balance and behavior [36,37]. Acidification of ocean surface water is a currently developing scenario that warrants a broadening of research foci in the study of acid–base physiology, ensuring a strong basis for the physiological interactions of ocean acidification with pollutants that may affect the same molecular and physiological pathways [38].

The cellular mechanisms of CO_2-induced changes in the physiology of mollusks are not yet fully understood but are likely to involve multiple pathways of metabolism, biomineralization and acid–base balance [39,40]. Metabolic effects of elevated pCO_2 vary between different species [41,42] and depend on the CO_2 concentration in seawater.

For decades, bivalve filter feeders have been used for environmental assessments [43–45]. Due to their sedentary habits, low metabolic transformation rates and their ability to bioconcentrate pollutants, bivalves have been used as bioindicators suitable for monitoring studies in coastal areas such as useful bioindicators of persistent pollutants [46]. Studies using clams [47–50], mussels [51–54] and oysters [55–57] have been performed worldwide.

It has been reported that the early life stages of marine invertebrates are more susceptible to environmental toxicants than are the adult forms [58]. However, the mechanism of action-oriented toxicity assays has been performed with adult organisms, and the results are sensitive to environmentally relevant concentration [59,60]. The use of biological endpoints (biomarkers) has been advocated as an important tool for assessing the bioavailability of contaminants and the general health of individual organisms [61]. These responses are taken into consideration for monitoring different sources of anthropogenic contamination in coastal areas using bivalves as bioindicator species [44,62].

The aim of this study is to assess adverse effects of CO_2-induced acidification on individuals of mussel *Perna perna* using a battery of sub-cellular effect biomarkers (lipid peroxidation, DNA primary damage and lysosomal membrane stability). Additionally, the mortality of the organisms was used to establish the toxicity related to the increase of proton concentration associated with the enrichment of CO_2. For this purpose, organisms were exposed over four days to different acidification scenarios with different pH values, and the influence of acidification associated with CO_2 enrichment was analyzed.

2. Materials and Methods

2.1. Laboratory Tests

To simulate the acidification process based on CO_2 enrichment event such as CCS, a CO_2 injection system was adapted from the experimental set up described by [63,64], with further detailed experiments in [65]. Briefly, the different acidification scenarios reached pH values of 8.0, 7.5, 7.0, 6.5, and 6.0, controlled using the injection system, and non-CO_2 injected treatment as control (pH 8.3). The experiment was developed using clean natural seawater from Enseada Beach in Guarujá, SP/Brazil [66]. Adult mussels *Perna perna* were acquired from a farming zone located at the Cocanha beach–Caraguatatuba municipality in São Paulo's coastal zone. The organisms were transported to the laboratory, where they were kept for 24 h prior to the assays under temperature (22 °C) and salinity (35 ppt) controlled in a 300 L aquarium with seawater.

Ten individuals of mussel *Perna perna* were introduced per 20 L chamber (two replicates per treatment). The laboratory conditions were controlled: seawater temperature at 20 ± 2 °C, salinity 35 ppt, 8.0 mg L^{-1} dissolved oxygen, and photoperiod 12:12. Water quality was maintained through renewing every 48 h.

After 48 and 96 h of exposure, survival was recorded, and ten organisms from each pH treatment (five organisms from each tank) had the hemolymph withdrawn before dissection gills were frozen (−80 °C). The hemolymph was collected to assess the lysosomal membrane stability through neutral red retention time assay (NRRT), while gills were used to assess DNA damage (strand breaks) and lipid peroxidation (LPO).

2.2. Chemical Characterization

The pH values and total alkalinity (TA) were measured using a potentiometric titration system (Metrohm 794 Basic Titrino, Switzerland, ref.6.0210.100) (pH scale calibration with 0.1 M HCl). Dissolved inorganic carbon (DIC) was determined via the experimental values and the seawater carbonate system speciation: HCO_3^-, CO_3^{2-}, CO_2, calcite saturation (Ω_{Cal}), aragonite saturation index (Ω_{Arag}) and partial pressure of carbon dioxide (pCO_2), using the CO2SYS v2.1 program as described in [65].

2.3. Neutral Red Retention Time Assay

The NRRT assay was carried out following the method described by [67]. This non-destructive method employed hemolymph withdrawn from the posterior adductor muscle of living mussels. The hemolymph was mixed to physiological saline solution (pH 7.3 containing HEPES 4.77 g L^{-1}, NaCl 25.48 g L^{-1}, $MgSO_4$ 13.06 g L^{-1}, KCl 0.75 g L^{-1}, $CaCl_2$ 1.47 g L^{-1}), spread on slides and transferred to a lightproof chamber for 15 min to allow cell attachment. Excess liquid was removed and 40 μL of the neutral red (NR) dye was added to the cell monolayer. A cover slip was added. After a 15 min incubation period, slides were examined every 15 min via optical microscopy (400×) for both structural abnormalities and NR dye loss from the lysosomes to the cytosol. The test was terminated when at least 50% of the examined cells exhibited these characteristics, and an NRRT mean value was calculated for each group. The same analyst carried out the assessment for all slides during the study.

2.4. Lipid Peroxidation and DNA Strand Breaks

For the LPO and DNA damage assays, mussel gills were homogenized with a buffer solution containing Tris-HCl (50 Mm), EDTA (1 mM), dithiothreitol (DTT) (1 mM), sucrose (50 mM), KCl (150 mM) and phenylmethylsulfonyl fluoride (PMSF) (100 mM).

LPO assay was carried out as demonstrated by [68]. LPO was determined in gill homogenates using thiobarbituric acid. Thiobarbituric acid reactants (TBARS) were determined via fluorescence at 530 nm for excitation and 630 nm for emission using a fluorescence microplate reader. Because the reagent could react with other aldehydes, the results were expressed as μg of TBARS mg^{-1} total protein.

The mitochondrial DNA damage was determined as outlined by [69], based on the K-SDS precipitation of DNA–protein crosslink. Fluorescence was used to quantify DNA strand breaks. DNA quantitation was achieved using Hoescht dye at a concentration of 100 nM in 200 mM Tris–HCl, pH 8.5, containing 300 mM NaCl and 4 mM sodium cholate. Salmon sperm DNA standards were used for calibration, and fluorescence readings were taken at 360 nm excitation and 460 nm emission. The results were expressed as μg of DNA·mg^{-1} total protein.

The Bradford methodology [70] was used to determine the total protein content (the calibration curve with bovine serum) and normalization.

2.5. Data Treatment

A normal distribution and equal variance among groups of ANOVA residues from the experiment were confirmed by means of the Shapiro–Wilk test and Levene's F-test, respectively. Two-way factorial ANOVA (pH values, exposure times and their combination) followed by the Dunnett's test identified significant differences between control group and treatments using the SPSS 15.0 statistical software program. In addition, the effect size estimated for ANOVAs (NRRT, DNA damage and LPO) were calculated to measure how much variance in the response variables (biomarkers) are accounted for by the explanatory variables (pH values and exposure times). Therefore, the "omega squared" (effect size, ES) method was used because it is considered a less-biased alternative, especially when sample sizes are small. Ref. [71] suggests the following interpretation of calculated effect sizes: 0–0.01 = very small; 0.01–0.06 = small; 0.06–0.14 = medium and >0.14 = large. The calculations were performed in the R Environment 4.0.4 [72] (using the software package "effectsize") [73].

Thereafter, the original data set including all biological responses (means) was analyzed through an integrative approach via factor analysis, employing principal component analysis as the extraction procedure. The data set was rearranged in a correlation matrix and two factors (or new variables) were extracted considering eigenvalues higher than 1.0 (Kaiser's criteria). For the factor analysis, the variables were auto scaled (Varimax normalized) to be treated with equal importance. The criterion for consideration of a variable as being associated with a particular factor was defined as it having a loading of 0.40 or higher. Besides the analysis of the variables aggregated by PCA, a representation of estimated factor scores from each studied area to the centroid of all cases for the original data was performed in order to confirm the factor descriptions and to characterize the studied stations. All analyses were performed using the PCA option of the multivariate exploratory technique procedure, followed by the basic set-up for factor analysis procedure from the STATISTICA software tool (Hamburg, Germany, Stat Soft, Inc., 2001; version 6.0).

3. Results and Discussion

Data had sufficient homogeneity of variance for ANOVA. There was no significant difference between mussels held in tanks at any treatment.

3.1. Chemical Analysis

The averaged values for the carbonate system speciation are presented in Table 1. This data refers to the carbon parameters measured and calculated at the end of the experiments. The salinity parameter used was 35, and the temperature was 23.5 ± 0.5 °C. As expected, the total inorganic carbon (TIC) was higher as pH was reduced, and the saturation index for calcite (Ω_{Cal}) and aragonite (Ω_{Arag}) decreased as pH decreased and pCO_2 increased.

Table 1. Carbonate system speciation in assays exposed to the different pH treatments for both bioassays of fertilization and embryo-larval development.

pH Treatment	TA (μmol/kg)	TCO_2 (μmol/kg)	TIC/DIC (μmol/kg)	HCO_3^- (μmol/kg)	CO_3^{2-} (μmol/kg)	CO_2 (μmol/kg)	pCO_2 (μatm)	Ω_{Cal}	Ω_{Arag}
8.3	1644	1404	1404	1249	147	8	262	3.53	2.31
8.0	1662	1511	1511	1397	99	14	472	2.38	1.56
7.5	1731	1718	1718	1629	39	50	1672	0.93	0.61
7.0	1682	1828	1828	1650	12	165	5451	0.29	0.19
6.5	2483	3249	3249	2469	6	774	25,996	0.14	0.09
6.0	3172	5996	5996	3167	3	2827	97,130	0.06	0.04

Mussel shells are made of up to 83% aragonite [74]. Aragonite is more soluble than calcite, therefore shell dissolution may be a considerable problem faced by mussels in acidified seawater [75]. The Ω_{Arag} remains favorable for calcification when >1 [73]. The present study shows a decrease in the Ω value related to an increase in protons H^+, being a critical value for calcification when pH $\leq$ 7.5 (Table 1). According to [76], $CaCO_3$ is of high ecological relevance, acting on the calcification kinetics and interfering in bivalve shell formation. Its reduction and dissolution depend on $\Omega_{Arag}/\Omega_{Cal}$ availability, which are changed via H^+ variance.

3.2. Biological Responses

The percentage of mussel mortality for each treatment assay is shown in Table 2. In general, the mortality increases as pH values decrease. For pH values up to 7.5, no mortality was registered, while for pH values below 7.0, mortality was recorded at a minimum of 20% after 96 h of exposure. These results concur with study [77], in which the clam *Scrobicularia plana* showed a mortality of 20% at the pH value of 7.1 at 96 h of exposure.

Table 2. *Perna perna* mortality percentage registered after 48-and 96 h exposure for the different pH levels (8.0, 7.5, 7.0, 6.5, and 6.0) and control (8.3 with no CO_2 added).

pH Value	Mortality (%)	
	48 h	96 h
8.3	0	0
8.0	0	0
7.5	0	0
7.0	0	20
6.5	10	20
6.0	0	30

Lysosomes are central to many key biological processes, including reproduction, digestion and immune response; thus, any deleterious changes to their function can have serious consequences for those processes. The general health status of the mussels was therefore considered in terms of changes to the lysosomal system of the blood cells [75].

The NRRT assay is based on the principle that only lysosomes in healthy cells take up and retain the vital dye neutral red. Lysosomal membrane damage caused by the impact of xenobiotics can decrease the NRRT times through inducing the leaking of lysosomal components. Lysosomal membrane damage caused by the impact of xenobiotics [78] or by CO_2-induced seawater acidification [75] can decrease the NRRT through inducing the leaking of lysosomal components [79].

The results of the biomarker of exposure NRRT assay for *Perna perna* involving acidified seawater are shown in Figure 1. The "exposure times" presented very small estimates ($F = 0.197$; df = 1; $p = 0.65$; ES = 0.001), while the isolated "pH values" ($F = 10.328$; df = 5; $p < 0.001$; ES = 0.37) and the combination of both explanatory variables ($F = 3.518$; df = 5; $p = 0.007$; ES = 0.15) also evidenced a large effect on NRRT responses. The range in NRRT values observed for lysosomes from control mussels ($n = 10$; no CO_2 added) at time zero, 48 h and 96 h of assay were 81 ± 26 min, 93 ± 23 min and 77 ± 24 min, respectively. In the first 48 h, the RT significantly ($p < 0.05$) decreased at pH 7.0 and 6.5. After 96 h, the organisms exposed to different pH values showed a significant decrease ($p < 0.01$) in the dye retention time in lysosomes only at pH 6.0. There was no significant difference between RT of the dye in the control group for the highest pH values (8.0 and 7.5).

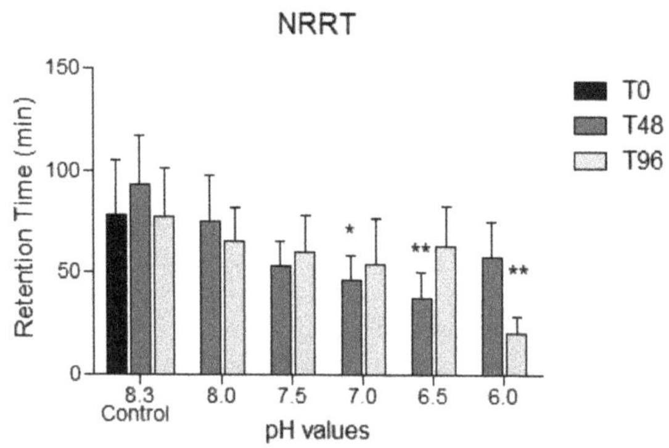

Figure 1. Mean (and standard error) of LMS assessed through NRRT assay in the hemocytes of *P. perna* exposed to different pH levels (Dunn's test: * $p < 0.05$; ** $p < 0.01$).

The results presented in this study for the control sample (81 ± 26 min) are reliable and fit with those obtained in previous studies reporting an NRRT ranging from 60 to 90 min in a study on healthy or non-exposed *P. perna* [80] and 60 min in a study using hemolymph of *Perna viridis* [81]. Furthermore, [82] reported NRRTs of 90 min using *Mytilus galloprovincialis* hemolymph, and 106 min in the [83] study using clam *Ruditapes philippinarum* hemolymph.

According to the criteria established by [84], in the cytochemical method, animals are considered to be stressed but compensating if NRRT were between 50 and 120 min. However, studies have demonstrated that lysosomal membrane stability differs in tropical zones compared to temperate zones, as higher temperatures in the tropics lead the neutral red dye to be retained in the lysosomes for a shorter period of time [85], agreeing with the results of this work and previous studies on NRRT of *P. perna* populations from the Brazilian coast [43,65,86,87]. Therefore, the present study may contribute to the development of a health index for tropical mussels.

Lysosomal integrity is directly correlated with physiological scope for growth and is also mechanistically linked with the processes of protein turnover [88]. Ref. [89] showed that lysosomal stability in adult oysters is also correlated with larval viability. Therefore, the integrity or stability of this membrane is considered an indicator of the cell "well-being", acting as an important cellular and nonspecific biomarker of stress [46] once its activity is directly related to ion transportation [59].

The significant reduction in lysosomal health observed for pH below 7.0 (48 h) may be related to the increase in Ca^{2+} concentrations in the hemolymph [75], which can cause lysosomal destabilization [90]. According to [22], decreasing pH promotes the dissolution of shells and, consequently, an increasing concentration of Ca^{2+} in the hemolymph, altering the cellular metabolism and reducing immune function. According to [91], extracellular calcium enters in the cells via pinocytosis, and once in the cytosol, Ca^{2+} can activate calcium-dependent phospholipase A2, destabilizing the lysosome membrane.

The compensation of acid–base imbalance is also of great importance, since it is related to ion transportation through the cell membrane. The CO_2 produced in the cells during routine metabolism is typically hydrated to form bicarbonate and H^+. The H^+ ions are buffered, while the bicarbonate is transported out of the cell in exchange for Cl^- via ion transport proteins. Refs. [92,93] indicated that mussels responded to maintain acid–

base balance through carbonic anhydrase activity, so that it was significantly higher in mussels exposed to low pH in comparison with control individuals. As observed in the present study, the increasing proton concentration in the environment, once more, may be associated with a decrease in the lysosome's capacity to retain the dye as a healthy cell.

DNA damage results evidenced that "exposure time" ($F = 0.298$; $df = 1$; $p = 0.58$; $ES = 0.001$) was not significant, presenting a very small influence on DNA-SB. The "pH values" variable, on the other hand, demonstrated a large significance ($F = 20.766$; $df = 5$; $p < 0.001$; $ES = 0.47$) and the combination of these explanatory variables showed a significantly medium effect ($F = 3.727$; $df = 5$; $p = 0.003$; $ES = 0.11$). Lipid peroxidation results showed that "exposure time" ($F = 34.953$; $df = 1$; $p < 0.001$; $ES = 0.24$), the "pH values" ($F = 4.940$; $df = 5$; $p < 0.001$; $ES = 0.15$) and the combination of these explanatory variables ($F = 5.603$; $df = 5$; $p < 0.001$; $ES = 0.17$) had a large statistical effect on LPO responses. The DNA strand break values and the LPO results are shown in Figure 2. In the present study, the DNA strand breaks exhibited significantly higher values ($p < 0.05$) than T0 in the gills of *Perna perna* mussels exposed to a pH value of 6.0 in both exposure times. Only LPO was significantly different ($p < 0.01$) at pH 6.0 after 96 h exposure.

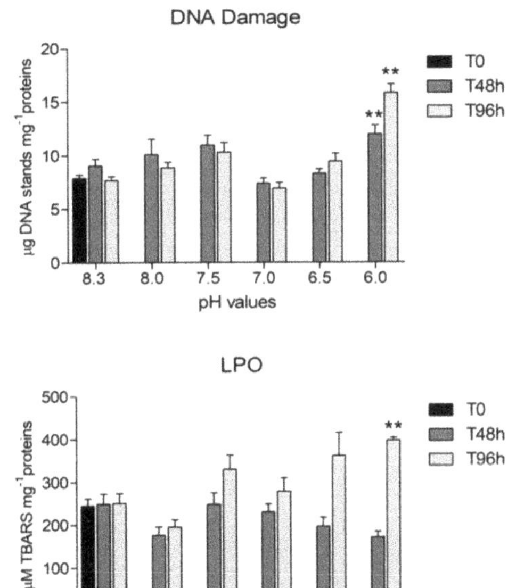

Figure 2. Mean (and standard error) of DNA damage (stand break) and LPO assessed in *P. perna* exposed to different pH levels for 96 h (** $p < 0.01$).

Oxidative stress occurs due to an imbalance between the production of reactive oxygen species (ROS) and their elimination via the antioxidant system defense, leading to protein degradation and enzyme inactivation [94]. If it is continuously generated, ROS can damage important biomolecules such as DNA, proteins and LPO [95]. Results obtained from this study may be related to the increase in the ROS production in acidified scenarios, agreeing with the study [96], where bivalve *Crassostrea gigas* showed a significant increase in the production of ROS in hemocytes, as well as DNA damage (comet assay) in different scenarios of ocean acidification (pH 7.6 and 7.8).

Ref. [97] reported that low pH caused a decrease in electron supply from NADH to the electron transport chain in the oyster *Crassostrea gigas*, and their response mechanism changed the cellular balance between resource supply and oxidative stress. As a protective system against the consequences of oxidative stress, lysosome autophagy consists of breaking down longer-lived proteins and organelles and recycling the products into protein-synthesis and energy production pathways [98].

3.3. Principal Component Analysis (PCA)

A multivariate analysis approach was applied to all data to help discriminate the main variables responsible for the variance of biological effects detected in mussels (Table 3, Figure 3). The application of PCA to the original 13 variables at the pH bioassay indicates that they can be grouped in two new factors. These new variables explain more than 96% of the total variance of the original data set. In the present study, we selected a group of variables associated with a component loading ≥ 0.40, corresponding to an associated explained variance of more than 30%.

Table 3. Sorted rotated factor loadings of 13 variables for the two principal factors.

Variables	Components	
	Factor 1	Factor 2
NRRT 48 h		−0.92
NRRT 96 h	−0.76	−0.52
DNA 48 h	0.69	
DNA 96 h	0.91	0.24
LPO 48 h	−0.76	
LPO 96 h	0.51	0.72
Mortality 48 h		
Mortality 96 h	0.55	0.68
TA	0.88	0.40
DIC	0.89	0.41
Ω_{Cal}	−0.26	−0.94
Ω_{Arag}	−0.26	−0.94
pCO_2	0.91	
Variance (%)	5.63	4.51
Cumulative (%)	0.43	0.34

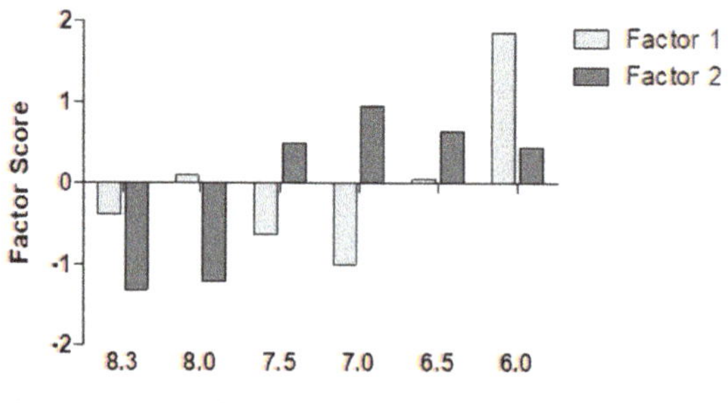

Figure 3. Estimated factor scores in relation to the pH treatments used in the study.

The F1 (61%) includes the lethal response and DNA damage of the mussel. According to [89], the positive LPO 96 h value may demonstrate a relationship between an increase of antioxidant enzymes and LPO levels as defense mechanisms due to possibly being overwhelmed by ROS produced in organism tissues. It was more representative for pH 6.0 with higher positive values (Figure 3).

The F2 (21%) showed the relationship between NRRT, DNA damage, LPO, mortality and pH reduction. The effects of this factor could be related with acidification by CO_2 enrichment because of the significant variance in lysosome membrane stability (NRRT) and the increase in DNA damage, the oxidative stress and the mortality.

The presented results corroborate the toxic effects observed from a decrease in pH value $\leq$ 6.5, while from pH values $\leq$ 7.5 a stress showed to be produced through activating enzymatic systems maintained at the lowest pH values. In summary, the stress produced via increasing proton concentration can be noticed in the enzymatic systems of the mussels (NRRT and LPO) in pH values lower than 7.5, being accentuated when the value of the concentration of protons is higher, which produces toxic effects related to pH 6.5 and 6.0.

A similar integration analysis was carried out for the same pH range, but with other toxicity species and endpoints (amphipods mortality, sea-urchin embryo-larval development and the benthic community) in the Bay of Santos [29]. Similar results were obtained; they determined that pH 6.0 was extremely toxic for marine life.

The responses of the mussels might be affected by the presence of other toxic compounds in the environment. The Santos Bay is affected by sewage discharges that carry illicit drugs, such as crack cocaine. This, in combination with acidification conditions (also below pH 7.5), provokes alterations in *Perna perna* reproduction [31].

4. Conclusions

The results obtained in this study suggest that mortality and reduced health status can occur when the tropical mussel *Perna perna* is exposed to pH levels lower than 7.5 for a short term (48 and 96 h). Lysosomal membrane stability of hemocytes showed the first signs of effects, followed by DNA damages and LPO in gills under acidified conditions. These findings should be considered when environmental monitoring approaches are performed in tropical marine areas receiving petroliferous or other activities employing CCS technologies.

Author Contributions: Conceptualization, L.d.S.S., C.D.S.P. and I.R.; methodology, L.d.S.S., E.B. and A.C.; software, L.d.S.S., E.B., I.R. and L.F.d.A.D.; validation, C.D.S.P., A.C., I.R. and L.F.d.A.D.; formal analysis, L.d.S.S. and L.F.d.A.D.; investigation, I.R., C.D.S.P. and A.C.; resources, C.D.S.P., I.R. and A.C.; data curation, L.d.S.S. and E.B.; writing—original draft preparation, L.d.S.S. and I.R.; writing—review and editing, all authors; visualization, E.B. and L.d.S.S.; supervision, C.D.S.P. and A.C.; project administration, C.D.S.P., I.R. and A.C.; funding acquisition, C.D.S.P., A.C. and I.R. All authors have read and agreed to the published version of the manuscript.

Funding: This study was funded by Fundação de Amparo à Pesquisa do Estado de São Paulo (FAPESP) through grants #2015/17329-0 and #2018/18456-4.

Institutional Review Board Statement: Not applicable.

Informed Consent Statement: Not applicable.

Acknowledgments: The first author would like to thank the Erasmus Mundus Program for the doctoral fellowship. Bonnail, E. thanks to ANID/CONICYT/FONDECYT 11180015. Pereira C.D.S. thanks CNPq for funding support (#409187/2016-0). Cesar, A., Pereira C.D.S., Ribeiro D.A. and T.A. DelValls thank CNPq (#305734/2018-0) for productivity fellowships. Riba, I. thanks FAPESP (#2017/25936-0) for her Visiting Researcher Program at Universidade Federal do Sao Paulo.

Conflicts of Interest: The authors declare no conflict of interest.

References

1. Schmittner, A.; Oschlies, A.; Matthews, H.D.; Galbraith, E.D. Future changes in climate, ocean circulation, ecosystems, and biogeochemical cycling simulated for a business-as-usual CO_2 emission scenario until year 4000 AD. *Glob. Biogeochem. Cycles* **2008**, *22*, 1–21. [CrossRef]
2. Caldeira, K.; Wickett, M.E. Ocean model predictions of chemistry changes from carbon dioxide emissions to the atmosphere and ocean. *J. Geophys. Res.* **2005**, *110*, C09S04. [CrossRef]
3. Feely, R.A.; Sabine, C.L.; Lee, K.; Berelson, W.; Kleypas, J.; Fabry, V.J.; Millero, F.J. Impact of anthropogenic CO_2 on the $CaCO_3$ system in the oceans. *Science* **2004**, *305*, 362–366. [CrossRef]
4. Barros, P.; Sobral, P.; Range, P.; Chícharo, L.; Matias, D. Effects of sea-water acidification on fertilization and larval development of the oyster *Crassostrea gigas*. *J. Exp. Mar. Biol. Ecol.* **2013**, *440*, 200–206. [CrossRef]
5. Harvey, B.P.; Gwynn-Jones, D.; Moore, P.J. Meta-analysis reveals complex marine biological responses to the interactive effects of ocean acidification and warming. *Ecol. Evol.* **2013**, *3*, 1016–1030. [CrossRef] [PubMed]

6. Rato, L.D.; Novais, S.C.; Lemos, M.F.L.; Alves, L.M.F.; Leandro, S.M. *Homarus gammarus* (Crustacea: Decapoda) larvae under an ocean acidification scenario: Responses across different levels of biological organization. *Comp. Biochem. Physiol. Part C Toxicol. Pharmacol.* **2017**, *203*, 29–38. [CrossRef] [PubMed]
7. Wei, L.; Wang, Q.; Wu, H.; Ji, C.; Zhao, J. Proteomic and metabolomic responses of Pacific oyster *Crassostrea gigas* to elevated pCO_2 exposure. *J. Proteom.* **2015**, *112*, 83–94. [CrossRef]
8. Fabry, V.J.; Seibel, B.A.; Feely, R.A.; Orr, J.C. Impacts of ocean acidification on marine fauna and ecosystem processes. *ICES J. Mar. Sci.* **2008**, *65*, 414–432. [CrossRef]
9. Rost, B.; Zondervan, I.; Wolf-Gladrow, D. Sensitivity of phytoplankton to future changes in ocean carbonate chemistry: Current knowledge, contradictions and research directions. *Mar. Ecol. Prog. Ser.* **2008**, *373*, 227–237. [CrossRef]
10. Rodríguez, A.; Hernández, J.C.; Brito, A.; Clemente, S. Effects of ocean acidification on juvenile sea urchins: Predator-prey interactions. *J. Exp. Mar. Biol. Ecol.* **2017**, *493*, 31–40. [CrossRef]
11. IPCC. *Summary for Policymakers, Climate Change 2014: Synthesis Report. Contribution of Working Groups I, II and III to the Fifth Assessment Report of the Intergovernmental Panel on Climate Change*; Cambridge University Press: Cambridge, UK, 2014. [CrossRef]
12. IEA. Carbon capture and storage: Model Regulatory Framework. *JAMA* **2010**, *303*, 1601. [CrossRef]
13. Kirchsteiger, C. Carbon capture and storage-desirability from a risk management point of view. *Saf. Sci.* **2008**, *46*, 1149–1154. [CrossRef]
14. Liu, X.; Godbole, A.; Lu, C.; Michal, G.; Venton, P. Source strength and dispersion of CO_2 releases from high-pressure pipelines: CFD model using real gas equation of state. *Appl. Energy* **2008**, *126*, 56–68. [CrossRef]
15. Dewar, M.; Wei, W.; McNeil, D.; Chen, B. Small-scale modelling of the physiochemical impacts of CO_2 leaked from sub-seabed reservoirs or pipelines within the North Sea and surrounding waters. *Mar. Pollut. Bull.* **2013**, *73*, 504–515. [CrossRef]
16. Leung, D.Y.C.; Caramanna, G.; Maroto-Valer, M.M. An overview of current status of carbon dioxide capture and storage technologies. *Renew. Sustain. Energy Rev.* **2014**, *39*, 426–443. [CrossRef]
17. Hall-Spencer, J.M.; Rodolfo-Metalpa, R.; Martin, S.; Ransome, E.; Fine, M.; Turner, S.M.; Rowley, S.J.; Tedesco, D.; Buia, M.C. Volcanic carbon dioxide vents show ecosystem effects of ocean acidification. *Nature* **2008**, *454*, 96–99. [CrossRef]
18. McGinnis, D.F.; Schmidt, M.; Delsontro, T.; Themann, S.; Rovelli, L.; Reitz, A.; Linke, P. Discovery of a natural CO_2 seep in the German North Sea: Implications for shallow dissolved gas and seep detection. *J. Geophys. Res. Ocean.* **2011**, *116*, 1–12. [CrossRef]
19. Canfield, D.E. Organic Matter Oxidation in Marine Sediments. In *Interactions of C, N, P and S Biogeochemical Cycles and Global Change*; Springer: Berlin/Heidelberg, Germany, 1993; pp. 333–363. [CrossRef]
20. Santana-Casiano, J.M.; González-Dávila, M.; Fraile-Nuez, E.; De Armas, D.; González, A.G.; Domínguez-Yanes, J.F.; Escánez, J. The natural ocean acidification and fertilization event caused by the submarine eruption of El Hierro. *Sci. Rep.* **2013**, *3*, 5–12. [CrossRef] [PubMed]
21. Leal, V.; Amonte, C.; Melián, G.V.; Meire Feijoo, A.; Di Nardo, D.; Pitti Pimienta, L.; Santana de León, J.M.; Rojas, S.; Barbero, L.; Pérez, N.M.; et al. Hydrogeochemical temporal variations related to the recent volcanic eruption at Cumbre Vieja volcano, La Palma, Canary Islands. In Proceedings of the EGU General Assembly 2022, Vienna, Austria, 23–27 May 2022. EGU22-8817. [CrossRef]
22. Bibby, R.; Widdicombe, S.; Parry, H.; Spicer, J.; Pipe, R. Effects of ocean acidification on the immune response of the blue mussel *Mytilus edulis*. *Aquat. Biol.* **2008**, *2*, 67–74. [CrossRef]
23. Blackford, J.; Stahl, H.; Bull, J.M.; Bergès, B.J.P.; Cevatoglu, M.; Lichtschlag, A.; Connelly, D.; James, R.H.; Kita, J.; Long, D.; et al. Detection and impacts of leakage from sub-seafloor deep geological carbon dioxide storage. *Nat. Clim. Chang.* **2014**, *4*, 1011–1016. [CrossRef]
24. Jones, D.G.; Beaubien, S.E.; Blackford, J.C.; Foekema, E.M.; Lions, J.; De Vittor, C.; West, J.M.; Widdicombe, S.; Hauton, C.; Queirós, A.M. Developments since 2005 in understanding potential environmental impacts of CO_2 leakage from geological storage. *Int. J. Greenh. Gas Control* **2015**, *40*, 350–377. [CrossRef]
25. Tait, K.; Stahl, H.; Taylor, P.; Widdicombe, S. Rapid response of the active microbial community to CO_2 exposure from a controlled sub-seabed CO_2 leak in Ardmucknish Bay (Oban, Scotland). *Int. J. Greenh. Gas Control* **2015**, *38*, 171–181. [CrossRef]
26. Widdicombe, S.; Beesley, A.; Berge, J.A. Impact of elevated levels of CO_2 on animal mediated ecosystem function: The modification of sediment nutrient fluxes by burrowing urchins. *Mar. Pollut. Bull.* **2013**, *73*, 416–427. [CrossRef] [PubMed]
27. Borrero-Santiago, A.R.; Carbú, M.; DelValls, T.A.; Riba, I. CO_2 leaking from sub-seabed storage: Responses of two marine bacteria strains. *Mar. Environ. Res.* **2016**, *121*, 2–8. [CrossRef] [PubMed]
28. Borrero-Santiago, A.R.; Ribicic, D.; Bonnail, E.; Netzer, R.; Koseto, D.; Ardelan, M.V. Response of bacterial communities in Barents Sea sediments in case of a potential CO_2 leakage from carbon reservoirs. *Mar. Environ. Res.* **2020**, *160*, 105050. [CrossRef]
29. Passarelli, M.C.; Riba, I.; Cesar, A.; Serrano-Bernando, F.; DelValls, T.A. Assessing the influence of ocean acidification to marine amphipods: A comparative study. *Sci. Total Environ.* **2017**, *595*, 759–768. [CrossRef]
30. Szalaj, D.; De Orte, M.R.; Goulding, T.A.; Medeiros, I.D.; DelValls, T.A.; Cesar, A. The effects of ocean acidification and a carbon dioxide capture and storage leak on the early life stages of the marine mussel *Perna perna* (Linneaus, 1758) and metal bioavailability. *Environ. Sci. Pollut. Res.* **2017**, *24*, 765–781. [CrossRef]
31. da Silva Souza, L.; Bonnail, E.; Luzzi, J.A.; Cesar, A.; DelValls, T.A.; Pereira, C.D.S. Could Acidified environments intensify illicit drug effects on the reproduction of marine mussels? *Appl. Sci.* **2022**, *12*, 11204. [CrossRef]

32. Bautista-Chamizo, E.; Sendra, M.; De Orte, M.R.; Riba, I. Comparative effects of seawater acidification on microalgae: Single and multispecies toxicity tests. *Sci. Total Environ.* **2019**, *649*, 224–232. [CrossRef]
33. Rodríguez-Romero, A.; Jiménez-Tenorio, N.; Basallote, M.D.; De Orte, M.R.; Blasco, J.; Riba, I. Predicting the impacts of CO_2 leakage from subseabed storage: Effects of metal accumulation and toxicity on the model benthic organism *Ruditapes philippinarum*. *Environ. Sci. Technol.* **2014**, *48*, 12292–12301. [CrossRef]
34. Sun, T.; Tang, X.; Jiang, Y.; Wang, Y. Seawater acidification induced immune function changes of haemocytes in *Mytilus edulis*: A comparative study of CO_2 and HCl enrichment. *Sci. Rep.* **2017**, *7*, 41488. [CrossRef]
35. Conradi, M.; Riba, I.; Almagro-Pastor, V.; DelValls, T.A. Lethal and sublethal responses in the clam *Scrobicularia plana* exposed to different CO_2-acidic sediments. *Environ. Res.* **2016**, *151*, 642–652. [CrossRef] [PubMed]
36. Ishimatsu, A.; Hayashi, M.; Kikkawa, T. Fishes in high-CO_2, acidified oceans. *Mar. Ecol. Prog. Ser.* **2008**, *373*, 295–302. [CrossRef]
37. Sokolova, I.I.; Matoo, O.B.; Dickinson, G.H.; Beniash, E. Physiological and ecological responses. In *Stressors in the Marine Environment*; Oxford Scholarship: Oxford, UK, 2016. [CrossRef]
38. Heuer, R.M.; Grosell, M. Physiological impacts of elevated carbon dioxide and ocean acidification on fish. *AJP Regul. Integr. Comp. Physiol.* **2014**, *307*, R1061–R1084. [CrossRef] [PubMed]
39. Fabry, V.J. Marine Calcifiers in a High-CO_2 Ocean. *Science* **2008**, *320*, 1020–1022. [CrossRef]
40. Pörtner, H.O. Ecosystem effects of ocean acidification in times of ocean warming: A physiologist's view. *Mar. Ecol. Prog. Ser.* **2008**, *373*, 203–217. [CrossRef]
41. Dupont, S.; Dorey, N.; Thorndyke, M. What meta-analysis can tell us about vulnerability of marine biodiversity to ocean acidification? *Estuar. Coast. Shelf Sci.* **2010**, *89*, 182–185. [CrossRef]
42. Hendriks, I.E.; Duarte, C.M.; Álvarez, M. Vulnerability of marine biodiversity to ocean acidification: A meta-analysis. *Estuar. Coast. Shelf Sci.* **2010**, *86*, 157–164. [CrossRef]
43. Cortez, F.S.; Seabra Pereira, C.D.; Santos, A.R.; Cesar, A.; Choueri, R.B.; Martini, G.D.A.; Bohrer-Morel, M.B. Biological effects of environmentally relevant concentrations of the pharmaceutical Triclosan in the marine mussel *Perna perna* (Linnaeus, 1758). *Environ. Pollut.* **2012**, *168*, 145–150. [CrossRef]
44. Maranho, L.A.; Pereira, C.D.S.; Choueri, R.B.; Cesar, A.; Gusso-Choueri, P.K.; Torres, R.J.; Abessa, D.M.D.S.; Morais, R.D.; Mozeto, A.A.; Delvalls, T.A.; et al. The application of biochemical responses to assess environmental quality of tropical estuaries: Field surveys. *J. Environ. Monit.* **2012**, *14*, 2608–2615. [CrossRef]
45. Moschino, V.; Delaney, E.; Da Ros, L. Assessing the significance of *Ruditapes philippinarum* as a sentinel for sediment pollution: Bioaccumulation and biomarker responses. *Environ. Pollut.* **2012**, *171*, 52–60. [CrossRef]
46. Moore, M.T.; Lizotte, R.E.; Smith, S. Responses of *Hyalella azteca* to a pyrethroid mixture in a constructed wetland. *Bull. Environ. Contam. Toxicol.* **2007**, *78*, 245–248. [CrossRef] [PubMed]
47. Liu, C.W.; Liang, C.P.; Lin, K.H.; Jang, C.S.; Wang, S.W.; Huang, Y.K.; Hsueh, Y.M. Bioaccumulation of arsenic compounds in aquacultural clams (*Meretrix lusoria*) and assessment of potential carcinogenic risks to human health by ingestion. *Chemosphere* **2007**, *69*, 128–134. [CrossRef] [PubMed]
48. Maranho, L.A.; DelValls, T.A.; Martín-Díaz, M.L. Assessing potential risks of wastewater discharges to benthic biota: An integrated approach to biomarker responses in clams (*Ruditapes philippinarum*) exposed under controlled conditions. *Mar. Pollut. Bull.* **2015**, *92*, 11–24. [CrossRef]
49. Riba, I.L.; DelValls, T.A.; Forja, J.M.; Gómez-Parra, A. The influence of pH and salinity on the toxicity of heavy metals in sediment to the estuarine clam *Ruditapes philippinarum*. *Environ. Toxicol. Chem.* **2004**, *23*, 1100–1107. [CrossRef] [PubMed]
50. Bonnail, E.; Borrero-Santiago, A.R.; Nordtug, T.; Øverjordet, I.B.; Krause, D.F.; Ardelan, M.V. Climate change mitigation effects: How do potential CO_2 leaks from a sub-seabed storage site in the Norwegian Sea affect *Astarte* sp. bivalves? *Chemosphere* **2021**, *264*, 128552. [CrossRef] [PubMed]
51. de Lafontaine, Y.; Gagné, F.; Blaise, C.; Costan, G.; Gagnon, P.; Chan, H. Biomarkers in zebra mussels (*Dreissena polymorpha*) for the assessment and monitoring of water quality of the St Lawrence River (Canada). *Aquat. Toxicol.* **2000**, *50*, 51–71. [CrossRef] [PubMed]
52. Gagné, F.; Blaise, C.; André, C.; Gagnon, C.; Salazar, M. Neuroendocrine disruption and health effects in *Elliptio complanata* mussels exposed to aeration lagoons for wastewater treatment. *Chemosphere* **2007**, *68*, 731–743. [CrossRef]
53. Parolini, M.; Pedriali, A.; Riva, C.; Binelli, A. Sub-lethal effects caused by the cocaine metabolite benzoylecgonine to the freshwater mussel *Dreissena polymorpha*. *Sci. Total Environ.* **2013**, *444*, 43–50. [CrossRef]
54. Pereira, C.D.S.; Abessa, D.M.D.S.; Bainy, A.C.D.; Zaroni, L.P.; Gasparro, M.R.; Bícego, M.C.; Taniguchi, S.; Furley, T.H.; De Sousa, E.C.P.M. Integrated assessment of multilevel biomarker responses and chemical analysis in mussels from São Sebastião, São Paulo, Brazil. *Environ. Toxicol. Chem.* **2007**, *26*, 462–469. [CrossRef]
55. Edge, K.J.; Johnston, E.L.; Roach, A.C.; Ringwood, A.H. Indicators of environmental stress: Cellular biomarkers and reproductive responses in the Sydney rock oyster (*Saccostrea glomerata*). *Ecotoxicology* **2012**, *21*, 1415–1425. [CrossRef]
56. Patterson, H.K.; Boettcher, A.; Carmichael, R.H. Biomarkers of dissolved oxygen stress in oysters: A tool for restoration and management efforts. *PLoS ONE* **2014**, *9*, e104440. [CrossRef] [PubMed]
57. Zanette, J.; Monserrat, J.M.; Bianchini, A. Biochemical biomarkers in gills of mangrove oyster *Crassostrea rhizophorae* from three Brazilian estuaries. *Comp. Biochem. Physiol. C Toxicol. Pharmacol.* **2006**, *143*, 187–195. [CrossRef] [PubMed]

58. His, E.; Beiras, R.; Seaman, M.N.L. The Assessment of Marine Pollution-Bioassays with Bivalve Embryos and Larvae. *Adv. Mar. Biol.* **1999**, *37*, 1–178. [CrossRef]
59. Marigómez, I.; Garmendia, L.; Soto, M.; Orbea, A.; Izagirre, U.; Cajaraville, M.P. Marine ecosystem health status assessment through integrative biomarker indices: A comparative study after the Prestige oil spill "mussel Watch". *Ecotoxicology* **2013**, *22*, 486–505. [CrossRef]
60. Tangwancharoen, S.; Burton, R.S. Early life stages are not always the most sensitive: Heat stress responses in the copepod *Tigriopus californicus*. *Mar. Ecol. Prog. Ser.* **2014**, *517*, 75–83. [CrossRef]
61. Viarengo, A.; Lowe, D.; Bolognesi, C.; Fabbri, E.; Koehler, A. The use of biomarkers in biomonitoring: A 2-tier approach assessing the level of pollutant-induced stress syndrome in sentinel organisms. *Comp. Biochem. Physiol. C Toxicol. Pharmacol.* **2007**, *146*, 281–300. [CrossRef]
62. Passarelli, M.C.; Bonnail, E.; Cesar, A.; DelValls, T.Á.; Riba, I. Integrative Assessment of Sediments Affected by CO_2 Enrichment: A Case Study in the Bay of Santos—SP, Brazil. *Appl. Sci.* **2021**, *11*, 11603. [CrossRef]
63. de Orte, M.R.; Sarmiento, A.M.; Basallote, M.D.; Rodríguez-Romero, A.; Riba, I.; DelValls, A. Effects on the mobility of metals from acidification caused by possible CO_2 leakage from sub-seabed geological formations. *Sci. Total Environ.* **2014**, *470–471*, 356–363. [CrossRef]
64. de Orte, M.R.; Sarmiento, A.M.; DelValls, T.A.; Riba, I. Simulation of the potential effects of CO_2 leakage from carbon capture and storage activities on the mobilization and speciation of metals. *Mar. Pollut. Bull.* **2014**, *86*, 59–67. [CrossRef]
65. da Silva Souza, L.; Bonnail, E.; Maranho, L.A.; Pusceddu, F.H.; Cortez, F.S.; Cesar, A.; Ribeiro, D.A.; Riba, I.; de Souza Abessa, D.M.; DelValls, Á.; et al. Sub-lethal combined effects of illicit drug and decreased pH on marine mussels: A short-time exposure to crack cocaine in CO_2 enrichment scenarios. *Mar. Pollut. Bull.* **2022**, *171*, 112735. [CrossRef] [PubMed]
66. CETESB. *Companhia Ambiental do Estado de São Paulo—Qualidade das Praias litorâneas no Estado de São Paulo*; CETESB: Sao Paulo, Brazil, 2016.
67. Lowe, D.M.; Pipe, R.K. Contaminant induced lysosomal membrane damage in marine mussel digestive cells: An in vitro study. *Aquat. Toxicol* **1994**, *30*, 357–365. [CrossRef]
68. Wills, E.D. Evaluation of lipid peroxidation in lipids and biological membranes. In *Biochemical Toxicology: A Practical Approach*; Oxford University Press: Oxford, UK, 1987; pp. 127–150.
69. Olive, P.L. DNA precipitation assay: A rapid and simple method for detecting DNA damage in mammalian cells. *Environ. Mol. Mutagen.* **1988**, *11*, 487–495. [CrossRef] [PubMed]
70. Bradford, M.M. A rapid and sensitive method for the quantitation of microgram quantities of protein utilizing the principle of protein-dye binding. *Anal. Biochem.* **1987**, *72*, 248–254. [CrossRef]
71. Field, A. *Discovering Statistics Using IBM SPSS Statistics*, 4th ed.; Sage Publications: Thousand Oaks, CA, USA, 2013; 915p.
72. Ihaka, R.; Gentleman, R. R: A language for data analysis and graphics. *J. Comput. Graph. Stat.* **1996**, *5*, 299–314. [CrossRef]
73. Ben-Shachar, M.; Lüdecke, D.; Makowski, D. Effectsize: Estimation of Effect Size Indices and Standardized Parameters. *J. Open Source Softw.* **2020**, *5*, 2815. [CrossRef]
74. Hubbard, F.; McManus, J.; Al-Dabbas, M. Environmental influences on the shell mineralogy of *Mytilus edulis*. *Geo-Mar. Lett.* **1981**, *1*, 267–269. [CrossRef]
75. Beesley, A.; Lowe, D.M.; Pascoe, C.K.; Widdicombe, S. Effects of CO_2-induced seawater acidification on the health of *Mytilus edulis*. *Clim. Res.* **2008**, *37*, 215–225. [CrossRef]
76. Thomsen, J.; Ramesh, K.; Sanders, T.; Bleich, M.; Melzner, F. Calcification in a marginal sea—Influence of seawater [Ca^{2+}] and carbonate chemistry on bivalve shell formation. *Biogeosciences* **2018**, *15*, 1469–1482. [CrossRef]
77. Freitas, R.; Almeida, Â.; Calisto, V.; Velez, C.; Moreira, A.; Schneider, R.J.; Esteves, V.I.; Wrona, F.J.; Soares, A.M.V.M.; Figueira, E. How life history influences the responses of the clam *Scrobicularia plana* to the combined impacts of carbamazepine and pH decrease. *Environ. Pollut.* **2015**, *202*, 205–214. [CrossRef]
78. Dailianis, S.; Domouhtsidou, G.P.; Raftopoulou, E.; Kaloyianni, M.; Dimitriadis, V.K. Evaluation of neutral red retention assay, micronucleus test, acetylcholinesterase activity and a signal transduction molecule (cAMP) in tissues of *Mytilus galloprovincialis* (L.), in pollution monitoring. *Mar. Environ. Res.* **2003**, *56*, 443–470. [CrossRef]
79. Binelli, A.; Cogni, D.; Parolini, M.; Riva, C.; Provini, A. In vivo experiments for the evaluation of genotoxic and cytotoxic effects of Triclosan in Zebra mussel hemocytes. *Aquat. Toxicol.* **2009**, *91*, 238–244. [CrossRef]
80. dos Santos Barbosa Ortega, A.; Maranho, L.A.; Nobre, C.R.; Moreno, B.B.; Guimarães, R.S.; Lebre, D.T.; de Souza Abessa, D.M.; Ribeiro, D.A.; Pereira, C.D.S. Detoxification, oxidative stress, and cytogenotoxicity of crack cocaine in the brown mussel *Perna perna*. *Environ. Sci. Pollut. Res.* **2019**, *26*, 27569–27578. [CrossRef] [PubMed]
81. Nicholson, S. Lysosomal membrane stability, phagocytosis and tolerance to emersion in the mussel *Perna viridis* (Bivalvia: Mytilidae) following exposure to acute, sublethal, copper. *Chemosphere* **2003**, *52*, 1147–1151. [CrossRef] [PubMed]
82. Martin-Diaz, L.; Franzellitti, S.; Buratti, S.; Valbonesi, P.; Capuzzo, A.; Fabbri, E. Effects of environmental concentrations of the antiepilectic drug carbamazepine on biomarkers and cAMP-mediated cell signaling in the mussel *Mytilus galloprovincialis*. *Aquat. Toxicol.* **2009**, *94*, 177–185. [CrossRef] [PubMed]
83. Aguirre-Martínez, G.V.; Buratti, S.; Fabbri, E.; DelValls, T.A.; Martín-Díaz, M.L. Using lysosomal membrane stability of haemocytes in *Ruditapes philippinarum* as a biomarker of cellular stress to assess contamination by caffeine, ibuprofen, carbamazepine and novobiocin. *J. Environ. Sci.* **2013**, *25*, 1408–1418. [CrossRef]

84. OSPAR. *Background Document and Technical Annexes for Biological Effects Monitoring*; Update 2013 Monitoring and Assessment Series; OSPAR: London, UK, 2013.
85. Pereira, C.D.S.; Martin-Díaz, M.L.; Zanette, J.; Cesar, A.; Choueri, R.B.; de Souza Abessa, D.M.; Catharino, M.G.M.; Vasconcellos, M.B.A.; Bainy, A.C.D.; de Sousa, E.C.P.M.; et al. Integrated biomarker responses as environmental status descriptors of a coastal zone (São Paulo, Brazil). *Ecotoxicol. Environ. Saf.* **2011**, *74*, 1257–1264. [CrossRef]
86. Abessa, D.M.D.S.; Zaroni, L.P.; De Sousa, E.C.P.M.; Gasparro, M.R.; Pereira, C.D.S.; De Figueredo Rachid, B.R.; Depledge, M.; King, R.S. Physiological and cellular responses in two populations of the mussel *Perna perna* collected at different sites from the Coast of São Paulo, Brazil. *Braz. Arch. Biol. Technol.* **2005**, *48*, 217–225. [CrossRef]
87. Souza, A.; Moreno, B.B.; Almeida, J.E.; Rogero, S.O.; Pereira, C.D.S.; Rogero, J.R. Cytotoxicity evaluation of Amoxicillin and Potassium Clavulanate in *Perna perna* mussels. *Ecotoxicol. Environ. Contam.* **2016**, *11*, 21–26. [CrossRef]
88. Allen, J.I.; Moore, M.N. Environmental prognostics: Is the current use of biomarkers appropriate for environmental risk evaluation? *Mar. Environ. Res.* **2004**, *58*, 227–232. [CrossRef]
89. Ringwood, A.H.; Hoguet, J.; Keppler, C.; Gielazyn, M. Linkages between cellular biomarker responses and reproductive success in oysters—*Crassostrea virginica*. *Mar. Environ. Res.* **2004**, *58*, 151–155. [CrossRef]
90. Marchi, B.; Burlando, B.; Moore, M.N.; Viarengo, A. Mercury- and copper-induced lysosomal membrane destabilisation depends on [Ca^{2+}] independent phospholipase A2 activation. *Aquat. Toxicol.* **2004**, *66*, 197–204. [CrossRef]
91. Jones, R.G.; Waufeh, I. Calcium-containing lysosomes in the outer mantle epithelial cells of Amblema, a fresh-water mollusc. *Anat. Rec.* **1982**, *203*, 337–343. [CrossRef]
92. Velez, C.; Figueira, E.; Soares, A.M.V.M.; Freitas, R. Native and introduced clams biochemical responses to salinity and pH changes. *Sci. Total Environ.* **2016**, *566–567*, 260–268. [CrossRef]
93. Freitas, R.; De Marchi, L.; Bastos, M.; Moreira, A.; Velez, C.; Chiesa, S.; Wrona, F.J.; Figueira, E.; Soares, A.M.V.M. Effects of seawater acidification and salinity alterations on metabolic, osmoregulation and oxidative stress markers in *Mytilus galloprovincialis*. *Ecol. Indic.* **2017**, *79*, 54–62. [CrossRef]
94. Halliwell, B.; Gutteridge, J.M.C. *Free Radicals in Biology and Medicine*, 4th ed.; Oxford University Press: New York, NY, USA, 2007.
95. Regoli, F.; Winston, G.W. Quantification of Total Oxidant Scavenging Capacity of Antioxidants for Peroxynitrite, Peroxyl Radicals, and Hydroxyl Radicals. *Toxicol. Appl. Pharmacol.* **1999**, *156*, 96–105. [CrossRef] [PubMed]
96. Cao, R.; Liu, Y.; Wang, Q.; Zhang, Q.; Yang, D.; Liu, H.; Qu, Y.; Zhao, J. The impact of ocean acidification and cadmium on the immune responses of Pacific oyster, *Crassostrea gigas*. *Fish Shellfish Immunol.* **2018**, *81*, 456–462. [CrossRef] [PubMed]
97. Timmins-Schiffman, E.; Coffey, W.D.; Hua, W.; Nunn, B.L.; Dickinson, G.H.; Roberts, S.B. Shotgun proteomics reveals physiological response to ocean acidification in *Crassostrea gigas*. *BMC Genom.* **2014**, *15*, 951. [CrossRef] [PubMed]
98. Cuervo, A.M. Autophagy: In sickness and in health. *Trends Cell Biol.* **2004**, *14*, 70–77. [CrossRef] [PubMed]

applied sciences

MDPI

Article

Assessment of the Accumulation of Trace Metals and Oxidative Stress Response Biomarkers in the Portunid *Portunus segnis*

Amel Dghim, Walid Ben Ameur and Ali Annabi *

Ecologie de la Faune Terrestre UR17ES44, Département des Sciences de la vie, Faculté des Sciences de Gabès, Université de Gabès, Gabès 6029, Tunisia; dsirine51@gmail.com (A.D.); benameurwalid@gmail.com (W.B.A.)
* Correspondence: medannabi@yahoo.fr or ali.annabi@isbm.rnu.tn

Abstract: The invasive blue crab *Portunus segnis*, which was collected from two sites on the Gulf of Gabès, is the subject of this work. This study is based on demonstrating the accumulation capacity of *P. segnis* by measuring the concentrations of cadmium, zinc, lead, and copper in the gills and hepatopancreas. The enzymatic activities of catalase, glutathione-S-transferase, reduced glutathione, and lipid peroxidase were assessed in this region for the first time. The main results show that the metals have high bioaccumulation potentials in *P. segnis* tissues between different sites. The possible adaptation of *P. segnis* in the Gulf of Gabès and the variations in the studied biomarkers and metal concentrations at different sites confirm the usefulness of the invasive blue crab as a sentinel species.

Keywords: aquatic pollution; biomarkers; heavy metals; *Portunus segnis*

1. Introduction

The aquatic environment is the ultimate container for pollutants. These include metal–elemental inorganic fertilizers that enter the aquatic environment, often through atmospheric deposition, the erosion of the geological matrix, or as a result of anthropogenic activities (e.g., the discharge of industrial, domestic, and mining wastes) and the use of pesticides and pesticides [1]. Metal source pollution is one of the greatest threats facing the world today. In fact, heavy metals pose serious ecological problems due to their toxicity and potential to accumulate in a variety of aquatic organisms, thereby causing damaging effects on the ecological balance of the aquatic environment. Therefore, this situation requires awareness of the vulnerability of these environments, regular monitoring, and the implementation of preventive measures.

However, the sound and scientific use of biological indicators, especially in environmental biological assessments (monitoring the state of the environment or the effectiveness of compensation or restoration measures) has not been recognized until recently. This discipline was originally based on the use of biological indicators, which can provide information about the contamination of the environment by chemicals. On the other hand, the identification of the bioavailability of contaminants and their accumulation in the tissues of these organisms requires the use of certain ecotoxicological tools such as biomarkers. The purpose of these tools is to provide information on the harmful effects of environmental exposures manifested at the individual level. However, due to the abundance of pollutants, the complexity of their effects, and the many interactions between them, environmental diagnosis must be based on the use of a multi-biomarker approach [2–4].

Like polluted aquatic environments, the Mediterranean marine environment is one of the most polluted semi-enclosed basins in the world, especially with respect to metallic pollutants. According to previous studies, about 49–82% of copper, 68–76% of zinc, 21–65% of lead, and 75–92% of cadmium are deposited in the Mediterranean Sea in soluble forms and can thus be easily accumulated by organisms [5].

One of the nations in the Mediterranean region in which the importance of metals to aquatic environments is growing is Tunisia, particularly in coastal regions where chemical

Citation: Dghim, A.; Ben Ameur, W.; Annabi, A. Assessment of the Accumulation of Trace Metals and Oxidative Stress Response Biomarkers in the Portunid *Portunus segnis*. *Appl. Sci.* **2023**, *13*, 7197. https://doi.org/10.3390/app13127197

Academic Editor: Ramaraj Boopathy

Received: 5 May 2023
Revised: 1 June 2023
Accepted: 13 June 2023
Published: 16 June 2023

manufacturing is most significant. Numerous studies revealed that across these coastal areas, the Gulf of Gabès constitutes an ecosystem under intense industrial stress and is therefore likely to be polluted with various contaminants such as metallic trace elements [6,7].

Numerous aquatic species have been applied in diverse ecotoxicological investigations to suggest the status of the pollution of ecosystems. Crustaceans, which are noted for their high accumulative potential, are among them [8]. The blue crab *Portunus segnis* (Forskl, 1775) was among the first lessepsian intruders to be identified in Egyptian waters 30 years after the Suez Canal was completed in 1868 [9]. It then established itself in coastal environments across the Levantine Basin, from Turkey to Egypt [10]. This crab species has been found in various Mediterranean locations such as the Aegean Sea, Sicily, and the northern Tyrrhenian Sea [11], as well as in Tunisian coastal waters and the surrounding region in recent years [10], such as Malta [12]. *P. segnis* was initially discovered in Tunisia in the Gulf of Gabès in 2014 [13]; it quickly spread throughout the Gulf, comprising Djerba Island and the northern beaches, namely, the Gulf of Hammamet, and it has also been found in Libya on occasion [14,15].

In this context, this work aims to use an invasive species of the Gulf of Gabès, *P. segnis*, as a bioindicator of the quality of the marine environment. This study aims to mirror the state of contamination of the aquatic surroundings in the Gulf of Gabès by pollutants through the determination of the heavy metal contents of the tissues of *P. segnis*, especially the gills and hepatopancreas, and the use of four biomarkers of pollution: catalase enzymatic activity, glutathione-S-transferase, reduced glutathione concentration, and an estimation of lipid peroxidation. To the best of our knowledge, this is the first study to biomonitor the two studied Tunisian aquatic ecosystems via combined chemical and biomarker analyses, using *Portunus segnis* as sentinel species.

2. Materials and Methods

2.1. Presentation of the Studied Sites

The Gulf of Gabès is located in the southeast of Tunisia and is characterized by a very wide continental shelf and a very gentle slope. It begins from Chebba (35.3° north latitude) in the north and ends at the Tunisia–Libya border in the south, with a total length of about 700 km. It is bordered by the Kerkennah Archipelago to the west, the shallow waters of the Kerkennah Islands to the south, and the mainland to the south. The Gulf basin is shallow, only 50 m deep at 110 km offshore.

The bottoms are generally sandy to muddy and are composed mainly of carbonate sediments [16]. The region is characterized by a pre-Saharan and arid to semi-arid climate, with generally high temperatures and a relatively low rainfall of around 200 mm per year [17]. In addition, in this gulf, the salinity of the seawater oscillates between 37.2 and 38 psu (practical salinity unit), with an average of 37.52 ± 0.29 psu. This is due to the low levels of precipitation, intense evaporation, and the low amount of runoff water [18]. Thus, as its geomorphological and climatic conditions are very favorable, the Gulf of Gabès is one of the most productive areas of the Mediterranean. Indeed, more than 42% of the production of the national fisheries occurs in this area [19].

2.2. Selection of Sampling Sites

The selection of sampling sites is a critical stage of any pollution impact assessment. For this purpose, two sites belonging to the Gulf of Gabès were selected, with different degrees of contamination (Figure 1):

- Ghannouch (respective geographical coordinates, 33°55′ N and 10°5′ E) is the area considered the most polluted due to the establishment of the chemical group, suggesting that the marine ecosystem and therefore the fauna and flora are threatened by liquid, solid, and gaseous industrial discharges [20].
- Sidi Salem (Djerba), the island of Djerba, is characterized by seasonal mass tourism, and the main sources of pollution are still anthropogenic waste.

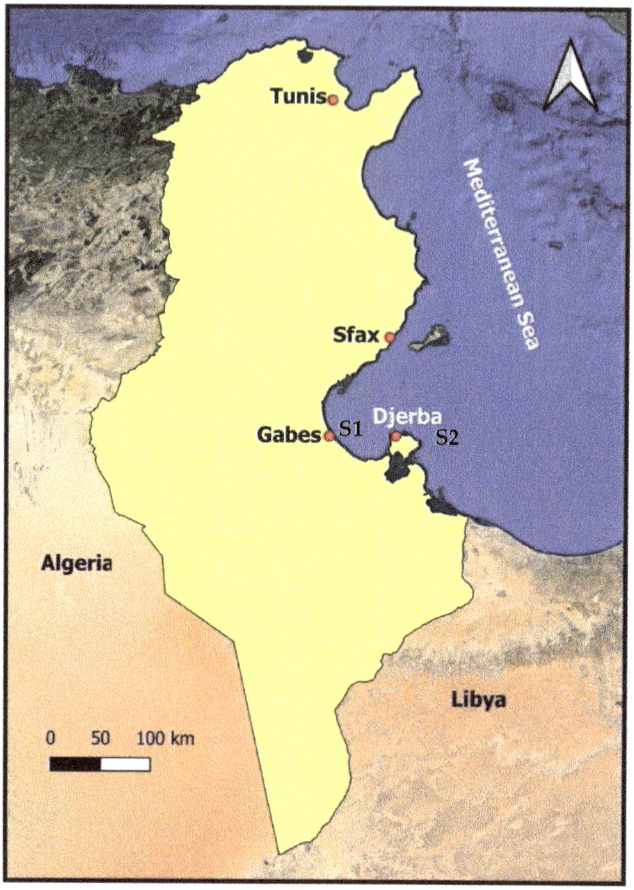

Figure 1. Location of *Portunus segnis* sampling sites in the Gulf of Gabès. S1: Ghannouch site (industrial area); S2: Djerba Site (touristic area).

Twenty adult specimens (n = 20) were collected from each area in November 2021. They were captured with the aid of neighborhood fishers around ten kilometers south of Gabès and around 20 km north of Djerba at a depth of roughly 1 m, which is the typical depth of a crab burrow. The live crabs were transported to the laboratory in buckets filled with aerated seawater as quickly as possible. Once the cephalothorax was opened on ice, the digestive glands and gills were quickly removed. The organs were dried and powdered for metal analysis, frozen at −20 °C, ground in a mortar with liquid N2, and stored at −80 °C until biochemical testing. Three pools were prepared from each sampling station for each collected tissue.

2.3. Chemical Analysis

The tissue samples were dried as the first phase of the technique by incubating them in a G-Therm115 oven at 65 °C for 24 h to maintain a constant dry weight. The samples were placed on spit for mineralization after the dehydration phase was finished, and the dry weight of each sample was determined.

The samples were placed on spit for mineralization after the dehydration phase was finished, and the dry weight of every specimen was determined. The method outlined by Annabi et al. [21] was used to analyze the chosen chemicals, Cd, Zn, Pb, and Cu. By using an acid attack, organic matter was removed during the mineralization step. After each sample had dried, not more than 10 mg was taken from it, and then 3 mL of concentrated nitric acid (HNO_3) was added. After being diluted to 30 mL with ultrapure water, the solution generated from the digestion process was maintained at 4 °C until it was analyzed via flame atomic absorption spectrometry, utilizing an atomic absorption spectrophotometer (Avanta GBC spectrometer, Australia) with an air–acetylene mixture.

Calibration curves were calculated using Agilent standard solutions for Cd, Pb, Cu, and Zn. They were prepared using serial dilutions of 3% nitric acid (HNO_3) and ultrapure water with a stock solution of 1000 mg L^{-1} of each metal. Working standards (Cd 0, 3, 5, and 15 µg/mL; Pb 0, 10, 15, and 20 µg/mL; Zn 0, 3, 5, 15, and 50 µg/mL; Cu 0, 15, 20, and 25 µg/mL) were used to calibrate the devices. Using external standards with coefficients for calibration curves greater than 0.99, the heavy metals of interest in this study were quantified. For every 10 samples analyzed, a blank solution was also analyzed. The samples were analyzed in triplicate (the variation coefficient was less than 10%), and the findings were computed using the mean and standard deviation.

The crab samples were spiked with known quantities of heavy metals to carry out the repeatability test and the verification of the analytical samples. The samples were examined in triplicate for each run. The same procedure utilized for the initial samples was employed to digest and evaluate the heavy-metal-spiked crab samples. Recoveries for the tested chemicals using the provided method ranged from 77 to 95%. Recovery correction was carried out for each sample.

Utilizing ultrapure water, aqueous solutions of the reagents and standards were prepared. Every chemical utilized was of the analytical reagent standard. Glass materials were cleaned and submerged for 24 h in a tank containing 3% nitric acid to prevent contamination. Each time a glass material was used, it was then cleaned with ultrapure water and dried in an oven.

2.4. Biomarker Analyses

Samples for enzymatic assays were weighed using a precision balance and then ground in a 0.1 M potassium phosphate buffer, pH 7.2, at a rate of 4 mL of buffer per gram of tissue. The cell suspension was then centrifuged (4000 rpm/min, 4 °C, 10 min), and the resulting supernatant was collected in tubes and stored at −20 °C until the biochemical assays were performed.

2.4.1. Tissue Preparation

Gill and hepatopancreas tissue samples were weighed for the oxidative biomarker analyses using a precision balance and then ground in a cold 0.1 M potassium phosphate buffer, pH 7.2, at a rate of 4 mL of buffer per gram of tissue. The centrifugation step of the cell suspension was performed at 4000 rpm/min and 4 °C for 10 min. After this step, the obtained supernatants were collected in tubes and stored at −20 °C until analysis.

2.4.2. Total Protein Quantification

The protein contents were determined using bovine serum albumin as a standard protein [22]. To perform this method, 2.5 mL of Bradford solution was mixed with 10 µL of enzyme extract. The mixture was then blended thoroughly and incubated at room temperature for 2 min. Finally, the absorbance was recorded at 595 nm, and the protein concentration is expressed in µg/µL.

2.4.3. Glutathione S-Transferase (GST) Determination

GST is an enzyme that aids in the detoxification of xenobiotics and/or protection against the toxic metabolites produced by the breakdown of macromolecules after exposure to oxidative stress. GSTs are phase II metabolic enzymes whose function is to conjugate to a glutathione molecule a wide variety of pollutants to allow for their elimination. These enzymes are therefore triggered by the presence of certain pollutants in the environment, such as metals, and are therefore considered biomarkers of pollution. In addition, the major role of GST is the mitigation of oxidative stress generated by trace metal elements (TMEs) [23]. The technique adopted used to assay GST activity was that of Habig et al. [24]. This consisted of presenting the enzyme contained inside the homogenate with a substrate (2,4-chloro dinitrobenzene (CDNB)) which reacts without problems with many forms of GST and glutathione. The conjugation reaction of these two products catalyzed by GST results in the formation of a new molecule, 1-glutathione-2,4-dinitrobenzene (GST-DNB), which absorbs light at a 340 nm wavelength. The value of the optical density measured is directly proportional to the amount of conjugate formed and is related to the intensity of GST activity. Assays were performed via the homogenization of 100 μL of the biological extract with 200 μL of 0.1 M NaPi (pH 6.5), 150 μL of 10 mM CDNB, and 900 μL of 10 mM glutathione. The obtained mixture was incubated at 37 °C, and then the absorbance was measured every minute at 340 nm with the extinction coefficient of 9.6 mM^{-1} cm^{-1}. The specific activity of GST is expressed in nanomoles of CDNB per minute per milligram of protein (nmol CDNB/min/mg protein).

2.4.4. Catalase (CAT) Determination

CAT is an enzyme specialized in the detoxification of hydrogen peroxide and its transformation into an oxygen and a water molecule. It is a biomarker of oxidative stress and is sensitive to certain contaminants that induce this type of stress, such as toxic metals [25]. The enzymatic activity of CAT was measured in the fractions of the gills and of the hepatopancreas, according to the method of Aebi [26], at 240 nm using a UV/visible spectrophotometer via the variation in the optical density, following the dismutation of hydrogen peroxide. The reaction mixture contained 950 μL of 0.05 M potassium phosphate buffer (pH 7), 500 μL of 30 mM H_2O_2, and 50 μL of enzyme extract. Every 20 s for 2 min, a spectrophotometer with an extinction value of 0.043 mM^{-1} cm^{-1} recorded the decline in absorbance at 240 nm. The specific activity of CAT is expressed in international units IU/mg of protein (μmoles of H_2O_2 destroyed/min/mg of protein).

2.4.5. Reduced Glutathione (GSH) Determination

The determination of reduced glutathione was carried via the colorimetric method using Ellman's reagent (DTNB) [27]. The principle is based on the oxidation reaction of GSH with 5,5-dithiodis-2-nitrobenzoic acid (DTNB), thus releasing thionitrobenzoic acid (TNB), which has an absorbance at 412 nm. To perform this method, a mixture of 500 μL of the homogenate and 500 μL of sulfosalicylic acid (4%) was centrifuged at 3500 rpm for 10 min. Subsequently, 200 μL of the supernatant was added to 1 mL of the phosphate buffer (0.1 M, pH 7.4) and 400 μL of 10 mM DTNB. The mixture was left for 5 min at room temperature, and the optical densities were read via a spectrophotometer at 412 nm with the extinction coefficient of 13.600 mM^{-1} cm^{-1}. The concentration of GSH is expressed in nanomoles per milligram of protein (nmol/mg of protein).

2.4.6. Malonedialdehyde Level (MDA) Determination

The MDA assay was carried out as described by Ohkawa [28], via the colorimetric method in the presence of thiobarbituric acid (TBA), which allows for the detection of small amounts of lipid peroxides and particularly free MDA. Thus, the detection of MDA present in biological samples was based on the reaction during which two molecules of TBA react with one molecule of MDA, which results in the formation of a red chromogen absorbing at 535 nm in an acidic medium. The intensity of the red coloration increases with the concentration of MDA. It should be noted that the MDA-TBA complex is only stable for 3 h in the dark and at room temperature. The operating mode was as follows: 100 mL of tissue extract was homogenized with 2 mL of butylated hydroxytoluene (BHT) in methanol, 50 mL of 0.37% TBA, 50 mL of 15% TCA, and 50 mL of 0.25 N HCl. Next, the obtained mixture was heated at 100 °C for 15 min. After the heating process, the mixture was chilled in ice for 10 min before being centrifuged at 3000 rpm for 10 min. At 535 nm, the optical density was read using the molar extinction coefficient (ε = 154 mM^{-1} cm^{-1}). The MDA concentrations are measured in nmol of MDA/mg of protein (nmol/mg).

2.5. Statistics

All biomarker determination data (GST, CAT, GSH, and MDA) and metal concentrations levels are shown as means ± SDs (standard deviations). The normality of the data was checked prior to statistical analysis. Using the StatView software tool (Statview 5.0), the findings were analyzed using a one-way ANOVA and Fisher's least significant difference technique (LSD). If $p < 0.05$, the value was judged statistically significant.

3. Results and Discussion

3.1. Metal Analysis

Figure 2a,b show the findings of the Cd, Cu, Zn, and Pb determinations in the gills and hepatopancreata of *P. segnis*. The studies confirmed the presence of these elements in the tissues of the target crab at heterogeneous levels. The Zn levels were higher at both study locations (Table 1). According to the findings, the heavy metal levels at the two research locations and in the two tissues (the gills and hepatopancreas) followed a concentration scale in the following decreasing order: Zn > Cd > Cu > Pb (Figure 2a,b). Statistical analyses tests for the levels of the four trace elements at the Ghannouch site revealed statistically significant differences in the *P. segnis* gills (F = 42.68; $p < 0.01$) and hepatopancreas (F = 42.68; $p < 0.01$) and in the hepatopancreas (F = 72.57; $p < 0.01$). In addition, the Zn, Cd, Cu and Pb contents at the Djerba site showed statistically significant differences for the gills (F = 20.31; $p < 0.01$) and hepatopancreas (F = 15.12; $p < 0.01$) (Table 1).

Table 1. Metal data (mg/kg dw) in *P. segnis* at the two studied sites for the gills and hepatopancreas.

Site	Tissues	Element	Mean ± SD	F	p
Ghannouch	Gills	Cd	19.21 ± 0.1	42.28	<0.01
		Cu	12.53 ± 0.078		
		Pb	10.15 ± 1.38		
		Zn	37.36 ± 12.57		
	Hepatopancreas	Cd	7.74 ± 0.14	72.57	<0.01
		Cu	6.96 ± 0.79		
		Pb	4.536 ± 0.24		
		Zn	18.14 ± 8.38		
Djerba	Gills	Cd	7.2 ± 0.049	20.31	<0.01
		Cu	6.45 ± 0.12		
		Pb	4.9 ± 0.01		
		Zn	10.72 ± 2.9		
	Hepatopancreas	Cd	3.18 ± 0.04	15.12	<0.01
		Cu	2.91 ± 0.08		
		Pb	2.21 ± 0.05		
		Zn	8.71 ± 1.03		

Figure 3a reveals that the highest concentrations in the gills were found in the individuals collected from Ghannouch, near the industrial zone of Gabès, for the four studied metals, Cd (19.21 ± 0.1 µg/g), Cu (12.53 ± 0.078 µg/g), Pb (10.15 ± 1.38 µg/g), and Zn (37.36 ± 12.57 µg/g) compared to the values noted in the individuals coming from Djerba, which were Cd (7.2 ± 0.049 µg/g), Cu (6.458 ± 0.122 µg/g), Pb (2.212 ± 0.054 µg/g), and Zn (10.722 ± 2.9 µg/g). For the values obtained from the hepatopancreas, the individuals from the Ghannouch site also presented the highest levels (Figure 3b) of metallic trace elements, Cd (7.74 ± 0.14 µg/g), Cu (6.96 ± 0.79 µg/g), Pb (4.536 ± 0.24 µg/g), and Zn (18.14 ± 8.38 µg/g), in comparison with the contents in the tissues of the individuals from Djerba, Cd (3.18 ± 0.04 µg/g), Cu (2.91 ± 0.087 µg/g), Pb (2.212 ± 0.054 µg/g), and Zn (8.71 ± 1.03 µg/g). In fact, the statistical analyses between the two study stations showed statistically significant differences in gill levels for Cd ($F = 23.32$; $p < 0.01$), Cu ($F = 92.35$; $p < 0.01$), Zn ($F = 62.24$; $p < 0.01$), and Pb ($F = 23.96$; $p < 0.01$), as well as for the levels in the hepatopancreas for Cd ($F = 74.39$; $p < 0.01$), Cu ($F = 17.11$; $p < 0.01$), Zn ($F = 54.31$; $p < 0.01$), and Pb ($F = 36.06$; $p < 0.01$) (Figure 3a,b).

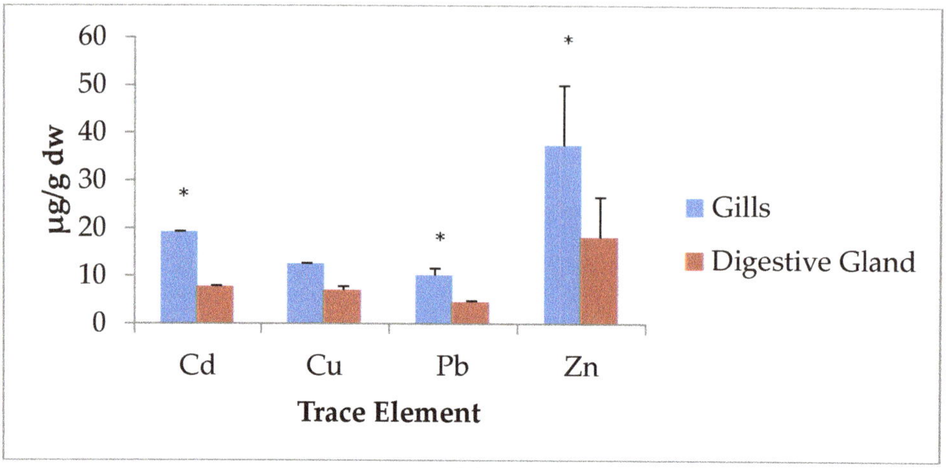

(a)

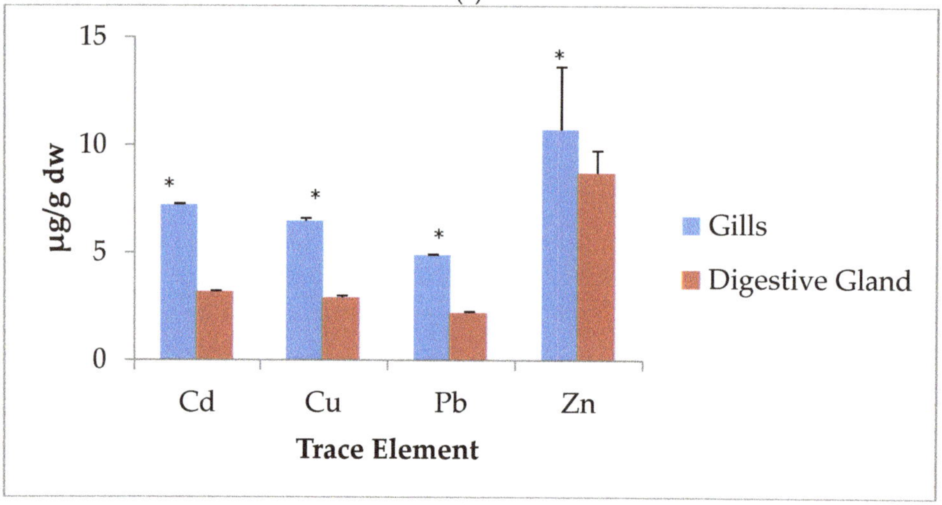

(b)

Figure 2. Heavy metal content s(Cd, Cu, Pb, and Zn) in *P. Segnis* tissues (n = 20) from Ghannouch site (**a**) and from Djerba site (**b**). *: statistically significant difference between the trace element contents in the gills and hepatopancreas ($p < 0.01$).

Only Italy [29] and Turkey [30,31] have provided data on metal levels in Mediterranean *Portunus segnis* populations. With a few exceptions, the majority of these research analyses were performed solely on soft tissues. The content of trace metal elements in crustaceans is estimated to be higher in the exoskeleton than in other tissues, as cadmium (with an ionic radius of 0.97 and 0.99 Å, respectively) and other elements such as manganese (Mn), lead, and zinc are gradually accumulated in the exoskeleton via calcium uptake routes [32]. Indeed, numerous investigations on brachyuran crustaceans have found that concentrations of metals in the exoskeleton are higher than in soft tissues [33,34].

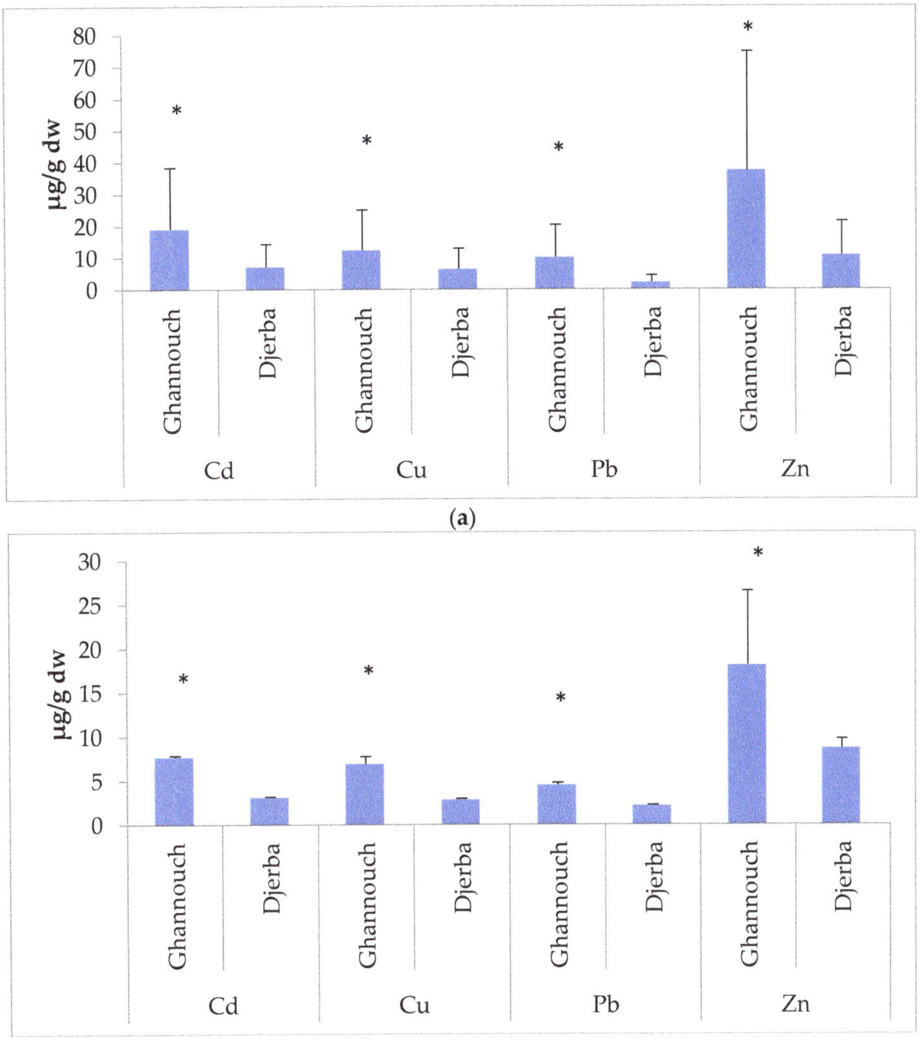

Figure 3. Trace element contents (μg/g dry weight) in the gills (**a**) and the digestive glands (**b**) between the two studied sites (n = 20). *: statistically significant difference for the metal contents between the two sites ($p < 0.01$).

Numerous environmental and biological factors, including molting, govern the trends of trace element bioaccumulation in crustaceans' soft tissues [35,36]. This is an important phase because it provides a detoxifying mechanism for heavy metals to be eliminated on a regular basis [37]. Metals have been shown to have distinct bioaccumulation patterns and can accumulate in various tissues at different concentrations. As an example, cadmium accumulated mostly in the midgut and gills, copper in the gills, and zinc in the gills/muscle of *Carcinus maenas* treated with metals for 32 days [38].

According to the results obtained in our study, the Cd and Pb levels were higher than the MOH, EC, and FAO standards [39–41], while for the other two elements (Cu and Zn), the values do not exceed the prescribed standards (Table 2). It should be noted that copper and zinc are essential for the metabolism and physiology of crustaceans, and their accumulation is generally independent of environmental concentrations [42,43]. However, when concentrations reach abnormally high levels, the ions accumulate in cytosolic proteins such as metallothionein [44]. The zinc concentrations (Table 2) in all studied samples in this study were higher than the concentrations of the other metals in the same samples of *P. segnis*. These high levels in crabs can be attributed to the biological interest in Zn for the organism. Indeed, Zn is an essential element for the metabolic process and is associated with certain Zinc transporter metalloproteins (such as metallothioneins). The physiological functions of these metalloproteins are related to their affinities for essential cations, especially copper and zinc, and to their antioxidant characteristics. They also play a major role in cellular homeostasis by controlling the bioavailability of certain essential metals.

Table 2. Data relating to some metal levels (mg/kg dw) in the tissues of aquatic organisms, according to standards [39–41].

Organism	Cu	Pb	Zn	Cd	References
Fish	-	0.3	-	0.05	[40]
Crustaceans	-	0.5	-	0.5	
Bivalves	-	1.5	-	1	
Fish	50	0.5	150	0.1	[39]
Crustaceans	50	0.5	150	0.5	
Bivalves	50	1.0	150	1.0	
Fish	30	0.5	30	0.05	[41]

Based on the data obtained in our study, we noted that the gill tissues at both studied sites had the highest concentrations of trace elements compared to the hepatopancreata (Table 3). These results are consistent with previous studies conducted on other crab species. A previous analysis [45] showed that *Potamonauteswarreni* crab gills contained more Cd and Cu than the hepatopancreas. Indeed, the work of Pan and Zhang [46] on the *Charybdis japonica* crab also revealed that the gills are more sensitive than the digestive glands to exposure to pollutants. Moreover, DNA strand breaks were positively correlated with Cd exposure only in the gill tissue. Our results are also consistent with the work performed by Corrêa et al. [47] on the *Ucides cordatus* crab. This work showed that metals accumulated in a concentration-dependent manner, with the highest levels found in the gills, followed by the hepatopancreas. Different species accumulate metals at different rates, depending on how they respond to metal exposure [48,49]. Within species that are closely related, as well as between genera, the processes of metal accumulation in aquatic crustaceans vary; some are net accumulators, whereas others are regulators. When an organism's rate of absorption exceeds its rate of excretion, net accumulation develops. The size of the individual organism can also have an impact on metal bioaccumulation in crabs. For both freshwater and marine crabs, smaller animals do indeed accumulate more metals [50,51].

Some work on the contamination of crab tissues by heavy metals has been reported in the literature. The values recorded in different regions of the world have been mentioned in Table 3. The concentrations of copper reported in our study in the gills of *P. segnis* individuals from Ghannouch and Djerba are lower than those reported in the samples of crabs collected from the Bizerte Lagoon, the Monastir Coast, and Gargour in Sfax [52,53]. The levels of the same element in the hepatopancreata of *P. segnis* in the two studied sites (Ghannouch and Djerba) are also lower than those recorded in crab samples from Canada [54], Australia [35], and the United States [55]. However, the concentration in the hepatopancreas samples of individuals from Ghannouch is similar to the concentration found in crab samples from Brazil [56].

The zinc contents in the gills of crabs from Ghannouch and Djerba are lower than those reported in the samples of crabs collected from the Bizerte Lagoon, Monastir Coast and Gargour in Sfax [52]. The zinc concentrations in the hepatopancreata of *P. segnis* from Ghannouch and Djerba are lower than those recorded in crab samples from Canada [54], Australia [35], and the United States [55]. However, the concentration of this element in the hepatopancreas samples of crabs from the Ghannouch site was similar to the concentration found in crab samples from Brazil [56].

For cadmium, the values noted in the gills of *P. segnis* from Ghannouch and Djerba are higher than those reported in the samples of crabs (*Carcinus maenas)* collected from the Lagoon of Bizerte, the Coast of Monastir, and Gargour in Sfax [52]. The cadmium content in the hepatopancreata of individuals from Ghannouch is higher than the values recorded in crabs from Canada [54], Australia [35], the United States [55], and Brazil [56], while its concentration in the hepatopancreata of the Djerba crabs is similar to that reported in samples from Canada [54], Australia [35], and the United States [55]. However, it is higher than in the samples from Brazil [56].

The concentrations of lead in the gills of crabs from the two sites Ghannouch and Djerba are lower than those observed in crab samples collected from the Bizerte Lagoon and Monastir Coast [52]. The content of this element in the gills of crabs from Djerba is close to that of crabs collected in the site of Gargour in Sfax [52]. In the hepatopancreata of Ghannouch individuals, the Pb contents are close to the concentrations reported in *C. maenas* from the Bizerte lagoon. However, for the values observed in the hepatopancreata of *P. segnis* individuals from both from Ghannouch and Djerba, the levels are higher than those obtained in the samples from the coast of Monastir and from Gargour in Sfax (Table 3). Regarding the data from the study by Ben-Khedher et al. [53] on *C. maenas* captured from Bizerte lagoon, the Pb concentrations are higher than those in the gills and hepatopancreata except for the levels in the gills of individuals from the Sfax site (Table 3). Comparing our data with other studies in other localities, we notice that the Pb levels in our samples are higher.

It has been shown that accumulation of trace elements in marine crabs might arise via saltwater or food [8]. The contribution of each route of uptake is linked to the bioavailability of these minerals. Cu and Zn concentrations in water are fairly low in the Gulf of Gabès region [21,57] for a Mediterranean-wide comparison, whereas sediments can be moderately to highly enriched in these two elements, as well as other heavy metals, based on the area [58,59]. As a result, as various previous investigations have shown, a transfer mechanism through sediments can be considered the primary predictor of reported accumulation patterns [60,61]. As a result, the contamination can be attributed to anthropogenic origins.

Table 3. Average contents of Cu, Zn, Cd, and Pb (mg/kg dw) in crab tissues from the literature.

Species	Site	Tissues	Cu	Zn	Cd	Pb	Reference
Scylla serrata	Canada	Digestive Gland	637	208	3.1	-	[59]
Pseudocarcinus gigas	Australia	Digestive Gland	52	108	2.2	-	[62]
Menippemer scenario	South Carolina-United States	Digestive Gland	308	46	4.6	-	[63]
Ucidescordatus	Brazil	Digestive Gland	6.6	8.89	0.16	-	[64]
Xenograpsustestudinatus	Taiwan	Whole tissue	53–290	119–610	0.49–1.29	1.83–2.64	[65]
Eriocheir sinensis	Poland	Whole tissue	3.17–21.9	1.15–26.6	0.016–0.025	0.29–0.61	[60]
Carcinusmaenas	Bizerte Lagoon, Tunisia	Gills	5.12–28.98	5.39–13.78	1.08–1.99	5.82–8.75	[13]
		Digestive Gland	15.16–31.93	5.3–25.11	0.67–2	5.01–8.56	
Carcinusmaenas	Bizerte Lagoon, Tunisia	Gills	857.3	627.1	0.61	21.1	[39]
	Monastir Coast, Tunisia		376.5	83.4	0.38	15.57	
	Sfax Coast, Tunisia		238.9	88.8	0.52	4.94	
	Bizerte Lagoon, Tunisia	Digestive Gland	863.5	985.1	21.5	4.5	
	Monastir Coast, Tunisia		396.9	102.4	0.62	0.67	
	Sfax Coast, Tunisia		205.1	99.9	3.04	0.14	
Portunussegnis	Gulf of Gabès (Zarrat), Tunisia	Muscle	206.45	590.04	0.21	0.4	[6]
		Gonad	584.85	36.46	0.23	0.14	
		Exoskeleton	65.78	11.4	0.14	0.06	

Table 3. *Cont.*

Species	Site	Tissues	Cu	Zn	Cd	Pb	Reference
Portunus pelagicus	Shuwaikh, Kuwait	Gills	79.1	29.1	0.28	3.29	[66]
		Hepatopancreas	68.6	18.9	0.43	4.69	
Chiromantes eulimene	Bhizolo Canal Richards Bay Harbour, South Africa	Gills	106.74	219.8	9.72	23.24	[67]
		Digestive Gland	153.6	140.57	7.12	8.94	
Scylla serrata	Punnakayal, Tuticorin, Southeast Coast of India	Gills	3.63	14.26	0.14	0.19	[68]
		Hepatopancreas	0.21	39.73	0.51	0.15	
Portunus pelagicus	Northern Bay of Bengal	Gills	21.06	-	1.01	1.67	[69]
		Hepatopancreas	12.46	-	2.82	2.07	
Cardisoma armatum	Kribi mangrove areas, Cameroon	Gills	3.87	64.77	0.18	0.09	[70]
		Hepatopancreas	7.88	51.34	0.35	2.07	
Portunus trituberculatus	Coast of Zhejiang Province, China	Gills	35.1	16.4	0.2	0.08	[71]
		Hepatopancreas	7.5	20.1	1.1	0.06	
Portunus segnis	Gulf of Gabès (Ghannouch), Tunisia	Gills	12.53	37.36	19.21	10.15	Present Study
		Digestive Gland	6.96	18.14	7.74	4.536	
	Djerba Island, Tunisia	Gills	6.45	10.72	7.2	4.9	
		Digestive Gland	2.91	8.71	3.18	2.21	

-: not determined.

3.2. *Biomarker Analysis*

3.2.1. Catalase Activity

A comparison of CAT activities at the two studied sites shows that this activity is more important at the Ghannouch site (Figure 4). Indeed, the results of the ANOVA indicate that there is a significant difference between the two studied sites for the hepatopancreas ($F = 71.25$; $p < 0.05$) and for the gills ($F = 17.9$; $p < 0.05$). The data also showed statistically significant differences between the two tissues within the same site (Djerba: $F = 114.68$; $p < 0.01$ and Ghannouch: $F = 24.8$; $p < 0.05$), with higher activities found in the gills. Depending on the type, concentration, and exposure duration of the metal, antioxidant enzymes may be stimulated or inhibited. These results may be in favor of an induction by pollution, caused by industrial discharges in Ghannouch, on the antioxidant activity of CAT. This is in agreement with several previous works that showed an increase in CAT enzymatic activities in *Carcinus maenas*, which were exposed to pollutants in the Monastir region [72] and in the Bizerte region [52], and in crab species exposed to metals [73,74].

These results can be explained by the fact that exposure to certain pollutants such as TMEs leads to an increase in the production of reactive oxygen species (ROS), such as the superoxide anion ($O_2^{\bullet-}$), hydrogen peroxide (H_2O_2), and the hydroxyl radical ($^{\bullet}OH$), which can alter the defense system of aquatic organisms [75]. In such a situation, increases in certain antioxidant enzymes may occur. These enzymes participate in the neutralization of these ROS and thus in the reduction of the damage caused by the contaminants. In this respect, Dellali et al. [76] affirmed that the enzymatic activity of CAT increases after exposure to certain pollutants.

In addition, our results also show a significant difference in CAT activity between the two studied tissues. Indeed, it is higher in the gills than in the hepatopancreas, with activity levels of about 4.7 μmol H_2O_2/min/mg protein and 3.49 μmol H_2O_2/min/mg protein, respectively, in Ghannouch, and 2.15 μmol H_2O_2/min/mg protein and 1.58 μmol H_2O_2/min/mg protein, respectively, in Djerba. In fact, our results show a profile in the same direction between the enzymatic activity of CAT and the accumulation of heavy metals at the level of the gills. These results are in agreement with several other scientific studies, such as the study by Jerome et al. [77], which showed an increase in this activity in the gills of *Callinectes amnicola* crabs exposed to industrial effluents.

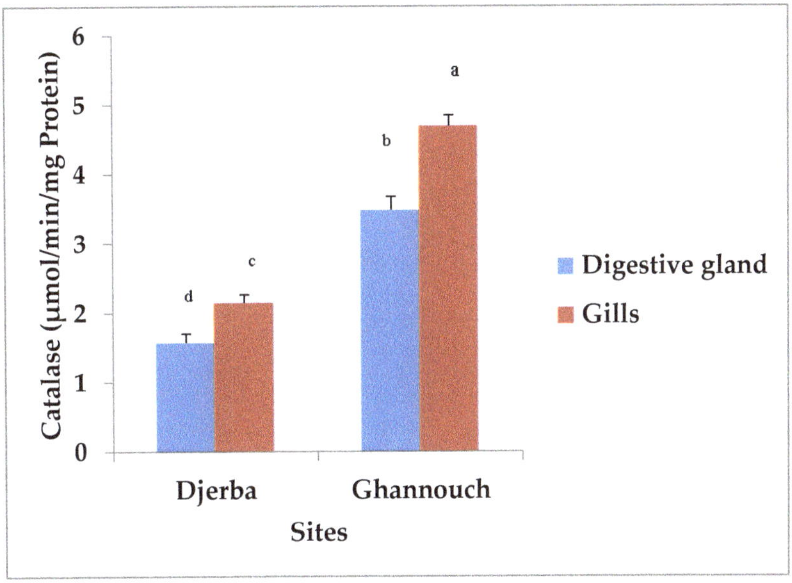

Figure 4. Catalase activity (µmol/min/mg Protein) in digestive gland and gill tissues of *P. segnis* from different localities. Different letters indicate statistically significant differences.

3.2.2. Glutathione S-Transferase Activity

Our results show the existence of a significant difference in GST activity, not only between the two studied sites but also between the two tissues examined in this study (Figure 5). GST showed important activity at the level of the gills compared to the hepatopancreas for the samples from Ghannouch (F = 20.87; $p < 0.01$) and Djerba (F = 37.21; $p < 0.01$). Between the two studied sites, GST activity was statistically more significant in the gills (F = 22.12; $p < 0.01$) and hepatopancreas (F = 37.21; $p < 0.05$).

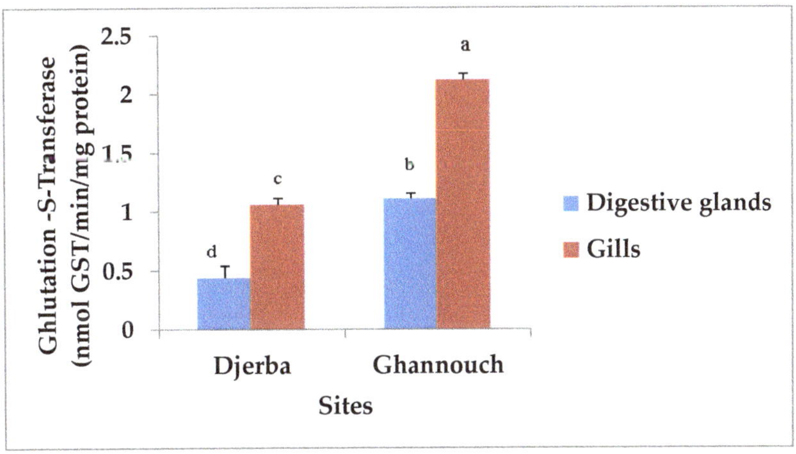

Figure 5. Glutathione S-transferase (GST) activity (nmol GST/min/mg Protein) in *P. segnis* (n = 20) from Ghannouch and Djerba. Different letters indicate statistically significant differences.

The induction of GST enzymatic activity in the Ghannouch individuals is suggested to be a response to environmental stress caused by the contamination of the marine environment by heavy metals that are accumulated in crab tissues, particularly the gills.

Indeed, our study shows an important activity of this enzyme in the individuals from the Ghannouch site, which present accumulations of Cd, Zn, Cu, and Pb at the level of the gills. Several studies have affirmed increases in the enzymatic activity of GST in crabs, fish, and mussels coming from sites contaminated by TMEs [65]. Furthermore, our data are in agreement with those of Barata et al. [64], who found an increase in GST activity in *Daphnia magna* following exposure to cadmium, and to those reported by Frías-Espericueta et al. [73] and by De Jesus et al. [74] for crab species exposed to metals.

The results of the present study can be partly explained by the bioavailability of heavy metals in the marine environment of Ghannouch, especially Cd, Zn, Cu, and Pb, which can generate disturbances that affect the homeostasis of individuals. To defend against and minimize this damage, an acceleration of the activities of certain enzymes, including GST, can protect the organisms and improve their overall health.

3.2.3. Malonedialdehyde Determination

The ANOVA results indicate an increase in MDA concentration in the crabs from Ghannouch. There is a significant difference between the two localities for both gill and hepatopancreas tissues, respectively (F = 54.22 and F = 42.51; $p < 0.01$) (Figure 6). Furthermore, according to our data for both sites, the concentration of MDA is higher in gills (Djerba: F = 51.46 and Ghannouch: F = 80.32; $p < 0.01$) than in the hepatopancreas. This increase in MDA levels was observed in individuals from the site with the highest levels of metals in the gills. In the same context, the studies of Bonneris et al. [78] confirm our results. They showed that following a 21-day exposure to Cd, MDA levels were elevated in the mussel *M. edulis*. This elevation was also demonstrated in *Pyganodon grandis* [79] and *Bathymodiolus azoricus* [80] captured from polluted environments and in crab species exposed to metals [73]. Therefore, we can suggest that an increase in MDA levels in individuals from Ghannouch is due to the bioavailability of heavy metals in the environment.

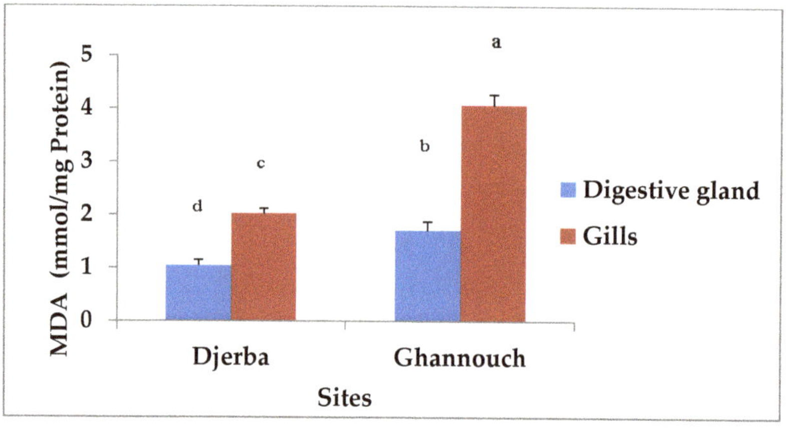

Figure 6. Comparison of MDA levels (nmol/mg Protein) in digestive glands and gills of *P. segnis* captured from the studied sites. Means that do not share the same alphabetic sign are significantly different.

Indeed, MDA is a biomarker generated secondarily after lipid peroxidation caused by an alteration of the plasma membrane through the attack of polyunsaturated fatty acids. This membrane lipoperoxidation seems to depend on the organ that accumulates more metals [81]. Lipid peroxidation is followed by structural changes in biological membranes [80] or other elements containing lipids [82]. This leads to a loss of permeability and the inactivation of receptors and membrane enzymes [83]. These functional disturbances can lead to cell death, and lipid peroxidation is an endogenous source of DNA damage [63].

A previous study by Giarratano et al. [84] revealed an increase in MDA levels in *Aulacomyaatra* mussels after exposure to iron, aluminum, and cadmium. The same is true for the work of Fahmy et al. [85], who revealed the induction of increased MDA levels in the hemolymph and soft tissues of snails treated with zinc. Radwan et al. indicated in their study [86] that there is a positive correlation between heavy metals and lipid peroxidation due to the presence of metal ions in the digestive gland of snails that catalyze the Fenton reaction and increase the risk of cell damage.

3.2.4. Determination of Reduced Glutathione

The results obtained in Figure 7 show an increase in GSH concentration in teh crabs from Djerba. There is a significant difference between the two localities ($p < 0.05$) (Figure 7). In addition, according to our results, we notice that for both sites, the concentration of GSH is higher in the hepatopancreas than in the gills.

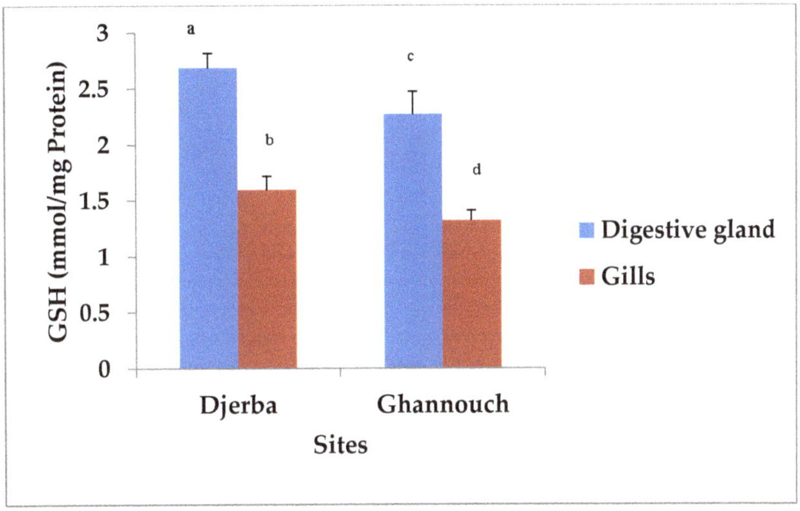

Figure 7. Glutathione content (mmol/mg Protein) in digestive glands and gills tissues of *P. segnis* from the studied sites. Means that do not share the same alphabetic sign are significantly different.

Indeed, GSH is the main intracellular non-enzymatic antioxidant agent [87]. The enzyme that catalyzes the conjugation of GSH to a wide variety of endogenous and exogenous electrophilic compounds is GST, which possesses the capacity of detoxification, thus playing a role in cellular protection from oxidative stress [88]. The conjugation of metribuzin metabolites with GSH, followed by the conversion of mercapturic acid derivatives, seems to play a major role in detoxification and excretion [89]. GSH presents the majority state of glutathione in cells; an increase in the oxidized form (GSSG) of GSH indicates the presence of an oxidative stress state [90]. The transformation of oxidized glutathione into reduced glutathione is achieved via glutathione reductase (GR) to activate GPx [91]. GSH plays a multifactorial role in the antioxidant defense mechanism. It is a direct scavenger of free radicals and a necessary co-substrate for GPx and glutathione-s-transferase activity. Therefore, changes in GSH levels can be considered particularly sensitive indicators of oxidative stress [92].

The decrease in GSH levels at the Ghannouch site, which is the site with the highest trace element levels, is in agreement with several other previous studies [93–95] which showed that GSH, in addition to other biochemical markers, seems to be a potential antagonist against metals. According to El-Shenawy et al. [86], the decrease in the GSH level is attributed to the increased consumption of this peptide for the synthesis of heavy-metal-binding proteins such as metallothioneins. In addition, Singaram et al. [96] reported that the decrease in the GSH level is also due to the increased utilization of GSH by glutathione peroxidase (GPx) to catalyze the reduction of H_2O_2. Leomanni et al. [97] also showed that the early induction of GPx activity plays a main role in the anticipated decrease in GSH.

4. Conclusions

The aim of this paper was to estimate the consequences of pollution on an invasive crab species, *Portunus segnis*, which is widely distributed on the Tunisian coasts.

The results of the analyses revealed a significant difference between the two studied populations in terms of both the biochemical parameters studied and the concentrations of the various trace elements measured in two tissues: the gills and hepatopancreas.

Results of the metal analysis revealed that *Portunus segnis* accumulates metallic trace elements in a variable manner. The general order of heavy metal bioaccumulation analyzed in the gills and hepatopancreas tissues in our study is: Zn > Cd > Cu > Pb. The bioaccumulation by the gills (37.36 ± 12.57 vs. 10.72 ± 2.9 for Zn, 19.21 ± 0.1 vs. 7.2 ± 0.049 for Cd, 12.53 ± 0.078 vs. 6.45 ± 0.12 for Cu, and 10.15 ± 1.38 vs. 4.9 ± 0.01 for Pb) is greater than that of the hepatopancreas (18.14 ± 8.38 vs. 8.71 ± 1.03 for Zn, 7.74 ± 0.14 vs. 3.18 ± 0.04 for Cd, 6.96 ± 0.79 vs. 2.91 ± 0.08 for Cu, and 4.536 ± 0.24 vs. 2.21 ± 0.05 for Pb) for the studied sites, Ghannouch and Djerba, respectively. Finally, the crabs living near the industrial zone, Ghannouch, accumulated more heavy metals in their tissues (gills and hepatopancreas) than those obtained from Djerba, the area furthest from the source of pollution.

The study of biochemical biomarkers in the gills and hepatopancreata of crabs has shown that crabs from Ghannouch, the most polluted area, have higher levels of biomarkers compared to those from Djerba. Indeed, the enzymatic activities of catalase and glutathione-S-transferase as well as the contents of malonedialdehyde and reduced glutathione all present significant differences between the two studied stations.

The variations in the levels of the studied biomarkers reflect the alarming state of this ecosystem and suggest that *Portunus segnis* is setting up a relatively effective detoxification system following metal contamination. As a result, this species could be used as a bioindicator of pollution in biomonitoring programs to assess the quality of aquatic environments in the Golf of Gabès.

Author Contributions: Formal analysis, W.B.A. and A.A.; Methodology, A.D., W.B.A. and A.A.; Resources, A.A.; Supervision, A.A.; Validation, A.A. All authors have read and agreed to the published version of the manuscript.

Funding: The authors declare that no funds, grants, or other support were received during the preparation of this manuscript.

Institutional Review Board Statement: Not applicable.

Informed Consent Statement: Not applicable.

Data Availability Statement: Not applicable.

Acknowledgments: The Life Sciences Department of the Faculty of Sciences of Gabès, University of Gabès, provided the reagents and technical scientific support for which the authors are very thankful. The authorsalso acknowledge the support of the study from Tunisia's Ministry of Higher Education and Scientific Research.

Conflicts of Interest: The authors declare no conflict of interest.

References

1. Reddy, M.S.; Mehta, B.; Dave, S.; Joshi, M.; Karthikeyan, L.; Sarma, V.K.S.; Basha, S.; Ramachandraiah, G.; Bhatt, P. Bioaccumulation of heavy metals in some commercial fishes and crabs of the Gulf of Cambay. *India Curr. Sci.* **2007**, *92*, 1489–1491.
2. Blaise, C.; Gagné, F.; Pellerin, J.; Hansen, P.-D.; Trottier, S. Molluscan Shellfish Biomarker Study of the Quebec, Canada, Saguenay Fjord with the Soft-Shell Clam, *Mya arenaria*: Molluscan Shellfish Biomarker StudyStudy. *Environ. Toxicol.* **2002**, *17*, 170–186. [CrossRef] [PubMed]

3. Galloway, T.S.; Brown, R.J.; Browne, M.A.; Dissanayake, A.; Lowe, D.; Jones, M.B.; Depledge, M.H. A Multibiomarker Approach To Environmental Assessment. *Environ. Sci. Technol.* **2004**, *38*, 1723–1731. [CrossRef]
4. Mayon, N.; Bertrand, A.; Leroy, D.; Malbrouck, C.; Mandiki, S.N.M.; Silvestre, F.; Goffart, A.; Thomé, J.-P.; Kestemont, P. Multiscale Approach of Fish Responses to Different Types of Environmental Contaminations: A Case Study. *Sci. Total Environ.* **2006**, *367*, 715–731. [CrossRef]
5. UNEP/MAP. Draft Transboundary Diagnostic Analysis for the Mediterranean Sea, 1997 (TDA/MED) (UNEP(OCA)MED IG.ll/inf.7). Presented at MEDITERRANEAN ACTION PLAN Tenth Ordinary Meeting of the Contracting Parties to the Convention for the Protection of the Mediterranean Sea against Pollution and Its Protocols Tunis, Tunis, Tunisia, 18–21 November 1997.
6. Hamza-Chaffai, A. Health Assessment of a Marine Bivalve *Ruditapes decussatus* from the Gulf of Gabès (Tunisia). *Environ. Internat.* **2003**, *28*, 609–617. [CrossRef]
7. Smaoui-Damak, W.; Hamza-Chaffai, A.; Berthet, B.; Amiard, J.C. Preliminary Study of the Clam *Ruditapes decussatus* Exposed In Situ to Metal Contamination and Originating from the Gulf of Gabès, Tunisia. *Bull. Environ. Contam Toxicol.* **2003**, *71*, 961–970. [CrossRef] [PubMed]
8. Rainbow, P.S. Trace Metal Concentrations in Aquatic Invertebrates: Why and so What? *Environ. Pollut.* **2002**, *120*, 497–507. [CrossRef]
9. Fox, H.M. The Migration of a Red Sea Crab through the Suez Canal. *Nature* **1924**, *113*, 714–715. [CrossRef]
10. Türeli, C.; Çelik, M.; Erdem, Ü. Comparison of meat composition and yield of blue crab (*Callinectes sapidus* RATHBUN, 1896) and sand crab (*Portunus pelagicus* Linne, 1758) caught in Iskenderun Bay, North-East Mediterranean. *Turk. J. Vet. Anim. Sci.* **2000**, *24*, 195–204.
11. Rabaoui, L.; Arculeo, M.; Mansour, L.; Tlig-Zouari, S. Occurrence of the lessepsian species *Portunus segnis* (Crustacea: Decapda) in the Gulf of Gabés (Tunisia): First record and new information on its biology and ecology. *Cah. Biol. Mar.* **2015**, *56*, 169–175.
12. Deidun, A.; Sciberras, A. A Further Record of the Blue Swimmer Crab *Portunus segnis* Forskal, 1775 (Decapoda: Brachyura: Portunidae) from the Maltese Islands (Central Mediterranean). *BIR* **2016**, *5*, 43–46. [CrossRef]
13. Rifi, M.; Ounifi-BenAmor, K.; Ben Souissi, J.; Zaouali, J. Première mention du crabelessepsien *Portunus segnis* (Forskål, 1775) (Décapode, Brachyoure, Portunidae) dans les eaux marines Tunisiennes. In Proceedings of the du 4ème Congrès Franco-Maghrébin et 5èmes Journées Franco-Tunisiennes de Zoologie, Korba, Tunisie, 13–17 November 2014; p. 9.
14. Bdioui, M. Premier signalement du crabe bleu Portunus segnis (Forskal, 1775) dans le Sud du Golfe de Hammamet (Centre-Est de la Tunisie). *Bull. Inst. Natn. Scien. Tech. Mer de Salammbô.* **2016**, *43*, 183–187.
15. Ounifi-Ben Amor, K.; Rifi, M.; Ghanem, R.; Draeif, I.; Zaouali, J.; Ben Souissi, J. Update of Alien Fauna and New Records from Tunisian Marine Waters. *Medit. Mar. Sci.* **2015**, *17*, 124. [CrossRef]
16. Burollet, P.F.; Clairefond, P.; Winnock, E. Géologie méditerranéenne, Tome VI, numéro 1. In *La Mer Pélagienne: Etude Sédimentologique et écologique du Plateau Tunisien et du Golfe de Gabès;* Département des sciences de la terre. Centre St Charles 13331 Marseille Cedex 3. Université de Provence: Marseille, France, 1979; p. 345.
17. Hamza, A. Le Statut du Phytoplancton Dans le Golfe de Gabès. Ph.D. Thesis, Faculté des Sciences de Sfax, Sfax, Tunisie, 2003; p. 297.
18. Drira, Z. Contribution à la Compréhension du Fonctionnement du Golfe de Gabès: Etude des Caractéristiques Dynamiques et Structurales des Communautés Phytozooplanctoniques. Ph.D. Thesis, Université de Sfax, Sfax, Tunisie, 2009; p. 229.
19. Hattab, T.; Ben RaisLasram, F.; Albouy, C.; Romdhane, M.S.; Jarboui, O.; Halouani, G.; Cury, P.; Le Loch, F. An Ecosystem Model of an Exploited Southern Mediterranean Shelf Region (Gulf of Gabes, Tunisia) and a Comparison with Other Mediterranean Ecosystem Model Properties. *J. Mar. Res.* **2013**, *128*, 159–174. [CrossRef]
20. Alaya-Ltifi, L.; Chokri, M.A.; Selmi, S. Breeding Performance of Passerines in a Polluted Oasis Habitat in Southern Tunisia. *Ecotoxicol. Environ. Safe* **2012**, *79*, 170–175. [CrossRef]
21. Annabi, A.; Bardelli, R.; Vizzini, S.; Mancinelli, G. Baseline Assessment of Heavy Metals Content and Trophic Position of the Invasive Blue Swimming Crab *Portunus segnis* (Forskål, 1775) in the Gulf of Gabès (Tunisia). *Mar. Pollut. Bull.* **2018**, *136*, 454–463. [CrossRef]
22. Bradford, M.M. A Rapid and Sensitive Method for the Quantitation of Microgram Quantities of Protein Utilizing the Principle of Protein-Dye Binding. *Anal. Biochem.* **1976**, *72*, 248–254. [CrossRef]
23. Almeida, J.R.; Gravato, C.; Guilhermino, L. Challenges in Assessing the Toxic Effects of Polycyclic Aromatic Hydrocarbons to Marine Organisms: A Case Study on the Acute Toxicity of Pyrene to the European Seabass (*Dicentrarchus labrax* L.). *Chemosphere* **2012**, *86*, 926–937. [CrossRef]
24. Habig, W.H.; Pabst, M.J.; Jakoby, W.B. Glutathione S-Transferases. *J. Biol. Chem.* **1974**, *249*, 7130–7139. [CrossRef]
25. Turdi, S.; Li, Q.; Lopez, F.L.; Ren, J. Catalase Alleviates Cardiomyocyte Dysfunction in Diabetes: Role of Akt, Forkhead Transcriptional Factor and Silent Information Regulator 2. *Life Sci.* **2007**, *81*, 895–905. [CrossRef]
26. Aebi, H. Catalase in vitro. *Methods Enzymol.* **1984**, *105*, 121–126. [CrossRef] [PubMed]
27. Ellman, G.L. Tissue Sulfhydryl Groups. Arch. *Biochem. Biophys.* **1959**, *82*, 70–77. [CrossRef]
28. Ohkawa, H.; Ohishi, N.; Yagi, K. Assay for Lipid Peroxides in Animal Tissues by Thiobarbituric Acid Reaction. *Anal. Biochem.* **1979**, *95*, 351–358. [CrossRef] [PubMed]

29. Catalano, D.; Torchia, G.; Pititto, F.; Greco, R. Contaminazione da metalli del "granchio americano", *Portunus pelagicus* (Linnaeus, 1758) nella Rada di Augusta (*Sicilia orientale*). *Biol. Mar. Mediterr.* **2006**, *13*, 696–699.
30. Ayas, D. The Effects of Season and Sex on the Nutritional Quality of Muscle Types of Blue Crab Callinectes sapidus (Rathbun, 1896) and Swimming Crab *Portunus segnis* (Forskal, 1775). *NESciences* **2016**, *1*, 1–14. [CrossRef]
31. Olgunoglu, M.; Olgunoglu, İ. Heavy Metal Contents in Blue Swimming Crab from the Northeastern Mediterranean Sea, Mersin Bay, Turkey. Pol. *J. Environ. Stud.* **2016**, *25*, 2233–2237. [CrossRef]
32. Rainbow, P.S.; Furness, R.W. Heavy Metals in the Marine Environment. In *Heavy Metals in the Marine Environment*; Furness, R.W., Rainbow, P.S., Eds.; CRC Press: Boca Raton, FL, USA, 2018; pp. 1–4. ISBN 978135107315831.
33. Gutiérrez, A.J.; Lozano, G.; Rubio, C.; Martín, V.; Hardisson, A.; Revert, C. Heavy Metals in Black Crabs in the Atlantic Coast (Tenerife, Spain)—Human Risk Assessment: Water. *Clean Soil Air Water* **2017**, 45. [CrossRef]
34. Nędzarek, A.; Czerniejewski, P.; Drost, A.; Harasimiuk, F.; Machula, S.; Tórz, A.; Masalski, P. The Distribution of Elements in the Body of Invasive Chinese Mitten Crabs (*Eriocheir sinensis* H. Milne-Edwards, 1853) from Lake Dąbie, Poland. *J. Food Compos. Anal.* **2017**, *60*, 1–9. [CrossRef]
35. Turoczy, N.J.; Mitchell, B.D.; Levings, A.H.; Rajendram, V.S. Cadmium, Copper, Mercury, and Zinc Concentrations in Tissues of the King Crab (*Pseudocarcinus gigas*) from Southeast Australian Waters. *Environ. Int.* **2001**, *27*, 327–334. [CrossRef] [PubMed]
36. Reichmuth, J.M.; Weis, P.; Weis, J.S. Bioaccumulation and Depuration of Metals in Blue Crabs (*Callinectes sapidus* Rathbun) from a Contaminated and Clean Estuary. *Environ. Pollut.* **2010**, *158*, 361–368. [CrossRef] [PubMed]
37. Bondgaard, M.; Bjerregaard, P. Association between Cadmium and Calcium Uptake and Distribution during the Moult Cycle of Female Shore Crabs, *Carcinus maenas*: An In Vivo Study. *Aquat. Toxicol.* **2005**, *72*, 17–28. [CrossRef]
38. Martín-Díaz, M.L.; Blasco, J.; Sales, D.; DelValls, T.A. Biomarkers Study for Sediment Quality Assessment in Spanish Ports Using the Crab Carcinus maenas and the Clam Ruditapes philippinarum. *Arch. Environ. Contam Toxicol.* **2007**, *53*, 66–76. [CrossRef]
39. *GB 5749-2006*; Standards for Drinking Water Quality. Ministry of Health of the People's Republic of China: Beijing, China, 2006.
40. European Commission. Commission Regulation (EC) No 1881/2006 of the European parliament and the council of 19 December 2006 setting maximum levels for certain contaminants in foodstuffs. *Off. J. Eur. Communities* **2006**, *18*, L364.
41. FAO. Compilation of legal limits for hazardous substances in fish and fishery products. *FAO Fish Circ.* **1983**, *464*, 5–100. Available online: http://trove.nla.gov.au/version/22206109 (accessed on 15 April 2023).
42. Baden, S.P.; Eriksson, S.P. Role, routes and effects of manganese in crustaceans. Oceanogr. Mar. Biol. *Annu. Rev.* **2006**, *44*, 61–83.
43. Mancinelli, G.; Papadia, P.; Ludovisi, A.; Migoni, D.; Bardelli, R.; Fanizzi, F.P.; Vizzini, S. Beyond the Mean: A Comparison of Trace- and Macroelement Correlation Profiles of Two Lacustrine Populations of the Crayfish *Procambarus clarkii*. *Sci. Total Environ.* **2018**, *624*, 1455–1466. [CrossRef]
44. Amiard, J.; Amiardtriquet, C.; Barka, S.; Pellerin, J.; Rainbow, P. Metallothioneins in Aquatic Invertebrates: Their Role in Metal Detoxification and Their Use as Biomarkers. *Aquat. Toxicol.* **2006**, *76*, 160–202. [CrossRef]
45. Schuwerack, P.-M.M.; Lewis, J.W.; Jones, P. The potential use of the South African river crab, Potamonauteswarreni, as a bioindicator species for heavy metal contamination. *Ecotoxicology* **2001**, *10*, 159–166. [CrossRef]
46. Pan, L.; Zhang, H. Metallothionein, Antioxidant Enzymes and DNA Strand Breaks as Biomarkers of Cd Exposure in a Marine Crab, Charybdis Japonica. Comp. *Biochem. Physiol. Part C Toxicol. Pharmacol.* **2006**, *144*, 67–75. [CrossRef] [PubMed]
47. Corrêa, J.D.; Da Silva, M.R.; Da Silva, A.C.B.; De Lima, S.M.A.; Malm, O.; Allodi, S. Tissue Distribution, Subcellular Localization and Endocrine Disruption Patterns Induced by Cr and Mn in the Crab Ucides Cordatus. *Aquat. Toxicol.* **2005**, *73*, 139–154. [CrossRef] [PubMed]
48. Rainbow, P.S. Biomonitoring of trace metals in estuarine and marine environments. *Aust. J. Ecotoxicol.* **2006**, *12*, 107–122.
49. Doherty, V.F.; Ogunkuade, O.O.; Kanife, U.C. Biomarkers of oxidative stress and heavy metal levels as indicators of environmental pollution in some selected fishes in Lagos, Nigeria. *Am. Eurasian J. Agric. Envrion. Sci.* **2010**, *7*, 359–365.
50. Bjerregaard, P.; Depledge, M.H. Trace Metal Concentrations and Contents in the Tissues of the Shore Crab *Carcinus maenas*: Effects of Size and Tissue Hydration. *Mar. Biol.* **2002**, *141*, 741–752. [CrossRef]
51. Peng, S.-H.; Hung, J.-J.; Hwang, J.-S. Bioaccumulation of Trace Metals in the Submarine Hydrothermal Vent Crab Xenograpsus testudinatus off Kueishan Island, Taiwan. *Mar. Pollut. Bull.* **2011**, *63*, 396–401. [CrossRef] [PubMed]
52. Ghedira, J.; Jebali, J.; Banni, M.; Chouba, L.; Boussetta, H.; López-Barea, J.; Alhama, J. Use of Oxidative Stress Biomarkers in Carcinus Maenas to Assess Littoral Zone Contamination in Tunisia. *Aquat. Biol.* **2011**, *14*, 87–98. [CrossRef]
53. Ben-Khedher, S.; Jebali, J.; Houas, Z.; Nawéli, H.; Jrad, A.; Banni, M.; Boussetta, H. Metals Bioaccumulation and Histopathological Biomarkers in *Carcinus maenas* Crab from Bizerta Lagoon, Tunisia. *Envrion. Sci. Pollut. Res.* **2014**, *21*, 4343–4357. [CrossRef] [PubMed]
54. Mortimer, M.R. Pesticide and Trace Metal Concentrations in Queensland Estuarine Crabs. *Mar. Pollut. Bull.* **2000**, *41*, 359–366. [CrossRef]
55. Reed, L.A.; Pennington, P.L.; Wirth, E. A Survey of Trace Element Distribution in Tissues of Stone Crabs (*Menippe mercenaria*) from South Carolina Coastal Waters. *Mar. Pollut. Bull.* **2010**, *60*, 2297–2302. [CrossRef]
56. Pinheiro, M.A.A.; Silva, P.P.G.E.; Duarte, L.F.D.A.; Almeida, A.A.; Zanotto, F.P. Accumulation of Six Metals in the Mangrove Crab UcidesCordatus (Crustacea: Ucididae) and Its Food Source, the Red Mangrove Rhizophora Mangle (Angiosperma: Rhizophoraceae). Ecotoxicol. *Environ. Safe* **2012**, *81*, 114–121. [CrossRef]

57. Bonanno, G.; Orlando-Bonaca, M. Trace Elements in Mediterranean Seagrasses: Accumulation, Tolerance and Biomonitoring. A Review. *Mar. Pollut. Bull.* **2017**, *125*, 8–18. [CrossRef]
58. Ayadi, N.; Aloulou, F.; Bouzid, J. Assessment of Contaminated Sediment by Phosphate Fertilizer Industrial Waste Using Pollution Indices and Statistical Techniques in the Gulf of Gabes (Tunisia). Arab. *J. Geosci.* **2015**, *8*, 1755–1767. [CrossRef]
59. El Zrelli, R.; Courjault-Radé, P.; Rabaoui, L.; Castet, S.; Michel, S.; Bejaoui, N. Heavy Metal Contamination and Ecological Risk Assessment in the Surface Sediments of the Coastal Area Surrounding the Industrial Complex of Gabes City, Gulf of Gabes, SE Tunisia. *Mar. Pollut. Bull.* **2015**, *101*, 922–929. [CrossRef] [PubMed]
60. Jara-Marini, M.E.; Soto-Jiménez, M.F.; Páez-Osuna, F. Trophic Relationships and Transference of Cadmium, Copper, Lead and Zinc in a Subtropical Coastal Lagoon Food Web from SE Gulf of California. *Chemosphere* **2009**, *77*, 1366–1373. [CrossRef]
61. Schneider, L.; Maher, W.A.; Potts, J.; Taylor, A.M.; Batley, G.E.; Krikowa, F.; Adamack, A.; Chariton, A.A.; Gruber, B. Trophic Transfer of Metals in a Seagrass Food Web: Bioaccumulation of Essential and Non-Essential Metals. *Mar. Pollut. Bull.* **2018**, *131*, 468–480. [CrossRef] [PubMed]
62. El-Shenawy, N.S.; Mohammadden, A.; Al-Fahmie, Z.H. Using the Enzymatic and Non-Enzymatic Antioxidant Defense System of the Land Snail Eobania vermiculata as Biomarkers of Terrestrial Heavy Metal Pollution. *Ecotoxicol. Environ. Safe* **2012**, *84*, 347–354. [CrossRef]
63. Marnett, L.J. Oxy Radicals, Lipid Peroxidation and DNA Damage. *Toxicology* **2002**, *181–182*, 219–222. [CrossRef]
64. Barata, C.; Varo, I.; Navarro, J.C.; Arun, S.; Porte, C. Antioxidant Enzyme Activities and Lipid Peroxidation in the Freshwater Cladoceran daphnia magna Exposed to Redox Cycling Compounds. *Comp. Biochem. Physiol. Part C Toxicol. Pharmacol.* **2005**, *140*, 175–186. [CrossRef]
65. Stegeman, J.J.; Marius, B.; Giulio, R.T.D.; Lars, F.; Fowler, B.A.; Sanders, B.M.; Veld, P.A.V. Molecular responses to environmental contamination: Enzyme and protein systems as indicators of chemical exposure and effect. In *Biomarkers. Biochemical, Physiological and Histological Markers of Anthropogenic Stress*; Huggett, R.J., Kimerle, R.A., Mehrle, P.M., Jr., Bergman, H.L., Eds.; Lewis Publishers: London, UK, 1992; pp. 235–335.
66. Karam, Q.; Guermazi, W.; Subrahmanyam, M.N.V.; Al-Enezi, Y.; Ali, M.; Leignel, V.; Annabi-Trabelsi, N. *Portunus pelagicus* (Linnaeus, 1758) as a Sentinel Species to Assess Trace Metal Occurrence: A Case Study of Kuwait Waters (Northwestern Arabian Gulf). *Toxics* **2023**, *11*, 426. [CrossRef]
67. Majola, N.; Mzimela, H.M.; Izegaegbe, J.I. Metal Bioaccumulation and Energy Biomarkers in Tissues of Two Populations of Chiromantes Eulimene from Richards Bay Harbour, South Africa. *Sci. Afr.* **2020**, *10*, e00558. [CrossRef]
68. Yogeshwaran, A.; Gayathiri, K.; Muralisankar, T.; Gayathri, V.; Monica, J.I.; Rajaram, R.; Marimuthu, K.; Bhavan, P.S. Bioaccumulation of Heavy Metals, Antioxidants, and Metabolic Enzymes in the Crab Scylla serrata from Different Regions of Tuticorin, Southeast Coast of India. *Mar. Pollut. Bull.* **2020**, *158*, 111443. [CrossRef]
69. Karar, S.; Hazra, S.; Das, S. Assessment of the Heavy Metal Accumulation in the Blue Swimmer Crab (*Portunus pelagicus*), Northern Bay of Bengal: Role of Salinity. *Mar. Pollut. Bull.* **2019**, *143*, 101–108. [CrossRef] [PubMed]
70. Ngo-Massou, V.M.; Kottè-Mapoko, E.F.; Din, N. Heavy Metal Accumulation in the Edible Crab Cardisoma armatum (Brachyura: Gecarcinidae) and Implications for Human Health Risks. *Sci. Afr.* **2022**, *16*, e01248. [CrossRef]
71. Liu, Q.; Liao, Y.; Xu, X.; Shi, X.; Zeng, J.; Chen, Q.; Shou, L. Heavy Metal Concentrations in Tissues of Marine Fish and Crab Collected from the Middle Coast of Zhejiang Province, China. *Environ. Monit. Assess* **2020**, *192*, 285. [CrossRef] [PubMed]
72. Jebali, J.; Banni, M.; Almeida, E.A.D.; Boussetta, H. Oxidative DNA Damage Levels and Catalase Activity in the Clam Ruditapes Decussatus as Pollution Biomarkers of Tunisian Marine Environment. *Environ. Monit. Assess* **2007**, *124*, 195–200. [CrossRef]
73. Frías-Espericueta, M.G.; Bautista-Covarrubias, J.C.; Osuna-Martínez, C.C.; Delgado-Alvarez, C.; Bojórquez, C.; Aguilar-Juárez, M.; Roos-Muñoz, S.; Osuna-López, I.; Páez-Osuna, F. Metals and Oxidative Stress in Aquatic Decapod Crustaceans: A Review with Special Reference to Shrimp and Crabs. *Aquat. Toxicol.* **2022**, *242*, 106024. [CrossRef]
74. De Jesus, W.B.; Mota Andrade, T.D.S.D.O.; Soares, S.H.; Pinheiro-Sousa, D.B.; De Oliveira, S.R.S.; Torres, H.S.; Protazio, G.D.S.; Da Silva, D.S.; Santos, D.M.S.; De Carvalho Neta, A.V.; et al. Biomarkers and Occurrences of Heavy Metals in Sediment and the Bioaccumulation of Metals in Crabs (*Ucides cordatus*) in Impacted Mangroves on the Amazon Coast, Brazil. *Chemosphere* **2021**, *271*, 129444. [CrossRef]
75. Khessiba, P.; Hoarau, M.; Gnassia-Bar, A. Biochemical Response of the Mussel Mytilus galloprovincialis from Bizerta (Tunisia) to Chemical Pollutant Exposure. *Arch. Environ. Contam. Toxicol.* **2001**, *40*, 222–229. [CrossRef]
76. Dellali, M.; GnassiaBarelli, M.; Romeo, M.; Aissa, P. The Use of Acetylcholinesterase Activity in Ruditapes Decussatus and Mytilus Galloprovincialis in the Biomonitoring of Bizerta Lagoon. *Comp. Biochem. Physiol. Part C Toxicol. Pharmacol.* **2001**, *130*, 227–235. [CrossRef]
77. Jerome, F.C.; Hassan, A.; Omoniyi-Esan, G.O.; Odujoko, O.O.; Chukwuka, A.V. Metal Uptake, Oxidative Stress and Histopathological Alterations in Gills and Hepatopancreas of Callinectes amnicola Exposed to Industrial Effluent. *Ecotoxicol. Environ. Safe* **2017**, *139*, 179–193. [CrossRef]
78. Géret, F. Influence of Metal Exposure on Metallothionein Synthesis and Lipid Peroxidation in Two Bivalve Mollusks: The Oyster (*Crassostrea gigas*) and the Mussel (*Mytilus edulis*). *Aquat. Living Resour.* **2002**, *15*, 61–66. [CrossRef]
79. Bonneris, E.; Perceval, O.; Masson, S.; Hare, L.; Campbell, P.G.C. Sub-Cellular Partitioning of Cd, Cu and Zn in Tissues of Indigenous Unionid Bivalves Living along a Metal Exposure Gradient and Links to Metal-Induced Effects. *Environ. Pollut.* **2005**, *135*, 195–208. [CrossRef] [PubMed]

80. Bebianno, M.; Company, R.; Serafim, A.; Camus, L.; Cosson, R.; Fialamedoni, A. Antioxidant Systems and Lipid Peroxidation in from Mid-Atlantic Ridge Hydrothermal Vent Fields. *Aquat. Toxicol.* **2005**, *75*, 354–373. [CrossRef] [PubMed]
81. Ferrat, L.; Pergent-Martini, C.; Roméo, M. Assessment of the Use of Biomarkers in Aquatic Plants for the Evaluation of Environmental Quality: Application to Seagrasses. *Aquat. Toxicol.* **2003**, *65*, 187–204. [CrossRef] [PubMed]
82. Al-Mutairi, D.A.; Craik, J.D.; Batinic-Haberle, I.; Benov, L.T. Induction of Oxidative Cell Damage by Photo-Treatment with Zinc MetaN -Methylpyridylporphyrin. *Free Radic. Res.* **2007**, *41*, 89–96. [CrossRef] [PubMed]
83. Pampanin, D.M.; Camus, L.; Gomiero, A.; Marangon, I.; Volpato, E.; Nasci, C. Susceptibility to Oxidative Stress of Mussels (*Mytilus galloprovincialis*) in the Venice Lagoon (Italy). *Mar. Pollut. Bull.* **2005**, *50*, 1548–1557. [CrossRef] [PubMed]
84. Giarratano, E.; Gil, M.N.; Malanga, G. Biomarkers of Environmental Stress in Gills of Ribbed Mussel Aulacomya atra atra (Nuevo Gulf, Northern Patagonia). *Ecotoxicol. Environ. Safe* **2014**, *107*, 111–119. [CrossRef]
85. Fahmy, S.R.; Abdel-Ghaffar, F.; Bakry, F.A.; Sayed, D.A. Ecotoxicological Effect of Sublethal Exposure to Zinc Oxide Nanoparticles on Freshwater Snail Biomphalaria alexandrina. *Arch. Environ. Contam. Toxicol.* **2014**, *67*, 192–202. [CrossRef]
86. Radwan, M.A.; El-Gendy, K.S.; Gad, A.F. Biomarkers of Oxidative Stress in the Land Snail, Theba Pisana for Assessing Ecotoxicological Effects of Urban Metal Pollution. *Chemosphere* **2010**, *79*, 40–46. [CrossRef]
87. Jacob, L. *L'insuffisance Rénale Aiguë*; Springer: Paris, France, 2007; pp. 1–342. [CrossRef]
88. Schwab, M. *Encyclopedia of Cancer*; Springer: Berlin/Heidelberg, Germany, 2011. [CrossRef]
89. Morgan, J. Evidence on the Developmental and Reproductive Toxicity of Metribuzin. DRAFT, California, 2001; p. 65. Available online: https://oehha.ca.gov/media/downloads/crnr/metribuzin.pdf (accessed on 15 April 2023).
90. Pelletier, É.; Campbell, P.G.C.; Denizeau, F. *Écotoxicologie Moléculaire: Principes Fondamentaux et Perspectives de Développement*, 1st ed.; Presses de l'Université du Québec: Québec, QC, Canada, 2004; pp. 1–502. [CrossRef]
91. Poortmans, J.R.; Boisseau, N. *Biochimie des Activités Physiques*, 2nd ed.; De BoeckSupérieur: Bruxelles, Belgium, 2003; p. 480.
92. Taleb-Senouci, D.; Ghomari, H.; Krouf, D.; Bouderbala, S.; Prost, J.; Lacaille-Dubois, M.A.; Bouchenak, M. Antioxidant Effect of Ajuga Iva Aqueous Extract in Streptozotocin-Induced Diabetic Rats. *Phytomedicine* **2009**, *16*, 623–631. [CrossRef]
93. Haq, Q.; Jan, A.; Ali, A. Glutathione as an Antioxidant in Inorganic Mercury Induced Nephrotoxicity. *J. Postgrad. Med.* **2011**, *57*, 72. [CrossRef]
94. Jozefczak, M.; Remans, T.; Vangronsveld, J.; Cuypers, A. Glutathione Is a Key Player in Metal-Induced Oxidative Stress Defenses. *Int. J. Mol. Sci.* **2012**, *13*, 3145–3175. [CrossRef] [PubMed]
95. Hore, M.; Saha, R.; Bhaskar, S.; Mandal, S.; Bhattacharyya, S.; Roy, S. Oxidative Stress Responses in Puntius sarana Collected from Some Environmentally Contaminated Areas of River Mahananda, Malda, West Bengal. *Ecotoxicology* **2023**, *32*, 211–222. [CrossRef] [PubMed]
96. Singaram, G.; Harikrishnan, T.; Chen, F.-Y.; Bo, J.; Giesy, J.P. Modulation of Immune-Associated Parameters and Antioxidant Responses in the Crab (*Scylla serrata*) Exposed to Mercury. *Chemosphere* **2013**, *90*, 917–928. [CrossRef] [PubMed]
97. Leomanni, A.; Schettino, T.; Calisi, A.; Gorbi, S.; Mezzelani, M.; Regoli, F.; Lionetto, M.G. Antioxidant and Oxidative Stress Related Responses in the Mediterranean Land Snail Cantareus Apertus Exposed to the Carbamate Pesticide Carbaryl. *Comp. Biochem. Physiol. Part C Toxicol. Pharmacol.* **2015**, *168*, 20–27. [CrossRef] [PubMed]

applied sciences

MDPI

Article

Application of Positive Matrix Factorization for Source Apportionment of Polycyclic Aromatic Hydrocarbons (PAH) in the Adriatic Sea, and the Evaluation of PAH-Related Carcinogenic Risks

Jelena Mandić *, Jere Veža and Grozdan Kušpilić

Laboratory of Chemical Oceanography and Sea Sedimentology, Institute of Oceanography and Fisheries, Šetalište Ivana Meštrovića 63, 21000 Split, Croatia; veza@izor.hr (J.V.); kuspe@izor.hr (G.K.)
* Correspondence: mandic@izor.hr

Abstract: Concentrations of polycyclic aromatic hydrocarbons (PAH) were studied in 36 sediment samples collected from 29 sites in the central Adriatic Sea. The total concentration of PAH showed high variability, ranging from 42.85 μg kg^{-1} in open sea sediments to 28,662 μg kg^{-1} in the sediments of Šibenik Bay. Unsubstituted, parent PAH were dominant PAH compounds in the entire area of Šibenik Bay and most of the stations in the coastal area. In Kaštela Bay, methyl PAH concentrations were higher than the concentrations of unsubstituted PAH, whereas in the sediments along the mid-Adriatic transect, similar concentrations of both PAH fractions were determined. PAH sources were investigated by applying the Positive Matrix Factorization software developed by the U.S. Environmental Protection Agency (USEPA). The results indicate the presence of four sources, two of which can be considered petrogenic and two pyrogenic. The contribution of the sources to the total concentration PAH at each station indicates the dominance of pyrogenic sources in Šibenik Bay and at most stations in the coastal area. In the sediments along the mid-Adriatic transversal, as well as in Kaštela Bay, PAHs originate mainly from petrogenic sources. The obtained results show that PMF can be a suitable tool for control of PAH pollution and thus, for the management of various activities in the Adriatic region. Seasonal differences between the contributions of sources in the Kaštela and Šibenik bays indicate that the contribution of traffic to the total concentration of PAH was higher during the tourist season. The carcinogenic risk of PAH determined by toxicity equivalent (TEQ) calculations showed an increased carcinogenic risk in the whole area of Šibenik Bay, regardless of the number of PAH compounds considered for the calculations. A low PAH-related carcinogenic risk was found for the rest of the study area.

Keywords: PAH; sediments; Adriatic Sea; Positive Matrix Factorization; source apportionment; toxic equivalent quantity

Citation: Mandić, J.; Veža, J.; Kušpilić, G. Application of Positive Matrix Factorization for Source Apportionment of Polycyclic Aromatic Hydrocarbons (PAH) in the Adriatic Sea, and the Evaluation of PAH-Related Carcinogenic Risks. *Appl. Sci.* **2023**, *13*, 6992. https://doi.org/10.3390/app13126992

Academic Editor: Veeriah Jegatheesan

Received: 21 April 2023
Revised: 6 June 2023
Accepted: 7 June 2023
Published: 9 June 2023

1. Introduction

Polycyclic aromatic hydrocarbons are a large class of organic compounds whose structure consists of fused aromatic rings [1]. They are emitted from numerous sources in the form of a complex mixture that includes a significant amount of heterocyclic aromatic compounds in addition to PAH [1]. Although PAH may originate from natural sources, most of these compounds are emitted into the environment from anthropogenic sources such as transportation, power generation, industrial furnaces, waste incineration, coke and carbon production, asphalt and petroleum cracking [2–4]. PAH of anthropogenic origin are usually divided into the class of petrogenic PAH, which are formed during the conversion of organic-rich materials at low temperatures, and pyrogenic PAH, which are formed during the high-temperature combustion of organic materials in industrial furnaces, vehicle engines, power plants, etc. [5–10]. PAH mixtures emitted from petrogenic sources usually contain a greater amount of low molecular weight PAH and alkyl-substituted PAH, while PAH mixtures from pyrogenic sources are enriched in unsubstituted PAH compounds with higher molecular weight. PAH are introduced into the marine environment mainly through rivers, atmospheric deposition, municipal and industrial runoff, and accidental spills of fresh petroleum derivatives. Due to their high hydrophobicity, PAHs adsorb to particles in the water column and sink through the water column. Sediments can accumulate large amounts of PAH, which degrade over long periods of time and pose a threat to ecosystem health [2,3].

PAHs have been shown to have disruptive effects on the lymphatic system and hematopoiesis [11,12], while chronic exposure to PAH causes mutagenicity due to their tendency to interact with DNA [13]. While pyrogenic PAHs are associated with greater environmental and health risks, they are primarily associated with chronic toxicity [11,13]. In contrast, petrogenic PAH are more soluble and bioavailable to marine organisms, and as such considered to cause acute toxicity. Due to their adverse effects on human health and the health of marine organisms, the U.S. Environmental Protection Agency (USEPA) has included 16 PAHs in a list of priority pollutants to be monitored in the environment [14]. Various approaches have been developed to assess the environmental risk associated with PAH, of which the use of Sediment Quality Guidelines (SQGs) and bioassays have been the most widely used [15,16]. Laboratory experiments have been conducted on a variety of marine animals, including microorganisms, algae, invertebrates, fish and marine mammals [11,13,17], and the results served as the basis for the carcinogen classification system for PAH and mixtures of PAH, and for the development of the Toxic Equivalent (TEQ) method for PAH-related environmental risk assessment [13,17–20]. This method assumes that all carcinogenic PAH are equitoxic to benzo[*a*]pyrene and that the carcinogenic effect of a complex PAH mixture can be calculated as the sum of the effects of the individual PAH compounds [17,20]. Although this approach may lead to overestimation or underestimation of the effects of PAH due to the inhibitory/synergetic properties of PAH compounds when present as a complex mixture, the TEQ method has been used in many studies to assess the PAH-related environmental risk [18–20].

Since the toxicity of a PAH mixture is determined by its composition and, consequently, by the nature of the source from which the PAH mixture originated, knowledge of the PAH sources at a particular site is important for the assessment of PAH-related environmental risk. Many approaches have been applied to constrain PAH sources, of which diagnostic ratios, statistical tools and modelling have been most commonly used. The use of statistical multivariate analyses, such as principal component analyses (PCA), cluster analyses and alternating least squares (ALS) are a cost-effective method for attributing PAH sources, although these methods require the identification of a larger number of PAH compounds at a larger number of locations [19,21,22]. To provide a reliable tool for addressing PAH pollution, the USEPA has developed a specific model based on the factor analysis method: the Positive Matrix Factorizations (PMF) [23]. The PMF results provide the number of factors (sources), the contributions of the factors to the concentration of each PAH compound, and a contribution of the factors to the total pollution PAH for each case [4,19,21,24].

This study aims to obtain detailed information on the most important PAH sources and their contributions in the overall PAH pollution by application of PMF software. Another goal was to determine the PAH-related sediment toxicity by calculation of toxic equivalents at each station. The obtained results will be used for the management of various economic activities in the Adriatic region, as well as for the development of strategies to mitigate the PAH pollution and PAH-associated environmental risk.

2. Material and Methods

2.1. Study Area

Figure 1 shows PAH concentrations that were analyzed in surface sediments of 29 stations along the Adriatic coast and along the mid-Adriatic transect. The coastal area includes 11 stations located from the island of Pag in the north, to the city of Dubrovnik in the south.

Stations were selected to cover urbanized areas along the Croatian coast as well as areas that were heavily industrialized in the past. Six stations along the horizontal transect from Split to Mt. Gargano in Italy were selected to identify PAH concentrations and the main PAH sources in this area. Due to the greater anthropogenic pressures, Kaštela Bay and Šibenik Bay were selected for more in-depth study. In these areas, sediment samples were obtained at a greater number of stations during winter and early autumn, in order to assess the impact of touristic activities on trends in PAH pollution. Additional information for each station is given in Tables S1 and S2 of Supplementary Information (SI).

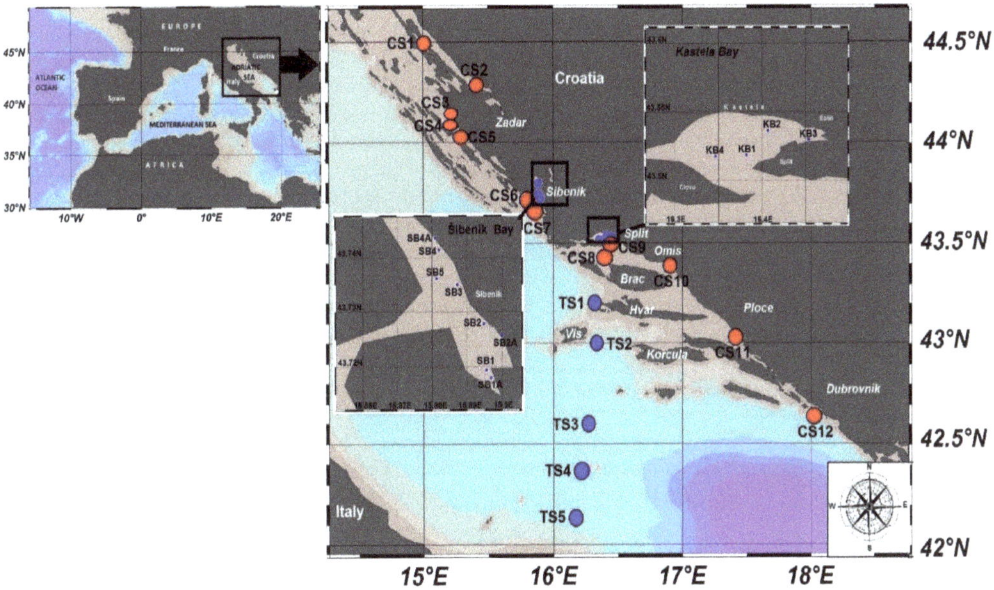

Figure 1. Sediment-sampling stations selected for source apportionment of polycyclic aromatic hydrocarbons and PAH-related toxicity evaluation in Adriatic Sea.

2.2. *Determination of Polycyclic Aromatic Hydrocarbon in Sediment Samples*

Polycyclic aromatic hydrocarbons were collected from the research vessels BIOS DVA using the Van Ween sampling device. Sediment samples were transferred in the laboratory for biogeochemistry of organic contaminants (*Biogéochimie des Contaminants Organique*) at the French institute for exploration of the sea: Ifremer (Institut Francais de Recherche pour l'exploitation de la Mer) in Nantes, for further analyses. Analytical procedure and instrumental analyses followed the protocol developed by Ifremer. Extraction of the sediment samples was performed using accelerated solvent extraction (ASE), with dichloromethane as the solvent. The sediment extracts were purified and fractionated by column chromatography, using activated SiO_2 and Al_2O_3 as adsorbents. After fractionation, the solvent was reduced by rotary evaporation and further concentrated under gentle nitrogen steam. Activated copper was added to the samples prior to instrumental analysis to remove sulphur traces, which may interfere with the chromatography column. Internal standards (phenanthrene-d_{10}, benzo[*e*]pyrene-d_{12}, fully deuterium-labelled compounds) were added to each sample prior to extraction to control the quality of the analyses. Instrumental analysis of PAHs was performed by high-resolution gas chromatography, using a HP 6890 gas chromatograph equipped with a 5973-mass selective detector (MSD). PAHs were separated on a DB5-MS capillary column made of fused silica (coated with 5% phenyl-methyl-silicone) with a length of 60 m and an inner diameter of 0.25 mm, using helium as the carrier gas. Injections were performed using a Combi-PAL CTC autosampler. Injection volume of 1–2 μL with an on-column port was used. The mass selective detector was operated in an electron impact mode at 70 eV and for quantitative analysis of individual PAHs in selected ion monitoring mode (SIM) in conjunction with perdeuterated calibration standards added prior to injection with the autosampler (acenaphthene-d10, fluorene-d10, pyrene-d10, benzo[*a*]anthracene-d12 and Indeno [1,2,3-cd]pyrene-d10). The list of analyzed PAH compounds, along with target ions, internal standards used for quantification, and abbreviations used in the text are listed in Table S3 of SI.

Identification of each compound was based on the single mass fragment ion characteristic for the compound and its retention time. The concentrations of the compounds were determined in the extracts using an eight-point calibration curve (linear regression) prepared for each compound and laboratory-prepared calibration solutions. The calibration curves were in the range of the concentrations of the analytes in the samples. A separate calibration curve was used for very low PAH concentrations in the extracts. Calibration solutions were prepared in iso-octane by appropriate dilutions (by weight). Prior to injection, deuterated surrogates for quantification were automatically added to each sample extract and to external calibration solutions. Example of characteristic chromatogram of PAH mixture in sediment samples, obtained by GC-MS operating under described conditions, is shown in Figure S1 of SI. Sediment samples with recoveries greater than 110% and lower than 70% were reanalyzed until recoveries for both internal standards were within the desirable range. Standard reference materials (SRM 1491 and SRM 2977) from the National Institute of Standards and Technology (NIST) were incorporated into a batch of 10 sediment samples for additional control of the analytical procedure. Internal standard recovery factors were used as an uncertainty matrix for the PMF model.

A detailed description of PAH analyses in sediment samples from the Adriatic Sea has been described in previous publications [22,25].

2.3. *PAH Source Apportionment by Positive Matrix Factorization*

Positive Matrix Factorization is a statistical method that reduces the large number of variables in datasets to combinations of species, called source types, and source contributions with weighted least squares regression [4,19,24].

The PMF model assumes that the concentration of each PAH compound in each sediment sample can be described as the sum of the concentration of the PAH compound emitted by the source and the source contributions for each sample, as shown in the following equation:

$$Xij = \sum_{j=1}^{p} g_{ik} f_{kj} + e_{ij}$$

where X_{ij} is the concentration of the ith PAH compound in the jth sample; g_{ik} is the concentration of the ith species in the kth source, f_{kj} is the contribution of the source kth to the sample jth and e_{ij} is the model error for the specie j determined in sample i.

The results are constrained to provide positive source contributions that make their interpretation physically meaningful. Dataset of PAH sediment concentrations at each station was used as model input, along with uncertainty data calculated using the following equation:

$$\text{Measurement uncertainty} - \text{M.U.} = |100 - \text{RF\%}|,$$

where RF is an average value of the recovery factor for two internal standards used for quality control of the analytical protocol.

Since the average recovery values ranged from 75 to 105%, the measurement uncertainties were within 0.25–0.05, depending on the sample (SI Tables S4–S7). The model was run for 4 to 6 factors and initialized with different starting points, changing the seed value from 1 to 20. The result with 4 factors, for which the Q/Q theoretical was 1.32, was considered satisfactory and adopted as the result. A more detailed description of the receptor model for source apportionment, as well as the guidelines used for model application in this work, were described in EPA guidelines [23].

2.4. Calculation of PAH-Related Toxic Equivalent-TEQ

Carcinogenic potential related with sediment PAH concentrations were calculated as summed toxic equivalent quantity for 9 EPA PAH compounds that are classified as known carcinogens, as well as for all 16 EPA PAH enlisted by USEPA as priority pollutants in the environment [14,18–20]. Toxic equivalent factors used for calculation of BaP equivalent toxicity for 16 EPA PAHs are given in Table 1.

Table 1. Toxic equivalent factors of 16 EPA PAHs (USEPA, 2014). * PAH compounds enlisted by EPA as known carcinogenic compounds.

PAH Compound	Toxic Equivalent Factor—TEF
Naphtalene *	0.001
Acenaphtene *	0.001
Acenaphtylene	0.001
Fluorene	0.001
Phenanthrene *	0.001
Anthracene	0.001
Fluoranthene *	0.001
Pyrene	0.001
Benzo[*a*]anthracene *	0.1
Chrysene *	0.001
Benzo[*b*]fluoranthene	0.1
Benzo[*k*]fluoranthene *	0.1
Benzo[*a*]pyrene *	1
Indeno [1,2,3-cd]pyrene	0.1
Dibenzo[*a,h*]pyrene	1
Benzo[*ghi*]perylene *	0.01

The original sediment concentrations of selected PAH compounds were multiplied by their toxic equivalent factor relative to the BaP, as follows:

$$TEQ = \sum C(n) \times (TEF)n$$

where *c* is the concentration of the PAH compound and *TEF* is a toxic equivalent factor relative to the Benzo[*a*]Pyrene of a specific PAH compound.

3. Results and Discussion

3.1. PAH Concentrations in the Sediments

Concentrations of all PAH compounds determined in the sediments of the study area, along with detection limits and recovery factors for two internal standards, are given in Tables S4–S7 of SI.

Concentrations of T-PAH and the characteristic groups of PAHs in sediments from 11 stations in the coastal area are shown in Figure 2. T-PAH concentration ranged from 180.71 μg kg^{-1} to 2793.18 μg kg^{-1}. The highest T-PAH concentration was measured in the sediment of the Brac-Split channel near the town of Omis (CS9), while the lowest PAH concentration was measured in the sediment near the island of Pag (CS1). High PAH concentrations in the coastal area are observed at the station near the town of Zadar (CS3). At this station, concentrations of parent-PAH were higher than those of C-PAH, indicating that PAH in this area originates mainly from pyrogenic sources, such as inland and marine traffic, residential heating, etc. C-PAH were the predominant PAH compounds only in the sediment at station CS2, whereas on the rest of stations, unsubstituted PAH were the predominant PAH compounds in the sediments. A high proportion of C-PAH fraction was also observed at stations CS4, CS9, CS10 and CS11. These stations are in the harbor (CS4) and in channel areas with dense marine traffic. Similar results with a dominance of pyrogenic PAH have been reported for sediments from coastal areas in the Mediterranean Sea, the Bay of Mexico and other coastal regions worldwide [2,26–31]. The dominance of

C-PAH in the sediment at station CS2 suggests that PAH in this area are mainly derived from unburned petroleum derivatives [10,32]. A petrogenic source of pollution at this station is also indicated by the relatively high amounts of S-PAH, which may be present in crude petroleum up to 5% [33,34].

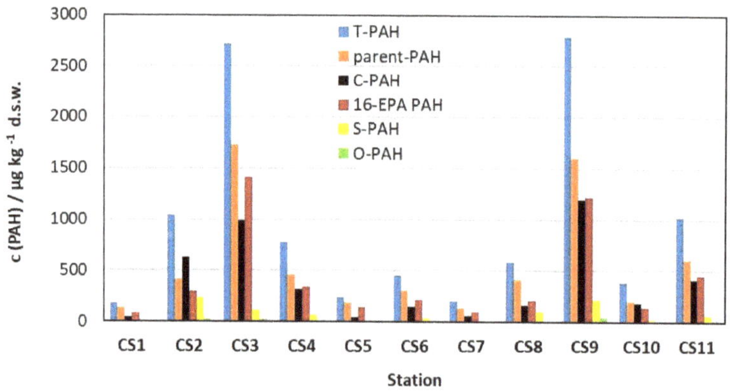

Figure 2. Concentration of total PAH (T-PAH), 16 EPA priority PAH (16 EPA PAH), unsubstituted PAH (parent-PAH), methyl-substituted PAH (C-PAH), sulphur-containing-PAH (S-PAH) and oxygen-containing-PAH (O-PAH) at stations in the coastal area.

PAH concentrations measured in the sediments of the stations along the mid-Adriatic transect are shown in Figure 3. Concentrations of T-PAH ranged from 42.859 µg kg^{-1} (TS2) to 963.517 µg kg^{-1} (TS6). PAH concentrations tend to decrease as the distance from the coast increases. Contributions of pyrogenic PAH were similar to the contribution of C-PAH at each station, except at TS1, where the concentration of parent-PAH was higher than the concentration of C-PAH fraction. Station TS1 is located near the island of Brač, where forest fires frequently occur in summer, which contributes to high levels of parent-PAH [35]. Similar results have been obtained for open sea regions around the world, where maritime traffic is the most pronounced anthropogenic pressure [2,27,29,36].

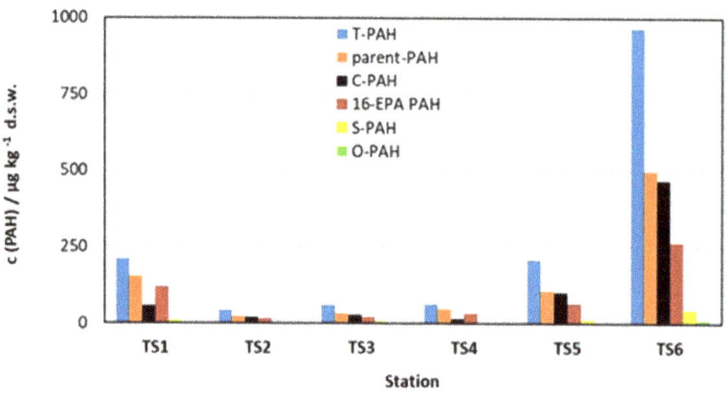

Figure 3. Concentration of total PAH (T-PAH), 16 EPA priority PAH (16 EPA PAH), unsubstituted PAH (parent-PAH), methyl-substituted PAH (C-PAH), sulphur-containing-PAH (S-PAH) and oxygen-containing-PAH (O-PAH) at stations along the mid-Adriatic transect.

PAH concentrations determined in the sediments of Kaštela Bay are shown in Figure 4. T-PAH concentration ranged from 278.64 μg kg^{-1} to 2888.9 μg kg^{-1}. The highest T-PAH concentrations during the two campaigns were measured at station KB3, which is located near the industrial port of Split and the shipyard area. C-PAH were the dominant PAH compounds, with concentrations ranging from 151.62 μg kg^{-1} to 1798.8 μg kg^{-1}. Predominance of C-PAH fraction at each station of Kaštela Bay indicate that PAH pollution in this area is mostly contributed by petrogenic sources [2]. PAH concentrations in Kaštela Bay are similar to the concentrations reported for the urban, coastal areas of the Mediterranean Sea, although they are lower than the PAH concentrations measured in the sediments of big harbour cities in the Mediterranean Sea, such as Toulon and Barcelona [8,26,27,30].

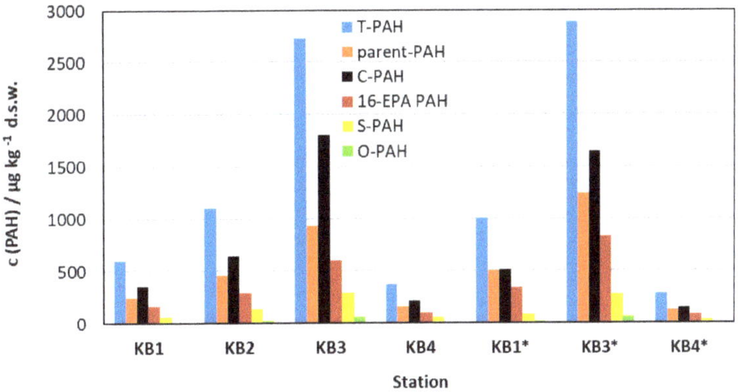

Figure 4. Concentration of total PAH (T-PAH), 16 EPA priority PAH (16 EPA PAH), unsubstituted PAH (parent-PAH), methyl-substituted PAH (C-PAH), sulphur-containing-PAH (S-PAH) and oxygen-containing-PAH (O-PAH) at stations in Kaštela Bay.

Concentrations of specific groups of PAH, determined in the sediment of Šibenik Bay, are shown in Figure 5. T-PAH concentrations ranged from 8280.8 μg kg^{-1} to 28,662 μg kg^{-1}. The concentrations of the unsubstituted parent compounds were higher than the concentrations of C-PAH at every station. The highest concentrations of PAH were measured at station SB4A, which is closest to the area where a ferroalloy plant operated from 1880 to 1990. The high concentrations of PAH observed at station SB4 during both sampling periods can also be attributed to past industrial activities, as this station is only 100 m away from SB4A. A high proportion of the C-PAH fraction in the total concentrations of PAH was observed at station SB1, in February. Since this station is located near a tourist marina, the elevated concentrations of C-PAH can be explained by accidental leakage of petroleum derivatives during ship maintenance, which usually occurs during the winter season. The lowest concentration of T-PAH was observed at station SB5, which is farthest from shore. No seasonal trend in the concentration of PAH was observed. The very high PAH concentrations measured at each station in Šibenik Bay are similar to concentrations measured in the sediments of highly industrialized areas, such as Venice, Toulon, Barcelona, Mexico Bay and other coastal areas worldwide [8,19,28,30,31].

3.2. PAH Sources in the Adriatic Sea

Factor profiles obtained by the PMF modelling are shown in Figure 6. The application of the PMF revealed four main factors contributing to the pollution of PAH in the studied area. Each factor is considered as one of the specific sources of polycyclic aromatic hydrocarbons and is further discussed as such.

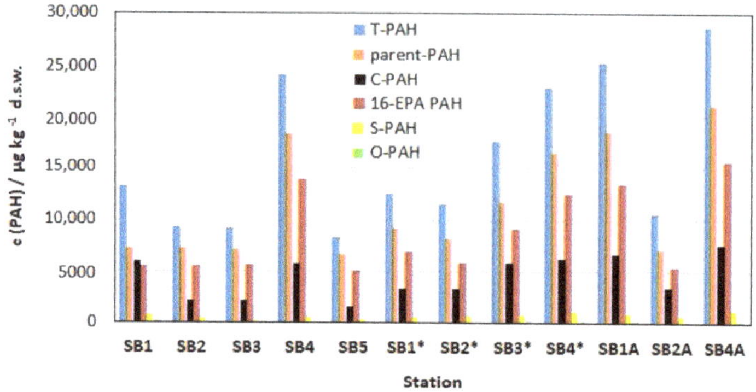

Figure 5. Concentration of total PAH (T-PAH), 16 EPA priority PAH (16 EPA PAH), unsubstituted PAH (parent-PAH), methyl-substituted PAH (C-PAH), sulphur-containing-PAH (S-PAH) and oxygen-containing-PAH (O-PAH) at stations in Šibenik Bay.

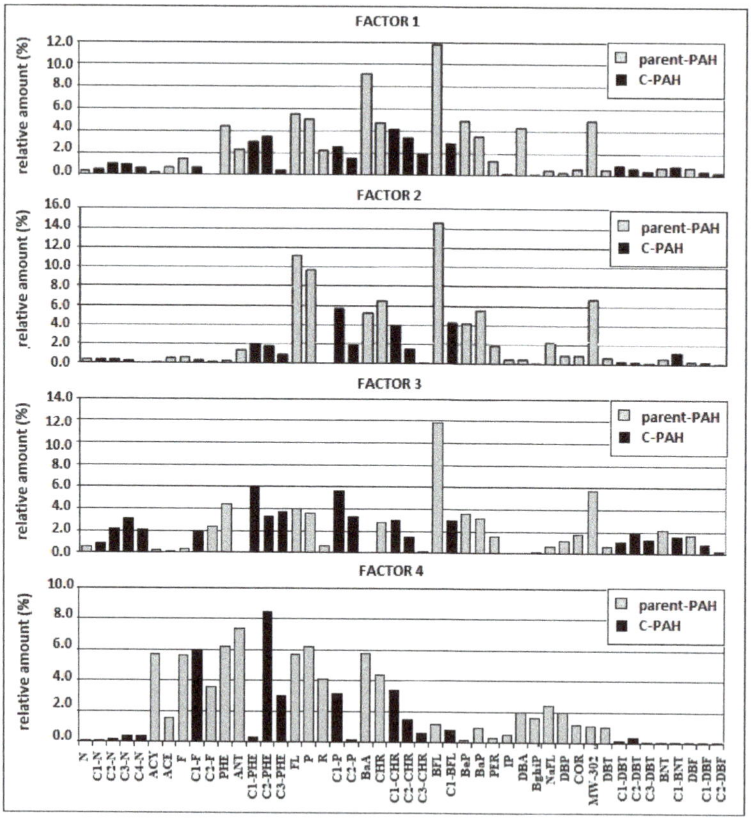

Figure 6. Factor (Source) profiles obtained by application of Positive Matrix Factorization.

Source 1 emits PAH mixtures enriched in unsubstituted PAH compounds. Dominant PAH compounds in the mixture are BFLs with a relative amount of 11.81% and BaA with a relative amount of 9.13%. Among the methyl-substituted compounds, C1-CHR (4.17%), C2-CHR (3.36%), C1-PHE (2.97%) and C2-PHE (3.43%) were the most abundant. The relative amounts of S-PAHs and O-PAH in the PAH mixture derived from this source were less than 1%.

Unsubstituted parent-PAH compounds were the most abundant fraction of the PAH mixture emitted from Source 2. The relative amounts of BFLs, FL and Py were 14.53%, 11.61% and 9.65%, respectively. The predominant C-PAH compounds in the mixture were C1-P (5.66%) and C1-BFLs (4.19%), while the predominant heterocyclic aromatic compounds were C1-BNTs, DBTs and DBFs, with relative amounts of 1.18%, 0.65% and 0.30%.

The unsubstituted parent-PAH compounds with more than three aromatic rings in the structure were the predominant compounds in the PAH mixtures emitted from both Source 1 and Source 2, indicating that both sources refer to high-temperature combustion [33,34]. However, the mixture emitted from Source 1 contained lower amounts of unsubstituted PAH and higher amount of phenanthrene than the mixture emitted from Source 2. BFLs, BaP and BaA are compounds typically formed during gasoline combustion, while mixtures formed during diesel combustion are enriched in phenanthrene, fluoranthene, pyrene and chrysene. Comparing the obtained results with the literature data, it is observed that the PAH mixture emitted by Source 1 is similar to the composition of the PAH mixtures produced by the combustion of liquid fossil fuels in vehicle engines. In contrast, Source 2 emits PAH mixtures in which the amount of low-molecular weight PAH is significantly lower than the amount of high-molecular-weight PAH. PAH mixtures enriched in unsubstituted parent-PAH compounds, such as BFLs, BaA, FL, P and CHR resemble PAH mixtures typically formed during high-temperature combustion of crude fossil fuels in industrial furnaces or forest fires [2,10,33,34]. The above-mentioned differences in the composition of the PAH mixtures emitted from these two sources indicate that Source 1 can be ascribed to inland and marine transport, while Source 2 represents high-temperature combustion of crude fossil fuels and biomass.

PAH mixture emitted from Source 3 also contained a high proportion, but the contribution of the C-PAH fraction to the total concentration of PAH was higher than the contribution of the parent-PAH compounds. The PAH mixture contained higher relative amounts of methyl phenanthrenes, naphthalenes and pyrenes, while the dominant heterocyclic compounds were C1-BNTs with a contribution of 1.18%. Comparison with the literature data suggests that Source 3 is a petrogenic PAH source, mostly associated with recent accidental leakage of unburned fossil fuels and other petroleum derivatives [2,32,37].

Source 4 emitted a PAH mixture enriched in C2-FEN (8.41%), A (7.33%) and PHE (6.17%). Compared to the other three sources, the mixture emitted from Source 4 was enriched in PHE, ANT, ACE and F, which is similar to the composition of PAH mixtures in the heavier petroleum fractions but may also indicate weathered petroleum contamination. Thus, Source 4 may also be considered as a petrogenic source, but more related to accidental leakage of heavier petroleum fractions during vessel maintenance and weathered petroleum pollution [37].

Figure 7 shows the contribution of sources (factors) to the total concentration of PAH in the sediment of each studied station. According to the model results, the two most important sources in the coastal area were transportation, both inland and marine (Source 1), and weathered petroleum products (Source 4). Transport activities were the predominant PAH sources at the stations near the island of Pag (CS1), near Biograd (CS5), in the channel of Šibenik (CS6 and CS7) and in the port of Split (CS8), where Source 1 contributed with more than 50% to the total PAH pollution.

At the other stations in the coastal zone, the main source of PAH pollution was Source 4, which can be attributed to accidental spills of fresh motor oil and other heavier petroleum distillates during vehicle maintenance. Source 4 was present at every site in the coastal zone, with contributions ranging from 27.02% to 96.06%. Source 3 was also present at stations CS2, CS9 and CS10, indicating recent petroleum pollution in these areas. These stations are in the canal areas (CS2 and CS9) and in the port area (CS10), which explains the higher contribution of recently spilled petroleum. The contribution of coke and biomass burning

(Source 2) was not observed at any station in the coastal area, which can be explained by the absence of industrial activities in these areas.

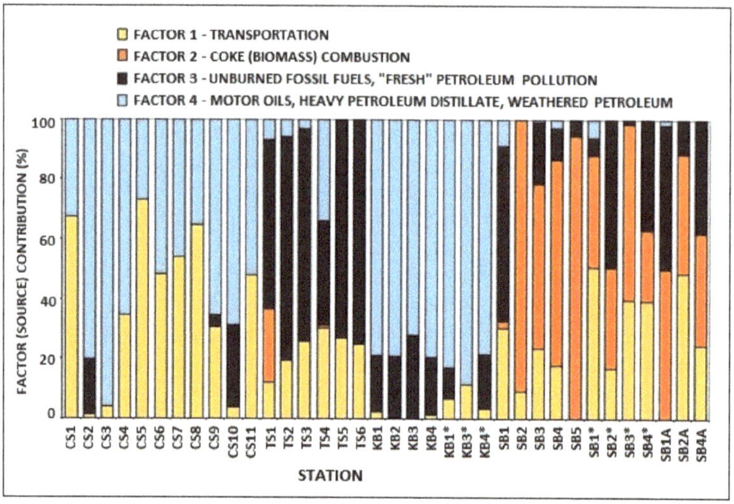

Figure 7. Factor (source) contributions (%) in the total PAH concentrations in the sediment of the investigated stations. * Refers to the sediments obtained in October of 2013.

The contribution of fresh petroleum pollution (Source 3) to the total PAH concentration was the highest in the sediments of the stations along the mid-Adriatic transect, ranging from 34.85%, as determined at station TS4, to 75.07%, as determined at station TS6. In the transect area, a high contribution of Source 1 was also observed, especially at station TS4, which is located along the high-density traffic route from the Strait of Otranto to the northern Adriatic Sea. These results indicate that accidental leakage of unburned fossil fuels and traffic are the main PAH sources in the transect area. This agrees well with the literature data for the open sea areas, where the main PAH sources are associated with maritime traffic. Elevated contributions of coke and biomass burning (Source 2) were determined in the sediment from station TS1 (24.67%), which is located near the island of Brač. Since this area is not heavily industrialized, these results can be explained by the frequent forest fires on the island of Brač, which occur frequently during the summer season [35].

In Kaštela Bay, the highest contribution to the sediment PAH concentration was determined for Source 4, ranging from 72.17% at KB3 in October, to 88.48% as observed at the same station in February. Seasonal differences in the contributions of each PAH source were observed only at station KB3, where the contribution of recently spilled petroleum was higher in February than in October. The highest contribution of Source 3 to the total concentrations of PAH in the sediment of Kaštela Bay was observed at station KB3 in February. This station is located near the industrial port of Split and the shipyard area, suggesting that the high contribution of Source 2 may be due to accidental spillage of unburned petroleum during ship maintenance [32,37]. The absence of Source 2, or minimal contributions from this source in the coastal area, Kaštela Bay and most of the open-sea stations are likely due to the absence of industrial activities in these areas.

In Šibenik Bay, a high contribution from Source 1 is observed at station SB1 in October. The contribution of Source 2 to the total PAH concentration was highest at each station in Šibenik Bay, especially at station SB5, where 94.57% of the PAH sediment concentration came from Source 2. High contributions from Source 2 were also observed in October at stations SB2 (91.07%) and SB4 (68.71%). These results can be explained by the influence of coke-burning in the ferroalloy factory that operated on the shores of the bay from 1880 to 1990. The high concentrations of PAH in the surface sediment are probably due to the diffusion of PAH from the deeper sediment layers that are oversaturated with PAH. The highest contributions from Source 3 were detected at station SB1 in February, where model results indicate that 58.72% of PAH originated from Source 3. Station SB1 is located near the tourist marina where most of the vessel maintenance is performed during the winter, suggesting that the high Source 3 contribution detected at this station in February is caused by accidental leaks of petroleum derivatives during vessel maintenance. An elevated contribution from Source 4 was also detected at this station, confirming the assumption that this station is greatly affected by vessel maintenance activities.

3.3. Benzo[a]pyrene Equivalent Toxicity of PAH Mixtures in the Adriatic Sea

Toxic equivalent quantity (TEQ) for 9 known carcinogenic PAH compounds and for 16 EPA compounds, calculated for the studied sediment samples are shown on Figure 8. It can be seen that summed TEQ for 9 and 16 PAHs exceeded 600 μg kg^{-1} only at the stations in Šibenik Bay. At the remaining stations, the TEQ values were below 200 μg kg^{-1}, regardless of the number of compounds considered. In contrast, TEQ values in Šibenik Bay were significantly higher, especially at the stations near the former ferroalloy factory and the tourist marina. When all 16 PAH compounds were considered, TEQ values at most stations exceeded 600 μg kg^{-1}, indicating a high carcinogenic risk associated with sediment PAH concentration in Šibenik Bay. Results obtained by calculation of TEQ are in good agreement with the results of PAH-related sediment toxicity obtained by applying the sediment quality guidelines, where a high environmental risk was observed for the total area of Šibenik Bay [38]. However, in the same study, the results of the bioassay application indicated poor environmental status associated with the PAH sediment concentration at stations outside Šibenik Bay. This suggests that assessing the toxicity of PAH solely on sediment concentration may significantly underestimate the environmental risk [38].

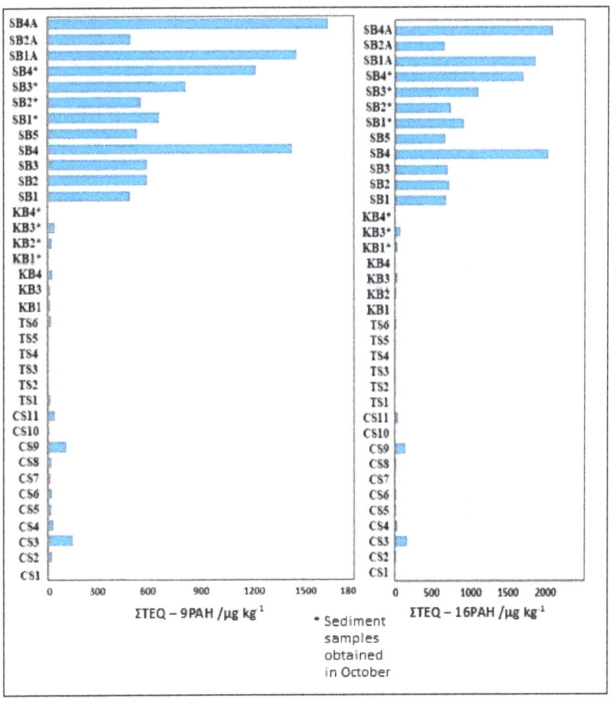

Figure 8. Toxic equivalent quantities for 9 known carcinogenic PAH and 16 EPA PAH in the sediments of investigated stations in the Adriatic Sea.

4. Conclusions

PAH concentrations in sediments of the mid-Adriatic Sea show great variability, ranging from very low concentrations in the open sea area to very high concentrations in urbanized, coastal areas. The application of the PMF model for PAH source apportionment revealed the presence of four main sources of PAH pollution in the investigated area. Among them, Source 1 and Source 4 can be considered as the most important ones since their contributions to the total PAH pollution were observed in each studied area. Positive Matrix Factorization provided detailed information about PAH sources and the contribution of each source to the total PAH in each location, which makes it a reliable tool for the management of various activities in the Adriatic region. To obtain more accurate information on pollution sources, the dataset used as model input should include a greater number of stations as well as other pollutants, such as trace metals, pesticides, dioxins, antifouling chemicals, etc. The PAH-toxicity assessment revealed a high cancer risk associated with PAH concentration in the sediments of Šibenik Bay, while in the rest of the study area the cancer risk associated with the sediment concentrations of PAH was low. While these results are in good agreement with the results obtained by applying the sediment quality guidelines, they differ greatly from the results of the bioassay used to test the acute toxicity of the sediments. It can be concluded that additional bioassays conducted on a larger number of different marine species are needed to accurately assess the environmental risk associated with PAH.

Supplementary Materials: The following supporting information can be downloaded at: https://www.mdpi.com/article/10.3390/app13126992/s1, Figure S1: Chromatogram of the PAH mixture in sediment sample separated by DB5-MS column and obtained by GC-MS operating in the SIM mode. Table S1. Sampling locations and descriptions of the sites in the Coastal zone and along the Mid Adriatic transect selected for the Source apportionment and toxicity assessment. Table S2. Sampling locations and descriptions of the sites in Kaštela and Šibenik Bay selected for the Source apportionment and toxicity assessment. Table S3. List of parent and alkyl-substituted PAH, sulphur heterocyclic aromatic compounds and oxygen heterocyclic aromatic compounds determined in sediment samples; abbreviation, molecular weight and internal standard used as a reference. Table S4. PAH concentration ($\mu g\ kg^{-1}$ dry sediment weight) with limits of detection (LOD) determined in the sediments of Coastal area and measurements uncertainties for each sample used as input data matrices for PMF. Table S5. PAH concentration ($\mu g\ kg^{-1}$ dry sediment weight) with limits of detection (LOD) determined in the sediments of transect area and measurements uncertainties (M.U.) for each sample used as input data matrices for PMF. Table S6. PAH concentration ($\mu g\ kg^{-1}$ dry sediment weight) with limits of detection (LOD) determined in the sediments of Kastela Bay are and measurements uncertainties (M.U.) for each sample used as input data matrices for PMF. Table S7. PAH concentration ($\mu g\ kg^{-1}$ dry sediment weight) with limits of detection (LOD) determined in the sediments of Šibenik bay and measurements uncertainties (M.U.) for each sample used as input data matrices for PMF.

Author Contributions: Conceptualization, J.M.; Methodology, J.M.; Software, J.M.; Formal analysis, J.M. and J.V.; Resources, G.K.; Writing—original draft, J.M.; Writing—review & editing, J.M. and J.V.; Visualization, J.M.; Project administration, G.K.; Funding acquisition, G.K. All authors have read and agreed to the published version of the manuscript.

Funding: This study was financially supported by Ifremer Centre Atlantique, Nantes, France; IOF-Split, Croatia and the Croatian Science Foundation (grant IP-2014-09-3606: "Marine plankton as a tool for assessment of climate and anthropogenic influence on the marine ecosystem—MARIPLAN"). The study was conducted within the EU FP7 N_287600 PERSEUS project, which provided additional financial support.

Institutional Review Board Statement: Not applicable.

Informed Consent Statement: Not applicable.

Data Availability Statement: Data supporting reported results can be found in the Doctoral thesis written by Jelena Mandić at the following link: https://dr.nsk.hr/islandora/object/morest%3A21 (accessed on 5 June 2023).

Acknowledgments: Authors wish to express their gratitude to Jacek Tronczynski, Celine Tixier and Nadege Bely for their assistance and support during the scientific visit at Ifremer institute laboratories. Many thanks to the crew of research vessel R/V "BIOS DVA" for their help during the fieldwork.

Conflicts of Interest: The author declare no conflict of interest.

References

1. Vollhardt, K.P.C.; Schore, N.E. *Organic Chemistry*, 3rd ed.; W. H. Freeman and Company: New York, NY, USA, 1999.
2. Neff, J.M. *Polycyclic Aromatic Hydrocarbons in the Aquatic Environment, Sources, Fates and Biological Effect*; Applied Science Publishers Ltd.: Essex, UK, 1979.
3. Tsapakis, M.; Stephanou, E.G.; Karakassis, I. Evaluation of atmospheric transport as a nonpoint source of polycyclic aromatic hydrocarbons in marine sediments of the Eastern Mediterranean. *Mar. Chem.* **2003**, *80*, 283–298. [CrossRef]
4. Duodu, G.O.; Ogogo, N.K.; Nanayakkara, M.; Sandya, W.; Harden, F.; Goonetilleke, A.; Ayoko, G. Source apportionment and risk assessment of PAHs in Brisbane River sediment, Australia. *Ecol. Indic.* **2017**, *73*, 784–799. [CrossRef]
5. Guo, J.; Wu, F.; Luo, X.; Liang, Z.; Liao, H.; Zhang, R.; Mai, B. Anthropogenic input of polycyclic aromatic hydrocarbons into five lakes in Western China. *Environ. Pollut.* **2010**, *158*, 2175–2180. [CrossRef] [PubMed]
6. Gschwend, P.M.; Hites, R.A. Fluxes of polycyclic aromatic hydrocarbons to marine and lacustrine sediments in the northeastern United States. *Geochim. Cosmochim. Acta* **1981**, *45*, 2359–2367. [CrossRef]
7. Sicre, M.A.; Marty, J.C.; Saliot, A.; Aparicio, X.; Grimalt, J.; Albaiges, J. Aliphatic and aromatic hydrocarbons in different sized aerosols over the Mediterranean Sea: Occurrence and origin. *Atmos. Environ.* **1987**, *21*, 2247–2259. [CrossRef]
8. Budzinski, H.; Jones, I.; Bellocq, J.; Pierard, C.; Garrigues, P. Evaluation of sediment contamination by polycyclic aromatic hydrocarbons in the Gironde estuary. *Mar. Chem.* **1997**, *58*, 85–97. [CrossRef]
9. Yunker, M.; MacDonald, R.; Vingarzan, R.; Mitchell, R.; Goyette, D.; Sylvestre, S. PAHs in the Fraser Rive Basin: A critical appraisal of PAH ratios as indicators of PAH source and composition. *Org. Geochem.* **2002**, *33*, 489–515. [CrossRef]
10. Tobiszewski, M.; Namiesnik, J. PAH diagnostic ratios for identification of pollution emission sources. *Environ. Pollut.* **2012**, *162*, 110–119. [CrossRef]
11. Agency for Toxic Substances and Disease Registry (ATSDR). *Toxicological Profile for Polycyclic Aromatic Hydrocarbons (PAHs)*; Department of Health and Human Services, Public Health Service: Atlanta, GA, USA, 1995.
12. Santodonato, J.; Howard, P.; Basu, D. Health and Ecological Assessment of Polynuclear Aromatic Hydrocarbons. *J. Environ. Pathol. Toxicol.* **1981**, *5*, 100–368.
13. IARC (International Agency for Research on Cancer). *Overall Evaluations of Carcinogenicity: An Updating of IARC Monographs Volumes 1 to 42*; IARC Monographs on the Evaluation of Carcinogenic Risk to Humans Suppl. 7; IARC: Lyon, France, 1987.
14. USEPA (U.S. Environmental Protection Agency). *Ambient Water Quality Criteria for Polynuclear Aromatic Hydrocarbons*; EPA 440/5-80-069. US NTIS PB81-117806; USEPA: Washington, DC, USA, 1980.
15. McGrath, J.A.; Joshua, N.; Bess, A.S.; Parkerton, T.F. Review of Polycyclic Aromatic Hydrocarbons (PAHs) Sediment Quality Guidelines for the Protection of Benthic Life. *Integr. Environ. Assess. Manag.* **2019**, *15*, 505–518. [CrossRef]
16. Bihari, N.; Najdek, M.; Floris, R.; Batel, R.; Zahn, R.K. Sediment toxicity assessment using bacterial bioluminescence: Effect of an unusual phytoplankton bloom. *Mar. Ecol. Prog. Ser.* **1989**, *57*, 307–310. [CrossRef]
17. USEPA. *Human Health Evaluation Manual, Supplemental Guidance: Update of Standard Default Exposure Factors*; OSWER Directive 9200; USEPA: Washington, DC, USA, 2014, pp. 1–120. Available online: https://www.epa.gov/sites/default/files/2015-11/documents/oswer_directive_9200.1120_exposurefactors_corrected2.pdf (accessed on 8 March 2023).
18. Deelaman, W.; Choochuay, C.; Pongpiachan, S.; Han, Y. Ecological and health risks of polycyclic aromatic hydrocarbons in the sediment core of Phayao Lake, Thailand. *J. Environ. Expo. Assess.* **2023**, *2*, 3. [CrossRef]
19. Cao, Y.; Lin, C.; Zhang, X.; Liu, X.; He, M.; Ouyang, W. Distribution, source, and ecological risks of polycyclic aromatic hydrocarbons in Lake Qinghai, China. *Environ. Pollut.* **2020**, *266*, 115401. [CrossRef] [PubMed]
20. Delistraty, D. Toxic equivalency factor approach for risk assessment of polycyclic aromatic hydrocarbons. *Toxicol. Environ. Chem.* **1997**, *64*, 81–108. [CrossRef]
21. Lin, T.; Qin, Y.; Zhen, B.; Li, Y.; Chen, Y.; Guo, Z. Source apportionment of polycyclic aromatic hydrocarbons in the Dahuofang Reservoir, Northeast China. *Environ. Monit. Assess.* **2013**, *185*, 945–953. [CrossRef] [PubMed]
22. Mandić, J.; Tronczynski, J.; Kušpilić, G. Polycyclic aromatic hydrocarbons in surface sediments of the mid- Adriatic and along the Croatian coast: Levels, distributions and Sources. *Environ. Pollut.* **2018**, *242*, 519–527. [CrossRef]
23. USEPA. *Positive Matrix Factorization (PMF) 5.0 Fundamentals & User Guide*; USEPA: Washington, DC, USA, 2014. Available online: http://www.epa.gov/heasd/research/pmf.html (accessed on 14 February 2023).
24. Joint Research Centre, Institute for Environment and Sustainability; Comero, S.; Gawlik, B.; Capitani, L. *Positive Matrix Factorisation (PMF)—An Introduction to the Chemometric Evaluation of Environmental Monitoring Data Using PMF*; Publications Office: Tokyo, Japan, 2011; Available online: https://data.europa.eu/doi/10.2788/2497 (accessed on 8 March 2023).
25. Tronczyński, J.; Munschy, C.; Moisan, K.; Guiot, N.; Truquet, I.; Olivier, N.; Men, S.; Furaut, A. Contamination of the Bay of Biscay by polycyclic aromatic hydrocarbons (PAHs) following the T/V "Erika" oil spill. *Aquat. Living Resour.* **2004**, *17*, 243–259. [CrossRef]

26. Lipiatou, E.; Tolosa, I.; Simo, R.; Bouloubassi, I.; Dachs, J.; Marti, S.; Sicre, M.A.; Bayona, J.M.; Grimalt, J.O.; Salliot, A.; et al. Mass budget and dynamics of polycyclic aromatic hydrocarbons in the Mediterranean sea. *Deep-Sea Res.* **1997**, *44*, 881–905. [CrossRef]
27. Wakeham, S.G. Aliphatic and polycyclic aromatic hydrocarbons in Black Sea sediments. *Mar. Chem.* **1996**, *53*, 187–205. [CrossRef]
28. Soclo, H.H.; Garrigues, P.H.; Ewald, M. Origin of polycyclic aromatic hydrocarbons (PAHs) in coastal and marine sediments: Case studies in Cotonou (Benin) and Aquitane (France) Areas. *Mar. Pollut. Bull.* **2000**, *40*, 387–396. [CrossRef]
29. Witt, G. Occurrence and transport of polycyclic aromatic hydrocarbons in the water bodies of the Baltic Sea. *Mar. Chem.* **2002**, *79*, 49–66. [CrossRef]
30. De Luca, G.; Furesi, A.; Leardi, R.; Micera, G.; Panzanelli, A.; Piu, P.C.; Sanna, G. Polycyclic aromatic hydrocarbons assessment in the sediments of the Porto Tores Harbor (Northern Sardinia, Italy). *Mar. Chem.* **2004**, *86*, 15–32. [CrossRef]
31. Ponce-Vélez, G.; Botello, A.V.; Díaz-González, G. Organic and inorganic pollutants in marine sediments from northern and southern continental shelf of the Gulf of Mexico. *Inter. J. Environ. Pollut.* **2006**, *26*, 295–311. [CrossRef]
32. Bjørseth, A.; Ramdahl, T. *Handbook of Polycyclic Aromatic Hydrocarbons*; Bjorseth, A., Ed.; Marcel Dekker: New York, NY, USA, 1985; Volume 2, p. 416.
33. Lima, A.L.; Farrington, W.J.; Reddy, C.M. Combustion-derived polycyclic aromatic hydrocarbons in the environment—A review. *Environ. Forensics* **2005**, *6*, 109–131. [CrossRef]
34. Khalili, N.R.; Scheff, P.A.; Holsen, T.M. PAH source fingerprints for coke ovens, diesel and gasoline engines, highway tunnels and wood combustion emissions. *Atmos. Environ.* **1995**, *29*, 533–542. [CrossRef]
35. Mifka, B.; Vučetić, V. Weather analysis during extreme forest fire on island of Brač from 14 to 17 July 2011. *Vatrog. I Upravlj. Požarima* **2012**, *1*, 13–25. Available online: https://hrcak.srce.hr/103617 (accessed on 22 May 2023).
36. Stout, S.A.; Uhler, A.D.; Emsbo-Mattingly, S.D. Comparative evaluation of background anthropogenic hydrocarbons in surficial sediments from Nine urban waterways. *Environ. Sci. Technol.* **2004**, *38*, 2987–2994. [CrossRef] [PubMed]
37. Wang, Z.; Fingas, M.; Blenkinsopp, S.; Sergy, G.; Landriault, M.; Sigouin, L. Comparison of oil composition changes due to biodegradation and physical weathering in different oils. *J. Chromatogr. A* **1998**, *809*, 89–107. [CrossRef]
38. Mandić, J.; Veža, J.; Kušpilić, G. Assessment of environmental risk related to the polycyclic aromatic hydrocarbons (PAH) in the sediments along the eastern Adriatic coast. *Acta Adriat.* **2022**, *63*, 135–150. [CrossRef]

applied sciences

MDPI

Article

An Integrated Approach for the Environmental Characterization of a Coastal Area in the Southern Atacama Desert

Estefanía Bonnail [1,*], Edgardo Cruces [1], Eva Rothäusler [1], Rómulo Oses [2], Ayón García [3], Christopher Ulloa [3], Nuria Navarro [4], Yesenia Rojas-Lillo [5], Álvaro Parra Valdivia [6], Ricardo Catalán Garrido [6] and Manuel Abad [4]

1 Centro de Investigaciones Costeras de la Universidad de Atacama (CIC-UDA), Av. Copayapu 485, Copiapó 1530000, Chile
2 Centro Regional de Investigación y Desarrollo Sustentable de Atacama (CRIDESAT), Universidad de Atacama, Av. Copayapu 485, Copiapó 1530000, Chile
3 Instituto de Investigaciones Científicas y Tecnológicas de la Universidad de Atacama (IDICTEC-UDA), Av. Copayapu 485, Copiapó 1530000, Chile
4 Departamento Biología y Geología, Física y Química Inorgánica, Universidad Rey Juan Carlos, Calle Tulipán s/n, 28933 Móstoles, Spain
5 Independent Researcher, Copiapó 1530000, Chile
6 SEREMI del Medio Ambiente Región de Atacama, Ministerio de Medio Ambiente, Gobierno de Chile, Calle Portales 830, Copiapó 1530358, Chile
* Correspondence: estefania.bonnail@uda.cl

Abstract: Desert areas in northern Chile are highly valuable ecosystems. While human activities are impacting the area in different ways, there are few environmental studies available. The current study analysed the ecological health status (water, sediment, biota) of a northern coastal area in the Atacama Region, including a national park (with a protected marine area), a tourist and benthic management area, and an industrial area. Results from the physical–chemical characterization and physiological state of organisms of ecological importance (macroalgae and microalgae) were integrated to determine pollution and toxic responses. The results identified high and moderate pollution levels for Bi, Ca, As, Ag and Cd in sediments. The As concentration in sediments is the leading environmental problem, with average values above the threshold effect level, associated with fine sediments. The stations showed increasing contamination and stress from north to south (national park > tourist and benthic management area > industrial area), associated with the proximity to the discharge of mining waste from the Salado River. The national park registered the poorest health status as demonstrated by high Cu bioaccumulation and high photosynthetic stress in the macroalgae and the lowest biomass concentration of the microalgae in water. The tourist and benthic management area demonstrated high As concentrations in sediments and Cd bioaccumulation. The industrial area was the least contaminated area, exhibiting lower photosynthetic stress and bioaccumulation.

Keywords: desert; marine pollution; ecosystem health status; algae monitoring

Citation: Bonnail, E.; Cruces, E.; Rothäusler, E.; Oses, R.; García, A.; Ulloa, C.; Navarro, N.; Rojas-Lillo, Y.; Parra Valdivia, Á.; Catalán Garrido, R.; et al. An Integrated Approach for the Environmental Characterization of a Coastal Area in the Southern Atacama Desert. *Appl. Sci.* **2023**, *13*, 6360. https://doi.org/10.3390/app13116360

Academic Editors: José A. González-Pérez and Abdeltif Amrane

Received: 23 March 2023
Revised: 4 May 2023
Accepted: 11 May 2023
Published: 23 May 2023

1. Introduction

Desert areas have a low level of environmental monitoring worldwide, as they are remote and spread out over a wide area and are therefore difficult to access. One such area is in northern Chile, including the Atacama Desert in the Atacama region, which occupies 10% of the national territory and is home to 1.6% of the national population. Here, the coastal waters are affected by human activities, such as industrial development, ports, desalination plants and tourism, among others [1]. One of the most important economic sectors in northern Chile is the mining of gold, silver, molybdenum, iron and lithium, and above all the extraction of copper (30% of global production). This region is also known for its vineyards as well as the cultivation of olives and other fruit crops. Other pressures on these coastal areas are intensive aquaculture, marine resource extraction and tourism. Vast amounts of water are needed due to these activities, which requires the installation of desalination plants. The largest desalinization plant in Latin America (1200 L s^{-1} treated water, [2]) will be installed in this region. Hence, human actions have a direct and indirect impact on the marine environment [3]. Despite these intensive human pressures in this desert area, [1] there is a considerable scientific knowledge gap about whether and how contaminants affect the health status of marine ecosystems.

One of the easiest ways to achieve environmental assessment is biomonitoring. For instance, the use of organisms to test an ecosystem's health status is an appealing tool for assessing pollution in aquatic ecosystems [4]. The target organisms are usually (semi-)sessile and can reflect the effects of the environment through biochemical, morphological or physiological responses. Bivalves, gastropods, lugworms, fish and macrophytes are commonly used. In the current study, a kelp of the genus *Lessonia* was selected as a target species because it is a suitable model for describing ecosystems. *Lessonia berteroana* is widely distributed on hard bottoms in northern Chile [5]. It is an economic and social resource species, common in the intertidal and shallow subtidal zone [6], and has been previously used for biomonitoring through bioaccumulation [7], abundance and morphometric features, fertility and sporophyte formation [8], and ecophysiological traits [9,10]. Microalga biomass is one of the most commonly used attributes in water quality monitoring studies, as it consistently responds to stressors [11] but also serves as an early warning system [12].

Integrating the environmental status (presence and abundance of contaminants) and biological responses of biota is often carried out using statistical tools. In particular, factor analysis allows the generated data to be grouped into new variables. The methodology is used in numerous areas of the world to determine the contamination and toxicity caused by different sources, using different target species and endpoints [13–17]. It is the main statistical method of research in the life sciences [18].

An earlier study [1] found that there was a lack of environmental studies and much data were obsolete for an adequate environmental assessment; these studies are the basis for creating national environmental legislation. The aim of the current study is to assess a sector of the Atacama Region at 26° S along a stretch of 80 km with an expected pollution gradient, where an industrial area, a tourism/fishing area and a national park are located. We analysed and integrated different matrices (sediment, water, macroalga tissue and phytoplankton) to determine the pollution level and the health status of this coastal environment.

2. Materials and Methods

2.1. Sampling Sites

Sampling was carried out over three matrices (water, sediment and macroalgae) at three sites in the northern coastal area (26° S) of the Atacama Region (Figure 1). Pan de Azúcar (PA) National Park is located in the north of the Atacama Region, near a public access beach and campsite off Pan de Azúcar Island. It is a zone of marine and ecological interest, as well as an exclusive and preferential conservation zone, as this area is home to endemic species. Flamenco (PB) is a small fishing village, with a port for artisanal fishing and a benthos management area; it is considered an industrial zone but is also used for tourism. The last and most southern site is Punta Totoralillo (PC), where there is industrial area and an important port that exports mainly iron materials.

At each site, we measured the physical–chemical parameters (temperature—T, electrical conductivity—EC, redox potential—Eh, pH, and total dissolved solid—TDS) using a Hanna (HI98195) multi-parameter meter.

Water samples were collected (in triplicate) in sterile plastic bottles, filtered (0.22 μm), and acidified (pH < 2) for metal analyses. Water samples were collected in amber bottles for nutrient analysis. Some samples were collected for alkalinity analysis. An amount greater than one litre per site was used for chlorophyll determination. Sediment samples (triplicate) were collected from the upper layer (5 cm) from the intertidal area for grain size characterisation and multi-element analyses. Photosynthetic activity was determined in fronds of *L. berteroana* using a pulse-amplitude modulated fluorometer (Junior-PAM).

Leaves of *L. berteroana* were sampled by hand at low tide for photosynthetic measurements. Additional fronds samples were preserved in the water from the collection site for transportation before further elemental analysis in the laboratory. All samples were refrigerated and transported to the laboratory in cold and dark conditions.

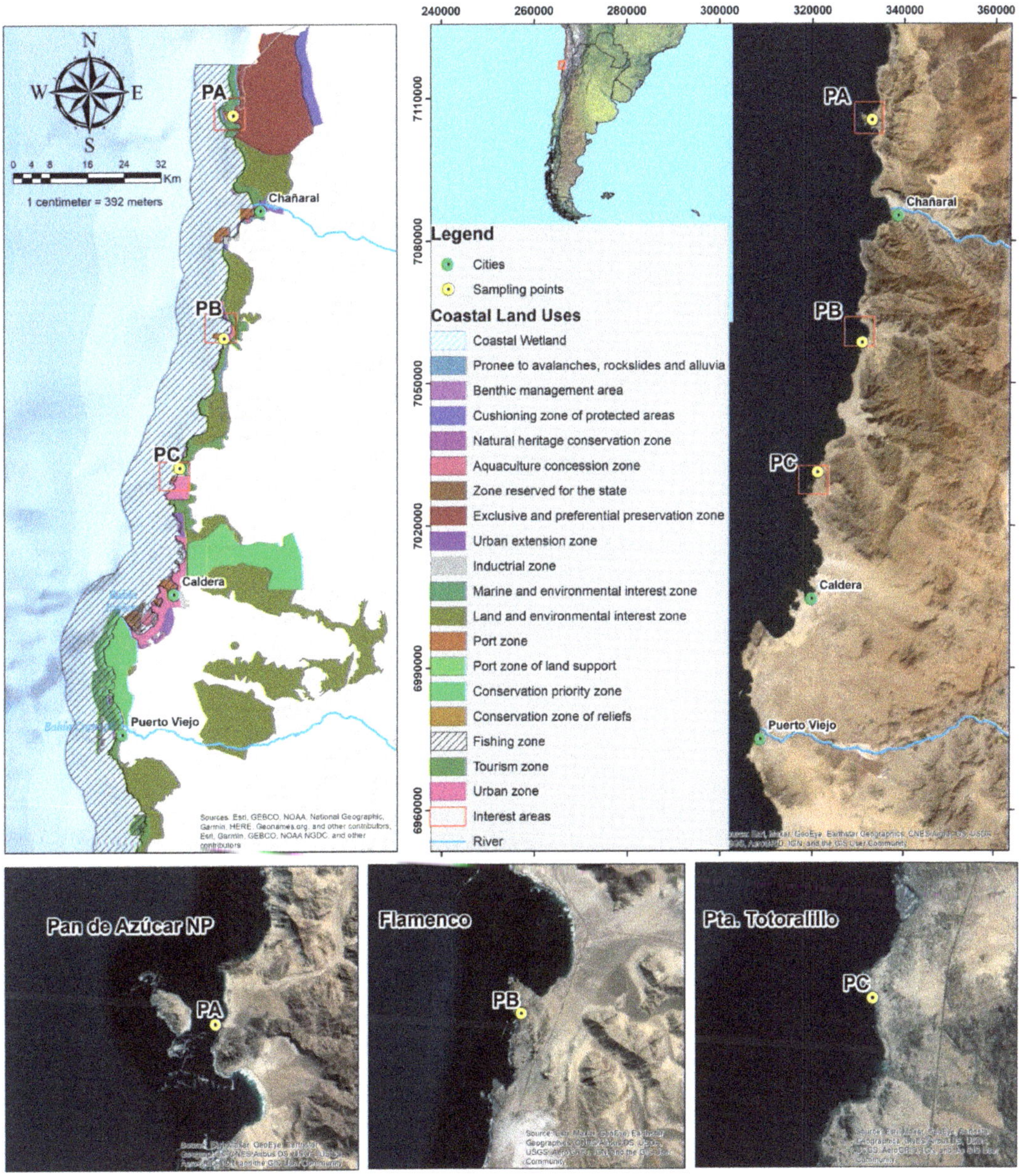

Figure 1. Location of the study area and the sampling stations along a coastal stretch of around 80 km: Pan de Azúcar National Park (PA), Flamenco (PB) and Punta Totoralillo (PC).

2.2. Chemical Analyses

The determination of carbonate and bicarbonate in the water was carried out with a titrimetric alkalinity test (MQUANT® Sigma-Aldrich, Schnelldorf, Germany). Grain size analysis was performed through sieve determination with Rotap instrumentation and following the standard classification [19]. Then, sediment samples were dried (40 ± 5 °C), sieved and homogenised. A subsample of 1 g was acid-digested (HNO_3, H_2O_2, HCl) over a thermo-block (EPA 3051/3050). Fronds of *L. berteroana* were dried in an oven at 40 ± 5 °C and ground before acid digestion (HNO_{3conc}, H_2O_2 and HCl_{conc}). Subsequently, the elemental analysis of water, sediment and alga extracts was carried out using atomic absorbance spectroscopy (AAS) and an inductively coupled plasma mass spectrometer (ICP-MS Agilent 7900) according to [20,21]. Certified reference standard materials were employed (SCL-22-598- Standard 2A Multi-element, and SCL-22-598- Standard 4 Multi-element) to certify the quality and accuracy of the analysis.

2.3. Photosynthetic Activity Determination

Photosynthetic measurements were taken with a pulse-amplitude modulated (PAM) chlorophyll fluorometer Junior-PAM (Walz GmbH, Effeltrich, Germany). After the samples were dark-adapted for 15 min, the maximum [$F_v/F_m = (F_m - F_0)/F_0$] quantum yields of photosystem II (PSII) were determined from the ground fluorescence (F_0) in the presence of a weak measuring light (<1 μmol photons m^{-2} s^{-1}) and maximum (F_m) fluorescence due to the application of a saturation light pulse (>9000 μmol photons m^{-2} s^{-1}, 0.8 s). Samples were exposed to a series of increasing actinic light intensities (E_{AL}) to determine the effective PSII-quantum yields [$\Phi_{PSII} = (F'_m - F_t)/F'_m$]. Every 30 s and before the intensity of the actinic light increased, a saturation pulse (>9000 μmol photons m^{-2} s^{-1}, 0.8 s) was applied, and the corresponding terminal (F_t) and maximal (F'_m) fluorescence yields were measured. The photosynthetic electron transport rates (ETR) were calculated as follows [22]:

$$ETR = \Phi_{PSII} \times E_{AL} \times A \times F_{II} \quad (1)$$

where Φ_{PSII} is the effective PSII-quantum yield, E_{AL} is the actinic light intensity, A is the thallus absorptance and F_{II} is the fraction of quanta absorbed by PSII (i.e., 0.8). The thallus absorptance A was measured using a cosine-corrected 2π PAR quantum sensor (Licor 192 SB; Li-COR Inc., Lincoln, Dearborn, MI, USA), and A was calculated as

$$A = 1 - (E_t \times E_0{}^{-1}) - R \quad (2)$$

where E_t and E_0 are the irradiances of the PAR-emitting light source used in the experiment, measured when the sensor was covered with one thallus of *L. berteroana* (transmitted light: E_t) and remained uncovered (incident light: E_0), respectively, and R is the reflectance (reflected fraction) of the thallus (i.e., 0.05). Values of ETR were plotted against the actinic light intensity (E_{AL}), and the photosynthetic parameters (ETR_{max}, α_{ETR}, E_k) were determined by non-linear curve fitting of ETR-E_{AL} curves after the model of [23]:

$$ETR = ETR_{max} \times \tanh(\alpha_{ETR} \times E_{AL}\ ETR_{max}{}^{-1}) \quad (3)$$

where ETR_{max} is the maximum ETR, tanh is the hyperbolic tangent function, E_{AL} is the actinic light intensity and α_{ETR} is the initial slope of the ETR-E_{AL} curve. The saturation irradiance for the photosynthetic electron transport (E_k) is the intercept between αETR and ETR_{max} ($E_k = ETR_{max}\ \alpha_{ETR}{}^{-1}$).

2.4. Chlorophyll Determination

The concentration of chlorophyll *a* (Chl-*a*) was determined in the collected water samples. Chl-*a* was extracted for 24 h in the dark in 90% acetone *v*/*v*, and data were calculated using the trichromatic equation of [24].

2.5. Data Treatment

The geoaccumulation index (I_{geo}; [25]) determined the degree of pollution in the collected sediments by comparison with the average of the upper crust concentrations [26].

2.6. Statistical Analyses

A one-way ANOVA (Kruskal–Wallis) followed by Dunn´s comparison test were applied to evaluate the differences in the concentrations of the elements in the sediments and algae, using the statistical program GraphPad Prism (version 5.03).

To determine the variable distribution of the environmental data (water $n = 5$, sediment $n = 8$) and algal responses (bioaccumulation $n = 7$, physiological responses $n = 5$), a multivariate factor analysis (FA) approach was applied following the methodology described in [14,27,28], with factor rotation using the varimax normalised procedure of the statistical package PAST3. A cut-off of 0.7 for the component loading was used to group the variables. The prevalence of the new factors per site was also calculated.

3. Results and Discussion

3.1. Water Chemistry

Samples from three matrices (water, sediment and biota) were collected in PA, PB and PC and analysed in the laboratory. The physical–chemical characterisation of water samples in the different stations is summarised in Table 1. Similar electrical conductivity (59 mS cm^{-1}), salinity and total dissolved elements (29 ppt) were found in the water samples. The pH ranged between 7.53 and 7.68, and the redox potential was lower for PA (140 mV) than PB and PC (204 and 217 mV, respectively). The highest total alkalinity was found in PB (229 mg $CaCO_3$), and the lowest in PC (185 mg $CaCO_3$). The concentrations of a few elements (Fe, Ag, As, Pb, Mn) in water were below the detection limits (Supplementary Material Table S1) in the National Park (PA). The Cu concentration in water was highest in PA: 15 µg L^{-1} Cu, compared to 1.7 and 3.5 µg L^{-1} in PB and PC, respectively. The Cu concentration in PA was above the marine reference exposure limits of NOAA [29].

Table 1. Physical–chemical parameters in the water at each sampled station (PA, PB, PC), average element concentration in seawater (N = 3) and marine reference exposure limits according to NOAA [29].

		PA	PB	PC	Marine Surface Water (Acute/Chronic)
T	°C	15.66	16.43	15.02	
pH		7.68 ± 0.01	7.65 ± 0.01	7.53 ± 0.005	
S	psu	39.84 ± 0.03	39.60 ± 0.14	39.93 ± 0.001	
EC	µS cm^{-1}	59,431 ± 40	59,093 ± 194	59,580 ± 10	
TDS	ppt	29.71	29.54	29.79	
Eh	mV	140 ± 1.14	204 ± 2.21	217 ± 0.21	
Alk	$mgCaCO_3$ L^{-1}	216 ± 32	229 ± 67	185 ± 17	
Element concentrations					
Ag	mg L^{-1}	bdl	bdl	bdl	
Al	µg L^{-1}	106	670	713	
As	µg L^{-1}	bdl	1.60	1.75	69/36
Ca	mg L^{-1}	377	326	371	
Cd	µg L^{-1}	0.22	0.37	0.21	40/8.8
Cu	µg L^{-1}	15.1	1.7	3.5	4.8/3.1
K	mg L^{-1}	359	310	355	-/100
Li	µg L^{-1}	130	163	166	
Mg	mg L^{-1}	1207	1016	1241	
Mn	µg L^{-1}	bdl	bdl	3.40	
Na	mg L^{-1}	9946	8722	10,200	
Pb	µg L^{-1}	bdl	8.3	40	210/8.1
Zn	µg L^{-1}	33.3	46.6	60.0	90/81

bdl: below detection limits.

3.2. Sediment Characterisation

The grain size of sediment samples (SI, Figure S1) was predominantly coarse and very coarse sand (>60%). The least representative fractions were silt and mud (<0.23% for PA, <0.05% for PB and PC). The chemical composition of sediments in the different stations, plus average values and previous data collected, are summarised in Table 2.

Cd concentration in sediments was below the detection limit for the northern stations, while 1.11 mg kg^{-1} was found in PC; this value was above the concentrations previously seen in the area (0.095 mg kg^{-1}) (Table 2). Al concentration in sediments ranged between 1850 and 2247 mg kg^{-1}; these concentrations were below values (4070 mg kg^{-1}) previously reported in the area. Concentrations of Fe in sediments were similar to those found along the coast (3813 $\pm$ 271 mg kg^{-1}). There was a significant accumulation of Cu in sediments from the National Park (PA, 9.31 mg kg^{-1}), which surpassed the average concentration in the area (Table 2), although there is a massive accumulation in Chañaral from historical mining residue discharges (7.20–985 mg Cu kg^{-1}) [30,31]. The Pb concentration in sediments varied in the study area, ranging between 7.67 and 27 mg kg^{-1}: PA (21.30) > PC (17.17) > PB (10.66). Nevertheless, these are normal concentrations in the area; a previous study [30] found values from 1.57 to 51 mg Pb kg^{-1}. The same tendency was found for As, where averaged concentrations (39.88 $\pm$ 18 mg kg^{-1}) agreed with previous studies [32,33], who found concentrations between 38.10 and 117 mg kg^{-1}. However, a significant difference ($p < 0.05$) was found between concentrations in sediments from PB (60.25 mg As kg^{-1}) and PC (19.73 mg As kg^{-1}). The concentrations of As surpassed the threshold effect levels (TEL), i.e., the concentrations had a certain probability of being toxic as tested through standard bioassays. The concentration of Zn in sediments from PC was significantly different ($p < 0.05$) for PA, but none of the concentrations surpassed the TEL values.

Table 2. Element concentrations in sediments collected in the sampled stations (PA, PB, PC), descriptive statistics and median values found in the study area by the Marine Sediment Quality Atacama project (MASEQATA) and the TEL (lower threshold levels) values for the NOAA [29].

		PA	PB	PC	av	sd	Min	Max	MASEQATA	TEL
Fines	%	0.23	0.05	0.05						
Al	mg kg^{-1}	1887	2043	2151	2027	152	1850	2247	4070	730
Ag	mg kg^{-1}	bdl	1.09	1.46	1.24	0.62	0.70	1.95	0.59	
As	mg kg^{-1}	46.43 ^{ab}	60.25 ^{a}	19.73 ^{b}	39.88	18.67	12.10	66.70	1.37	7.24
Bi	mg kg^{-1}	18.13	12.77	18.67	16.52	8.80	3.71	26.30	0.01	
Ca	g kg^{-1}	279	194	199	224	42	186	288	14.60	
Cd	mg kg^{-1}	bdl	bdl	1.11	1.11				0.095	680
Cu	mg kg^{-1}	9.31	6.24	5.59	7.05	1.95	4.74	10.30	9.14	18.7
Fe	mg kg^{-1}	3887	3705	3848	3813	271	3517	4408	4686	
K	mg kg^{-1}	1135	1172	1252	1186	86	1034	1362		
Li	mg kg^{-1}	2.74	2.12	2.38	2.37	0.38	2.02	3.11		
Mg	mg kg^{-1}	4150	3436	3634	3740	353	3302	4238	3811	
Mn	mg kg^{-1}	43.57	43.33	44.37	43.76	2.09	40.00	46.00	55.11	
Na	mg kg^{-1}	4755	4544	4795	4698	196	4402	4982		
Pb	mg kg^{-1}	21.30	10.66	17.17	15.76	6.16	7.68	27.10	1.27	30.24
Zn	mg kg^{-1}	11.53 ^{a}	8.18 ^{ab}	7.58 ^{b}	9.10	2.03	7.14	13.00	7.78	124

bdl: below detection limits. Different letters indicate significant differences ($p < 0.05$) by Dunn´s multiple comparison test. The absence of a letter means no significant difference.

According to the calculated I_{geo} (Figure 2), the studied stations were classified from heavily to extremely polluted ($I_{geo} > 4$) by Bi and Ca, and moderately to heavily polluted ($2 < I_{geo} < 4$) by As, Ag and Cd. The study area was not polluted ($I_{geo} < 1$) for the rest of the studied elements (Mg, Na, Fe, Pb, Cu, Al, Li, Zn, Mn).

3.3. Algal Responses

The light curve and photosynthetic parameters of *L. bertoreana* showed latitudinal patterns for the study zone (Figure 3), with lower F_v/F_m values in PA and higher values towards PC. In contrast, ETR_{max} and E_K presented the highest importance in PA and lower values in PB and PC. The PB and PC zones did not show significant differences in photosynthetic parameters with PC; however, their light curves were apparently different from that of PA.

The results of surface phytoplankton biomass measured as extractable Chl-*a* concentration (Table 3) showed an increase in chlorophyll concentration from the PA site to the PC site. The PA site showed the lowest average ($p \leq 0.05$) surface chlorophyll relative to the PB and PC sites, which were not significantly different.

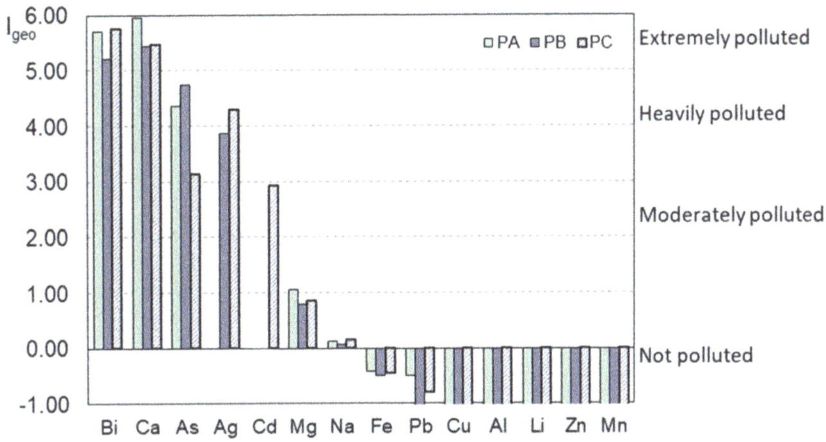

Figure 2. Geoaccumulation index (I_{geo}) calculated for Ag, Al, As, Cu, Fe, Mn, Pb and Zn for the studied stations (PA, PB, PC).

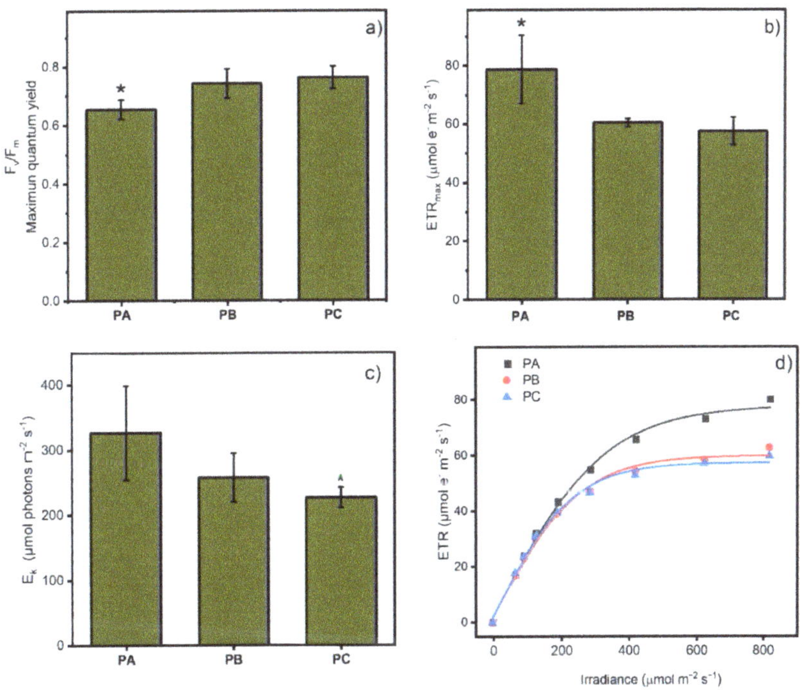

Figure 3. Changes in photosynthetic activity indicated by the maximum quantum yield (**a**) photosynthetic electron transport rate (**b**), light saturation index (**c**), and light curve (**d**). These Chl-*a* fluorescence parameters were calculated from the P-I light saturation curve in *L. bertoreana* of the different sampling areas. Data are presented as mean ± standard deviation ($n = 5$). Asterisks indicate significant differences ($p \leq 0.05$).

Table 3. Chlorophyll *a* concentration of the upper stratum of the water column of three study sites (Asterisk indicates significant difference ($p \leq 0.05$)).

Station	Chlorophyll *a* (μg mL^{-1})
PA	0.006675 ± 0.0017 *
PB	0.013502 ± 0.0047
PC	0.016339 ± 0.0001

3.4. Bioaccumulation in Algal Tissue

Different bioaccumulation behaviour was observed in algae from the three stations. Some of the studied elements, such as Ag, Li and Mn, were found in low concentration in tissues of *L. bertoreana.* No significant differences were found in concentrations in tissues of most of the elements (Al, As, Bi, Ca, Mg, Pb and Zn) between the studied stations. Despite significant differences among As concentrations in sediments (Table 2), no significant difference was found in As concentration of algal tissue (Table 4). Significantly ($p < 0.05$) greater Cu, K and Na bioaccumulation were found in organisms from PA than from PC. Significant bioaccumulation of Fe was found in PB algae.

Table 4. Concentration of elements in the tissues of *Lessonia berteroana* in sampling stations PA, PB and PC.

		PA	PB	PC
Ag	mg kg^{-1}	0.10	bdl	0.20
Al	mg kg^{-1}	3.33	3.67	3.00
As	mg kg^{-1}	12.10	15.92	6.32
Bi	mg kg^{-1}	3.31	2.53	2.75
Ca	mg kg^{-1}	3087	2952	2086
Cd	mg kg^{-1}	1.90 ^{ab}	3.94 ^{a}	1.68 ^{b}
Cu	mg kg^{-1}	2.28 ^{a}	1.28 ^{ab}	0.69 ^{b}
Fe	mg kg^{-1}	6.10 ^{ab}	7.47 ^{b}	3.03 ^{a}
K	mg kg^{-1}	14,787 ^{a}	11,430 ^{ab}	10,777 ^{b}
Li	mg kg^{-1}	bdl	bdl	bdl
Mg	mg kg^{-1}	2389	2189	1588
Mn	mg kg^{-1}	0.80	0.80	0.35
Na	mg kg^{-1}	8271 ^{a}	7106 ^{ab}	6209 ^{b}
Pb	mg kg^{-1}	1.80	1.45	6.25
Zn	mg kg^{-1}	2.47	1.83	0.55

bdl: below detection limits. Different letters mean significant differences ($p < 0.05$) with Dunn´s multiple comparison test. Absence of letter means no significant difference.

3.5. Multivariate Analysis Approach

Two principal factors were obtained from the FA (Table 5) run over the matrix of elements (Cu, Fe, Al, As, Pb, Zn) in water (W), sediments (S) and algae (A), percentage of fine sediments (fines) and the biological responses of the macroalga *L. berteroana* (ETR, E_k, α, F_v/F_m) and the microalga Chl-*a*. The relationship between the components and the factors is plotted as an estimated score per station in Figure 4.

Factor 1 accounted for 82.88% of the total variance, including environmental concentrations of As and its mobility in water, sediments and bioaccumulation, the effect over the microalga population, and the rest of the biological effects. It is also linked to contaminants associated with fine particulate matter of the sediment, As, Pb and Cd, and bioaccumulation of the elements (except Mn) with significant toxic effects. Therefore, Factor 1 is representative of PA and PB, and it is associated more with environmental degradation caused by As (Figure 4, Table 2), which is also reflected by As bioaccumulation in alga tissue and the biomass of microalgae (Table 4). The concentration of As in sediments from PB was eight times greater than the TEL value from the NOAA. However, the concentrations of other metals in alga tissues included in this factor, such as Fe, Cd, Mn and Zn, are also associated with the granulometry.

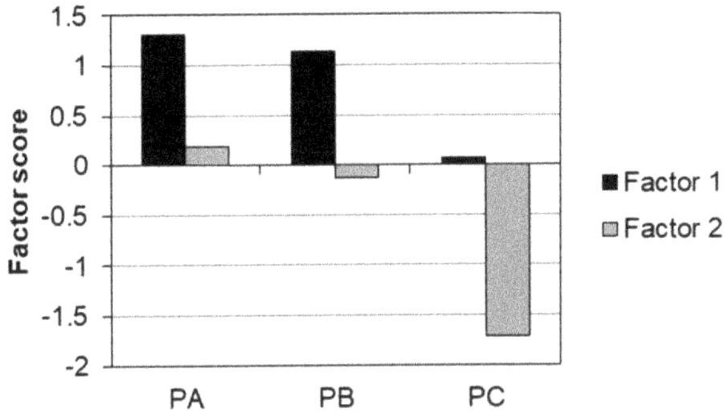

Figure 4. Scores of each factor for the sampling stations.

Table 5. Sorted rotated factor loading (varimax normalised) of 25 variables on the two main factors. For interpretation, the loading cut-off < 0.7.

	Factor 1	Factor 2
%variance	82.88	12.45
Fines	0.735	
AsW	0.817	
PbW		−0.977
ZnW		−0.754
CuS		−0.986
FeS		−0.713
AsS	0.904	
PbS	0.833	
CdS	0.957	
MnS	0.807	
ZnS		−0.981
CuA	0.830	
FeA	0.894	
AsA	0.956	
PbA	0.950	
CdA	0.944	
MnA		−0.915
ZnA	0.869	
F_v/F_m	0.963	
ETR	0.987	
E_k	0.812	
α	0.882	
Chl-*a*	0.894	

Factor 2, which accounted for 12.45% of the total variance, resulted in the combination of contamination by certain metals; only relations with the partition in the sediment of the Zn were included, as the rest had no influence (or minimum) with metal concentrations. There was a slight Cu accumulation, which is not appreciable compared to the bioaccumulation for the rest of the elements (except for Mn, which is not important from the point of view of contamination and the geochemical matrix). The biological effects were not significant in Factor 2. Therefore, Factor 2 is not associated with the toxicity of metals in water when there are negative values for PB and PC, and possibly also for PA, due to low values (Figure 4).

The PA site, at the northern border, is within the Pan de Azúcar National Park. Despite being a protected area, the results showed critical concentrations of Cu, Pb and Zn in sediments (Table 2), which might result from historical mining disposal in Chañaral Bay. The As concentration in sediments was also detected as dangerous in the area. These are responsible for most of the significant stress shown by the macroalgae, and contamination in the water is associated with the low chlorophyll activity of the phytoplankton (Table 3). PB is located in a tourism and fishing area, also influenced by components grouped in Factor 1; i.e., contamination of elements in sediments, such as As and Cu, reflects the specific stress on the physiology of *L. bertoreana*. Finally, PC is close to a factory; this site recorded low concentrations of toxic elements in water and sediments.

The superficial phytoplankton biomass measured as extractable chlorophyll-*a* in the three studied sites showed greater sensitivity to water contamination; thus, the biomass decreased with rising contamination. The PA biomass was significantly lower than PB and PC (Table 3). Although photosynthetic parameters of photoinhibition (F_v/F_m) of *L. bertoreana* showed the same pattern in terms of physiological responses, that is, the algae from PA appeared under higher stress than those from PB and PC, the ETR_{max} and the light saturation index were higher in PA and lower in PC. Due to their cell wall, these photosynthetic adaptations (Figure 3) and the high bioaccumulation capacity of these organisms (Table 4) allow them to survive in these conditions of environmental contamination without threshold. The stations serve as a pollution gradient from north to south; the possible pollution source is Chañaral Bay, due to the nature of the elemental contamination. This anthropogenic source was previously reported by [30]; the Cu tailing disposal in Chañaral, with an estimated mining discharge of 550 kg per year, is located at a beach between PA and PB [34]. Other previous studies detected Cu as responsible for chronic and sub-lethal effects and community distribution alterations [35], and different stress responses (antioxidant-induced) and bioaccumulation in molluscs promoted by Cu, As and Zn [36].

The consequences of contamination in marine environments, such as the physiological stress of algae or the distribution of phytoplankton, are just some environmental responses to the presence of harmful materials. Therefore, the need for a marine regulatory framework for coastal sediments in Chile [1] is also highlighted here. Furthermore, considering the different ecosystems in northern and southern Chile, the recommendation of international organizations to develop sediment quality guidelines for different regions is of utmost importance.

4. Conclusions

There is a large gap in available environmental data in the coastal area of the Atacama region. Even though the region serves as an excellent natural laboratory because of meteorological and oceanic dynamics (Humboldt Current, flooding, tsunamis, presence of wetlands and protected areas), historical and contemporary anthropogenic impacts have not yet been studied in detail. Waste discharges from mining activities for more than 50 years, through the Salado River, and the influence on the surrounding areas, such as Pan de Azucar National Park (PA) and, to a lesser extent, Flamenco Bay (PB), determines the contamination associated with each area studied. Thus, the results of the current study showed that the coastal sediments within a short stretch of 80 km are polluted by Bi, Ca, As, Ag and Cd. The average As concentration in the sediments (39.88 mg kg^{-1}) was above the threshold effect level, with the highest concentrations found in PB (a tourist and benthic management area). The algal biomonitoring revealed a gradient of physiological stress from south to north, with PA (the national park), despite being an exclusive conservation zone, registering the poorest health status (greater Cu bioaccumulation in algae tissue, higher photosynthetic stress in macroalgae and lower biomass of microalgae). Although As concentrations were significantly higher in sediments from PB, an important amount of Cd was found in the leaves of *L. bertoreana* at this site. A significant concentration of toxic elements, such as As, Zn, Cu, Pb and Fe, were found in PC, but the results showed that these were not responsible for the toxicological effects in algal species. The current study highlights the need for further research, including more marine organisms and sites, as well as the application of national guidelines for the quality of marine sediments. Future research recommendations would also include oceanographic studies to determine marine currents´ influence.

Supplementary Materials: The following supporting information can be downloaded at: https://www.mdpi.com/article/10.3390/app13116360/s1, Figure S1: Granulometry results from accumulated retention per particle size for samples from (a) PA, (b) PB, (c) PC, (d) total; Figure S2: Grain-size sediment distribution of collected sediments samples (PA, PB, PC, *n* = 3); Table S1: Detection limit (DL) and quantification limit (QL) for water, sediment and algal samples from the current study.

Author Contributions: Conceptualisation, E.B. and E.C.; methodology, all authors; software, E.B., E.C. and Y.R.-L.; validation, all authors; formal analysis, E.B., E.C. and Y.R.-L.; investigation, all authors; resources, all authors; data curation, E.B., E.C., Y.R.-L. and A.G.; writing—original draft preparation, all authors; writing—review and editing, all authors; visualisation, E.B. and E.C.; supervision, E.B.; project administration, E.B.; funding acquisition, E.B. All authors have read and agreed to the published version of the manuscript.

Funding: This project was funded by the Agencia Nacional de Investigación y Desarrollo of Chile (ANID/FOVI 210060).

Institutional Review Board Statement: Not applicable.

Informed Consent Statement: Not applicable.

Data Availability Statement: Not applicable.

Acknowledgments: The authors are grateful for the project ATA20992 from MINEDUC. E. Bonnail is grateful for ANID/FONDECYT 11180015 and M. Abad is grateful for the M2616 Project and TSUARA Project M2615, funded by Universidad Rey Juan Carlos (Spain).

Conflicts of Interest: The authors declare no conflict of interest.

References

1. Bonnail, E.; Diaz, A.; García, A.; Cruces, E.; Borrero-Santiago, A. Coastal uses and contaminants spread in the Desert coastal region of Atacama. *Chemosphere* **2022**, *288*, 132519. [CrossRef] [PubMed]
2. ECONSA Chile. Empresa Concesionaria de Servicios Sanitarios, S.A. Ley de Transparencia Active. Available online: https://www.econssachile.cl/proyecto-pda/28-planta-desalinizadora-de-agua-de-mar-para-atacama (accessed on 15 March 2023).
3. Campero, C.; Harris, L.M.; Kunz, N.C. De-politicising seawater desalination: Environmental Impact Assessments in the Atacama mining Region, Chile. *Environ. Sci. Pollut.* **2021**, *120*, 187–194. [CrossRef]
4. Zhou, Q.; Zhang, J.; Fu, J.; Shi, J.; Jiang., G. Biomonitoring: An appealing tool for assessment of metal pollution in the aquatic ecosystem. *Anal. Chim Acta.* **2008**, *606*, 135–150. [CrossRef]
5. Campos, L.; Berrios, F.; Oses, R.; González, A.; Bonnail, E. Unravelling *Lessonia trabeculata* management in coastal areas of the Atacama Region of northern Chile through a DPSIR approach: Insights for sustainable plans. *Mar. Pol.* **2021**, *133*, 104737. [CrossRef]
6. Berrios, F.; González, J.E.; Campos, L.; Campbell, D.E.; Bonnail, E. Brown algae (*Lessonia nigrescens*, species complex) fisheries of northern Chile evaluated through the DPSIR approach: Social and economic impact of management strategies. *Ocean Coast. Manag.* **2022**, *224*, 106198. [CrossRef]
7. Sáez, C.A.; Lobos, M.G.; Macaya, E.C.; Oliva, D.; Quiroz, W.; Brown, M.T. Variation in Patterns of Metal Accumulation in Thallus Parts of *Lessonia trabeculata* (Laminariales; Phaeophyceae): Implications for Biomonitoring. *PLoS ONE* **2012**, *7*, e50170. [CrossRef]
8. Oyarzo-Miranda, C.; Latorre, N.; Meynard, A.; Rivas, J.; Bulboa, C.; Contreras-Porcia, L. Coastal pollution from the industrial park Quintero bay of central Chile: Effects on abundance, morphology, and development of the kelp *Lessonia spicata* (Phaeophyceae). *PLoS ONE* **2020**, *15*, e0240581. [CrossRef]
9. Koch, K.; Thiel, M.; Tellier, F.; Hagen, W.; Graeve, M.; Tala, F.; Laeseke, P.; Bischof, K. Species separation within the *Lessonia nigrescens* complex (Phaeophyceae, Laminariales) is mirrored by ecophysiological traits. *Bot. Mar.* **2015**, *58*, 81–92. [CrossRef]
10. Celis-Pla, P.S.; Trabal, A.; Navarrete, C.; Troncoso, M.; Moenne, F.; Zuñiga, A.; Figueroa, F.L.; Sáez, C.A. Daily changes on seasonal ecophysiological responses of the intertidal brown macroalga *Lessonia spicata*: Implications of climate change. *Front. Plant Sci.* **2022**, *13*, 3619. [CrossRef]
11. Stevenson, J. Ecological assessments with algae: A review and synthesis. *J. Phycol.* **2014**, *50*, 437–461. [CrossRef]

12. Torres, M.A.; Barros, M.P.; Campos, S.C.; Pinto, E.; Rajamani, S.; Sayre, R.T.; Colepicolo, P. Biochemical biomarkers in algae and marine pollution: A review. *Ecotox. Environ. Saf.* **2008**, *71*, 1–15. [CrossRef]
13. Lu, X.; Liu, W.; Zhao, C.; Chen, C. Environmental assessment of heavy metal and natural radioactivity in soil around a coal-fired power plant in China. *J. Radioanal. Nucl. Chem.* **2013**, *295*, 1845–1854. [CrossRef]
14. Cesar, A.; Lia, L.R.B.; Pereira, C.D.S.; Santos, A.R.; Cortez, F.S.; Choueri, R.B.; De Orte, M.R.; Rachid, B.R.F. Environmental assessment of dredged sediment in the major Latin American seaport (Santos, São Paulo—Brazil): An integrated approach. *Sci. Total Environ.* **2014**, *497–498*, 679–687. [CrossRef]
15. Wang, J.; Fu, Z.; Qiao, H.; Liu, F. Assessment of eutrophication and water quality in the estuarine area of Lake Wuli, Lake Taihu, China. *Sci. Total Environ.* **2019**, *650*, 1392–1402. [CrossRef]
16. Bonnail, E.; Riba, I.; de Seabra, A.A.; DelValls, T.A. Sediment quality assessment in the Guadalquivir River (SW, Spain) using caged Asian clams: A biomarker field approach. *Sci. Total Environ.* **2019**, *650*, 1996–2003. [CrossRef]
17. Cunha Pasarelli, M.; Bonnail, E.; Cesar, A.; DelValls, T.A.; Riba, I. Integrative assessment of sediments affected by CO_2 enrichment: A case study in the Bay of Santos—SP, Brazil. *App. Sci.* **2021**, *11*, 11603. [CrossRef]
18. Kaplunovsky, A.S. Factor analysis in environmental studies. *HAIT J. Sci. Eng. B* **2005**, *2*, 54–94.
19. Wentworth, C.K. A Scale of Grade and Class Terms for Clastic Sediments. *J. Geol.* **1922**, *30*, 377. [CrossRef]
20. U.S. EPA. *Method 3050B: Acid Digestion of Sediments, Sludges, and Soils*; Revision 2; U.S. EPA: Washington, DC, USA, 1996. Available online: https://www.epa.gov/sites/production/files/2015-06/documents/epa-3050b.pdf (accessed on 15 September 2022).
21. *EPA 6010D*; Method 6010D (SW-846): Inductively Coupled Plasma-Atomic Emission Spectrometry. Revision 4; U.S. EPA: Washington, DC, USA, 2014.
22. Baker, N.R. Chlorophyll Fluorescence: A Probe of Photosynthesis In Vivo. *Annu. Rev. Plant Biol.* **2008**, *59*, 89–113. [CrossRef]
23. Jassby, A.D.; Platt, T. Mathematical formulation of the relationship between photosynthesis and light for phytoplankton. *Limnol. Oceanogr.* **1976**, *21*, 540–547. [CrossRef]
24. Jeffrey, S.W.; Humphrey, G.F. New spectrophotometric equations for determining chlorophylls a, b, c_1 and c_2 in higher plants, algae and natural phytoplankton. *Biochemie und Physiologie der Pflanzen* **1975**, *167*, 191–194. [CrossRef]
25. Müller, G. Schwermetalle in den sedimenten des Rheins—Veränderungen seit 1971. *Umschau* **1979**, *79*, 778–783.
26. McLennan, S.M. Relationships between the trace element composition of sedimentary rocks and upper continental crust. *Geochem. Geophys. Geosyst.* **2001**, *2*, 1021. [CrossRef]
27. DelValls, T.A.; Chapman, P.M. Site-Specific sediment quality values for the Gulf of Cádiz (Spain) and San Francisco Bay (USA), using the sediment quality triad and the multivariate analysis. *Cienc. Mar.* **1998**, *24*, 313–336. [CrossRef]
28. Bonnail, E.; Sarmiento, A.M.; DelValls, T.A. The use of a weight-of-evidence approach to address sediment quality in the Odiel River. *Ecotox Environ. Saf.* **2016**, *133*, 244–251. [CrossRef]
29. Buchman, M.F. *NOAA Screening Quick Reference Tables*; NOAA OR&R Report 08-1; Office of Response and Restoration Division, National Oceanic and Atmospheric Administration: Seattle WA, USA, 2008; 34p.
30. Ramirez, M.; Massolo, S.; Frache, R.; Correa, J. Metal speciation and environmental impact on sandy beaches due to El Salvador copper mine, Chile. *Mar. Pollut. Bull.* **2005**, *50*, 62–72. [CrossRef]
31. Bonnail, E.; Cruz-Hernández, P.; Galleguillos, S.; Izquierdo, T.; Abad, M. Metal contamination in Chañaral Bay (North of Chile)/La contaminación metálica en la Bahía de Chañaral (norte de Chile): Retrospección, prospección y proyección. *Geogaceta* **2020**, *67*, 59–62. Available online: http://www.sociedadgeologica.es/archivos/geogacetas/geo67/Geo67_p59_62.pdf (accessed on 10 May 2022).
32. Tapia, J.; González, R.; Townley, B.; Oliveros, V.; Álvarez, F.; Aguilar, G.; Menzies, A.; Calderón, M. Geology and geochemistry of the Atacama Desert. *Antonie Van Leeuwenhoek* **2018**, *111*, 1273–1291. [CrossRef] [PubMed]
33. Tapia, J.; Davenport, J.; Townley, B.; Dorador, C.; Schneider, B.; Tolorza, V.; von Tümpling, W. Sources, enrichment, and redistribution of As, Cd, Cu, Li, Mo, and Sb in the Northern Atacama Region, Chile: Implications for arid watersheds affected by mining. *J. Geochem. Explor.* **2018**, *185*, 3–51. [CrossRef]
34. Andrade, S.; Moffett, J.; Correa, J.A. Distribution of dissolved species and suspended particulate copper in an intertidal ecosystem affected by copper mine tailings in Northern Chile. *Mar. Chem.* **2006**, *101*, 203–212. [CrossRef]
35. Jara, C.; Gaete, H.; Lobos, G.; Hidalgo, M.E. Oxidative stress in the mollusk *Echinolittorina peruviana* (Gasteropoda: Littorinidae, Lamarck, 1822) and trace metals in coastal sectors with mining activity. *Ecotoxicology* **2014**, *23*, 1099–1108. [CrossRef] [PubMed]
36. Navarro, N.; Abad, M.; Izquierdo, T.; Bonnail, E. The arid coastal wetlands of northern Chile: Towards an integrated management of highly threatened systems. *J. Mar. Sci. Eng.* **2021**, *9*, 948. [CrossRef]

applied sciences

MDPI

Article

Risk Assessment of a Coastal Ecosystem from SW Spain Exposed to CO_2 Enrichment Conditions

Marina Cunha Passarelli [1,2], Estefanía Bonnail [3], Augusto Cesar [2,*] and Inmaculada Riba [1,2]

1 Department of Ecotoxicology, Santa Cecília University (Unisanta), Santos 11045-907, Brazil
2 Department of Marine Sciences, Federal University of São Paulo (Unifesp), Santos 11045-907, Brazil
3 Department of Coastal Research, Atacama University (CIC-UDA), Copiapó 1530000, Chile
* Correspondence: acesar@unifesp.br

Abstract: The Weight-of-Evidence (WOE) approach uses multiple lines of evidence to analyze the adverse effects associated with CO_2 enrichment in two stations from the Gulf of Cádiz (Spain) with different contamination degrees. Sediment contamination and metal (loid) mobility, toxicity, ecological integrity, and bioaccumulation from the samples exposed to different acidification scenarios (pH gradient from 8.0 to 6.0) were used in the WOE. The experiments were conducted under laboratory conditions using a CO_2-bubbling system. Different integration approaches such as multivariate analyses were used to evaluate the results. The results indicated that the adverse biological effects under pH 6.5 were related to the mobility of dissolved elements (As, Fe, Cu, Ni, and Zn). Furthermore, the pH reduction was correlated to the increase of bioaccumulation of As, Cr, Cu, Fe, and Ni in the tissues of mussels at pH 7.0. The noncontaminated sediment showed environmental degradation related to the acidification at pH values of 7.0; whereas the sediment moderately contaminated showed both environmental risks, caused by acidification and the presence and the increase of the bioavailability of contaminants. The WOE approach supposes an effective tool to identify and distinguish the causes of adverse effects related to the enrichment of CO_2 in marine environments.

Keywords: integrated analysis; CO_2 enrichment; mesocosms tests; metal contamination; sediment toxicity; ecological integrity

Citation: Passarelli, M.C.; Bonnail, E.; Cesar, A.; Riba, I. Risk Assessment of a Coastal Ecosystem from SW Spain Exposed to CO_2 Enrichment Conditions. *Appl. Sci.* **2023**, *13*, 5805. https://doi.org/10.3390/app13095805

Academic Editors: Ramaraj Boopathy, Mauro Marini and Angel DelValls

Received: 11 March 2023
Revised: 2 April 2023
Accepted: 28 April 2023
Published: 8 May 2023

1. Introduction

Understanding the impacts of climate change and global warming is one of the greatest challenges faced by the scientific community, as are promoting mitigation and adaptation strategies [1,2]. The release of carbon dioxide (CO_2) from human activities is considered a significant contributor to observed changes in global climate patterns and the acidification of the oceans [3]. In addition, ocean acidification associated with global warming is projected to amplify the adverse effects of warming (further at 2 °C), provoking impacts on the growth, development, survivorship, calcification, and richness/abundance of species [4]. Therefore, in the last years, many studies analyzed the potential effects of acidification and CO_2 enrichment in marine environments [5–9] among natural processes [10]. They demonstrated a relationship between adverse effects on marine organisms promoted by acidification in seawater and the increase in the bioavailable forms of toxic elements. However, not many of them were designed or executed from an integrative point of view integrating multiple lines of evidence (LOEs): contamination, toxicity, and bioavailability.

The integrative methods based on the weight-of-evidence (WOE) were applied to assess the environmental quality of aquatic ecosystems in the last decades [11–15]. This approach uses a different set of data obtained within the application of multiple lines of evidence (LOEs) that usually implies (1) sediment physical characteristics (grain-size distribution); (2) chemical factors (contaminants in specific phases); (3) ecotoxicologically based on the use of acute and chronic toxicity tests and biomarker responses in key organisms; (4) ecological factors through the analysis of the structure and function of benthic communities; and (5) the bioaccumulation assessment of contaminants in living organisms.

The risk assessment based on the WOE approach to assessing the pollution derived from CO_2 enrichment in aquatic environments was one of the adopted methods. It integrates the different responses, whilst it reduces the uncertainty associated with the complexity [10].

A recent study used the WOE to assess the environmental risks related to the enrichment of CO_2 in the coastal zones in Brazil [9]. Results showed that the elemental bioavailability increased from pH 7.0, being responsible for the adverse effects caused in marine organisms. In this sense, the current study uses the WOE integrating LOEs to understand the potential adverse effects associated with the enrichment of CO_2 in marine ecosystems, and to identify the source of contamination responsible for the pollution.

The main goal of the current study was to determine the environmental risk of CO_2 enrichment in marine ecosystems using the WOE approach. Therefore, different LOEs from two sampling sites with different contamination degrees were employed under the influence of acidification (pH gradient from 8.0 to 6.0).

2. Materials and Methods

2.1. Study Area

Two sediment samples were collected from the Bay of Cádiz (Figure 1). The selected areas were previously studied and were classified according to the metal contamination levels [5,16,17]. The Rio San Pedro (36°31′52.90″ N; 6°12′48.43″ W) sampling site is classified as low metal concentration (control site), while the Trocadero (36°31′15.5″ N; 6°12′25.1″ W) is considered a sampling site with a moderate metal concentration (contaminated site) [16,18].

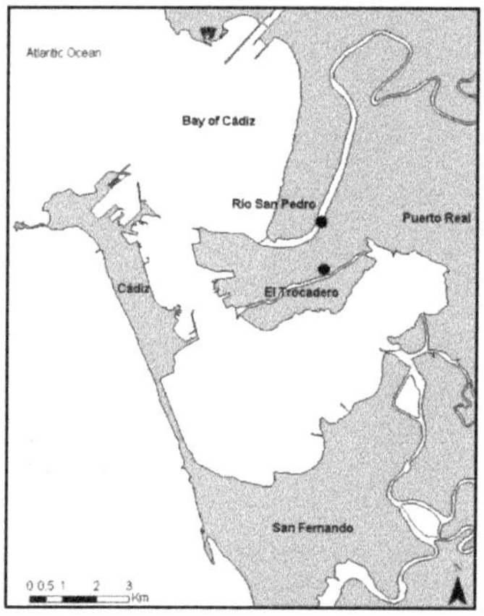

Figure 1. Sediment sampling sites located in the Bay of Cádiz (SW, Spain): Río San Pedro (RSP) and El Trocadero (TRO).

Water used in toxicity tests was collected from the surface (1 m depth) of the Rio San Pedro (RSP), during high tide, transported to the laboratory using 50 L gallons, and placed in a 400 L tank (pH 7.9 ± 1; Salinity 34 psu, dissolved oxygen >81%, and temperature 18 ± 1 °C). As previously mentioned, the RSP has been used as a reference point in other works since no contaminants are present.

2.2. Experimental Design

Figure 2 shows, in detail, the schematic design used for each experiment included in the WOE: (i) design to assess the benthic integrity; (ii) design used to elutriate assays employed as samples prior to acidification; (iii), and (iv) acute toxicity tests with amphipods and mussels design, respectively. In addition, it was adapted (iv) to be used for the bioaccumulation and NRRT assays.

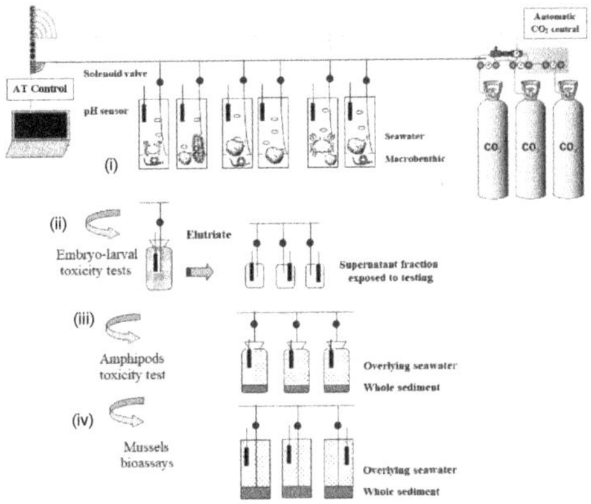

Figure 2. Schematic design of the CO_2 injection system used in this study for the different LOEs: (**i**) the benthic integrity; (**ii**) the elutriate assays for chronic toxicity; (**iii**,**iv**) acute toxicity tests with amphipods and mussels, respectively, and (**iv**) it was also applied for bioaccumulation and NRRT assays.

The laboratory-scale injection systems employed (Figure 2) for ecotoxicological and ecological integrity assessments were described by [16] and chemical analyses, acute and chronic toxicity tests, ecological integrity, and metal bioaccumulation are further described in [18,19]. The CO_2 injection system proposed mimic acidification by CO_2 enrichment in marine ecosystems.

2.3. *Chemical Analyses*

Sediment and water samples were collected from each station (RSP and TRO, Figure 1) at the beginning and the end of the toxicity and mesocosms tests. The metal and metalloid concentrations (Al, As, Cr, Cu, Fe, Ni, and Zn) in the sediment were analysed in the total fraction. The dissolved metal concentration in overlying and elutriate seawater were determined in filtered (0.45 mm) and acidified (pH < 2 with ultrapure HNO_3) subsamples for metal(loid) analyses.

2.4. *Toxicity Tests*

Different biological responses were selected: (i) acute toxicity tests: amphipod and mussel mortalities; (ii) chronic toxicity tests: embryo–larval development success of sea urchins and mussels; (iii) metal bioaccumulation in the soft bodies of mussels; (iv) macrobenthic community alteration and; (v) biomarker response of cellular stress: neutral red retention time (NRRT) in the hemolymph of mussels.

The pH values of the treatments ranged between 8.0 and 6.0 (8.0, 7.5, 7.0, 6.5±, and 6.0) and were conducted in replicates. Table 1 summarises the description of the different toxicity tests performed in the study using different species and conditions and the ecological integrity assessment conditions.

2.4.1. Acute-Toxicity Tests

Individuals of amphipods *Ampelisca brevicornis* were collected from a noncontaminated area at the Bay of Cádiz. These are used as environmental quality indicators and are recommended by the specific protocol for dredged-sediment characterization [20]. Individuals of mussels *Mytilus galloprovincialis* were obtained from an aquaculture farm (Cetária, Cádiz, Spain). Both organisms were acclimated for a week before testing. Parameters were adjusted according to the optimal conditions established for each species employed in this study and kept until the end of the experiments (Table 1). The experiments were adapted from [21,22].

Table 1. Summarized description of the biological tests and the ecological integrity assessment tested in this study.

Test Organism	pH Range	T (°C)	DO (%)	Duration	Endpoint	Type
A. brevicornis	8.0–6.0	19 ± 1	>80	10 d	Mortality	Acute
P. lividus	8.0–6.0	18 ± 1	>80	48 h	Plateaus stage	Chronic
M. galloprovincialis	8.0–6.0	19 ± 1	>80	48 h	D-phase	Chronic
M. galloprovincialis	8.0–6.0	19 ± 1	>80	10 d	Mortality	Acute
M. galloprovincialis	8.0–6.0	19 ± 1	>80	14 d	Metal concentration	Bioaccumulation
M. galloprovincialis	8.0–6.0	19 ± 1	>80	24/96 h, 10 d	LMS	Acute
Macrobenthic	8.0–6.0	19 ± 1	>80	21 d	Biological indices	Macro-benthic community Alteration

T: temperature; DO: dissolved oxygen; LMS: lysosomal membrane stability.

2.4.2. Chronic Toxicity Using Elutriates

Specimens of sea urchin *Paracentrotus lividus* were collected from a rocky intertidal platform from a protected area in the Gulf of Cádiz (Southwestern Spain); whereas the specimens of mussels *M. galloprovincialis* were obtained from an aquaculture farm (Cetária, Cádiz, Spain).

Sea urchin gametes were obtained from direct extraction from the gonads. On the other hand, mussels were induced to spawn by thermal stimulation according to the protocol recommended by [23,24] and exposed to acidified sediment elutriate following [22].

2.4.3. Metal Bioaccumulation

M. galloprovincialis mussels were exposed to acidified sediments for 14 days. Assays were performed using whole sediment with a sediment-to-water ratio of 1:2 v/v as reported by [25]. The concentrations of metals Zn, Cu, Al, Fe, Mn, Ni, Cr, and the metalloid As in the soft tissue were determined using an inductively coupled plasma mass spectrometer (ICP-MS) (Thermo Elemental Series-X, Santa Clara, CA, USA).

2.4.4. Ecological Integrity

After the acclimation time, mesocosms were maintained for 21 days, natural photoperiod and constant aeration and temperature. Then, the macrobenthic community structure (total abundance, richness, and diversity) was determined using Margalef's and Shannon's methods, respectively.

2.4.5. Neutral Red Retention Time Assay

The NRRT assay was based on the principle that only lysosomes in healthy cells take up and retain the neutral red vital stain [26]. Lysosomal membrane damage caused by xenobiotics can decrease the NRRTs [27]. The experiments were performed following the method described by [28].

2.5. *Statistical Analysis (WOE Approach)*

Figure 3 represents the four LOEs integration used to characterize and distinguish the impacts related to the contamination levels and acidification degrees.

The obtained results from the application of each LOE were linked for each component assessed using three different methods (WOE): (a) tabular matrix based on statistical significance ($p < 0.05$) compared to the reference case (RSP sediment, pH 8.0); (b) pie charts, showing statistical differences ($p < 0.05$ significantly different and positive effect; $0.01 \geq p > 0.05$ significantly different but moderate effect; and $p > 0.01$ not significantly different) among the samples sites and pH values compared to the reference case (RSP sediment, pH 8.0); and (c) linking the results using a multivariate analysis using the factor analysis principal component as extraction method [29].

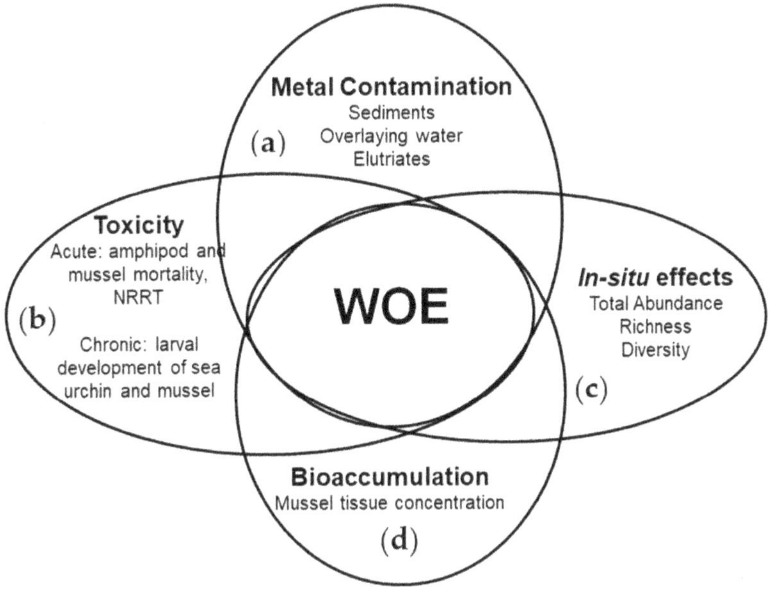

Figure 3. Integration of the LOEs used to assess the environmental risks in CO_2-enriched aquatic environments: (**a**) contamination type and degree; (**b**) laboratory toxicity tests; (**c**) in situ conditions alterations; (**d**) metal bioaccumulation.

Significant differences found between the reference site and the pH treatments were determined by an ANOVA and Dunnet's tests (TOXSTAT software, version 3.5) to develop the pie charts and tabular matrix.

The indexes of contamination, toxicity, bioaccumulation, and ecological integrity were calculated as outlined by [11], considering the ratio-to-reference (RTR) values. Results from each LOE and each pH treatment were referred to the reference case and then derived as the minimum value. The RTRs were averaged for contamination in sediments and mobility, acute and chronic toxicity, bioaccumulation, and ecological integrity.

The contamination index was obtained regarding the metal concentration in sediments ($RTR_{sediment}$), and concentration in seawater and elutriates ($RTR_{mobility}$). The elements As, Cr, Cu, Ni, and Zn were considered to determine the indexes. The acute toxicity index was derived from the average of RTR calculated for amphipod and mussel mortalities and NRRT assays. The values of NRRT were transformed as their inverse to be treated with equal importance. The chronic toxicity index was estimated from the average RTR for larval development inhibition results of mussels and sea urchins. Regarding the index of macrobenthic community structure (ecological integrity), the values were also inverted to be treated with equal importance, and the average of RTR was calculated based on the results of total abundance, diversity, and richness. The bioaccumulation index was derived from the average RTR estimated for the metals bioaccumulated in the tissues of mussels.

Furthermore, to develop an integrated and more precise interpretation of the risks associated with CO_2 enrichment in the marine environment, a multivariate analysis (principal-component analysis, PCA-with-varimax rotation) was carried out. Therefore, the correlation between increasing pollution, pH reduction, and biological effects measured in the laboratory was further explained. Factor scores for each sampling station were calculated following the protocols outlined by [11,30] and using the STATISTICA® software package (version 13.2). The principal factors were extracted and the eigenvalues above 1.0 were taken into account. Variables having loadings $\geq$0.40 to a particular factor were considered associated with the respective factor, following [31].

3. Results and Discussion

The results obtained for each of the LOEs used in this study are summarized in Table 2. The concentrations of most of the studied metal(loid)s in the TRO sediment sample were higher than in RSP sediment samples. Concentrations of Al, Cr, and Fe showed little variation among the pH values for both sediment samples (RSP and TRO). The Cu concentration in sediment decreased with pH for both sampling sites, whereas an increase in Zn concentration was observed in the TRO sediment at pH 6.0. However, greater variability in the concentration of metals was observed in the samples of overlying water and elutriates when compared with the sediment samples. Overall, the metal concentrations were higher in the elutriate treatment than in the overlying seawater. The concentrations of the studied elements were higher in sediment from TRO elutriates, while RSP sediments exhibited higher concentrations for most of the metals than for the overlying seawater samples.

Table 2. Summarized results for the different lines of evidence (LOEs) used in this study: a) chemical concentrations in sediment (sed, in μg g^{-1} dry weight), overlying water (sw, in mg L^{-1}), and elutriate (elu, in mg L^{-1}), and the bioaccumulation of elements in the soft tissue of mussels (bio, in μg g^{-1} dry weight); b) acute and chronic toxicity tests and neutral red assays (mortality of amphipods (Mort_A), and mortality of mussels (Mort_M); inhibition of larval development of mussels (Inh_M) and sea urchins (Inh_SU); neutral red retention time (NRRT) for 24 h, 96 h, and 10 d of exposure are expressed in minutes); and d) ecological integrity (total abundance, richness, and diversity) exposed to the pH treatments (8.0, 7.5, 7.0, 6.5, and 6.0) using the RSP and TRO sample sites.

Variable	pH Treatments									
	RSP					TRO				
	Control	pH 7.5	pH 7.0	pH 6.5	pH 6.0	Control	pH 7.5	pH 7.0	pH 6.5	pH 6.0
Contamination										
Cr sed	1.60	1.80	2.00	2.30	2.00	5.00	5.00	5.20	5.20	5.30
Cr sw	bdl	bdl	bdl	bdl	bdl	bdl	bdl	bdl	bdl	bdl
Cr elu	bdl	bdl	bdl	bdl	bdl	bdl	bdl	bdl	bdl	bdl
As sed	bdl	bdl	bdl	bdl	bdl	bdl	bdl	bdl	bdl	bdl
As sw	0.16	0.16	0.20	0.20	0.18	0.18	0.18	0.18	0.18	0.18
As elu	bdl	bdl	bdl	bdl	bdl	bdl	bdl	bdl	0.078	0.084
Cu sed	5.40	5.40	5.50	5.60	5.30	6.50	6.40	6.40	6.20	6.10
Cu sw	0.04	0.02	0.03	0.03	0.02	0.02	0.03	0.03	0.03	0.03
Cu elu	0.03	0.03	0.03	0.06	0.06	0.07	0.07	0.07	0.09	0.07
Fe sed	592	587	535	706	657	1456	1433	1413	1471	1566
Fe sw	0.02	0.02	0.07	0.03	0.02	0.01	0.01	0.01	0.02	0.02
Fe elu	0.01	0.04	0.04	0.04	0.07	0.05	0.06	0.14	0.19	1.07
Ni sed	bdl	bdl	bdl	bdl	bdl	bdl	bdl	bdl	bdl	bdl
Ni sw	13.3	22.9	48.6	55.7	54.0	52.6	55.9	60.7	72.3	71.2
Ni elu	bdl	bdl	bdl	bdl	bdl	bdl	bdl	bdl	bdl	bdl
Zn sed	2.40	2.50	2.70	2.60	2.50	6.20	6.20	6.20	6.70	7.50
Zn sw	0.02	0.02	0.03	0.05	0.11	0.02	0.03	0.04	0.06	0.07
Zn elu	0.01	0.02	0.01	0.01	0.01	0.02	0.02	0.02	0.02	0.02
Al sed	1139	1205	1259	1329	1279	2992	2987	2930	2929	3178
Al sw	0.04	0.02	0.03	0.02	0.02	0.03	0.06	0.04	0.04	0.05
Al elu	0.04	0.06	0.06	0.09	0.09	0.10	0.10	0.08	0.08	0.10
Concentrations in organisms										
Al bio	4.60	6.50	8.70	115	n.a.	6.20	7.60	6.00	n.a.	n.a.
As bio	972	776	1094	943	n.a.	1200	945	1072	n.a.	n.a.
Fe bio	12.3	18.8	21.1	127	n.a.	17.2	16.5	25.9	n.a.	n.a.
Ni bio	242	254	273	455	n.a.	390	266	256	n.a.	n.a.
Cu bio	993	895	987	966	n.a.	587	627	814	n.a.	n.a.
Cr bio	630	299	663	373	n.a.	400	84	480	n.a.	n.a.
Zn bio	11,200	13,700	18,600	14,600	n.a.	20,000	22,500	19,200	n.a.	n.a.
Toxicity										
% Mort_A	14	20	57	100	100	20	30.3	64	100	100
% Mort_M	16.6	17	36.6	33.3	100	35	35	73.3	83.3	100
% Inh_M	8.3	9.5	14.6	75	100	20	67	87.3	98.3	100
% Inh_SU	9.0	24	91	100	100	11.3	19	94	100	100
NRRT 24 h	70	72	36	20	7.50	54	50	38	28	6.0
NRRT 96 h	68	60	30	1.5	0.0	40.5	38	40.7	20	0.0
NRRT 10 d	68.3	72	50	0.0	0.0	35	38	18	0.0	0.0
Ecological Integrity										
Abundance	196	190	172	170	101	24	22	20	21	16
Richness	0.96	1.1	1.16	1.16	0.21	1.22	1.21	1.3	0.67	0.73
Diversity	4.01	3.90	2.45	3.93	1.75	1.12	0.98	0.6	0.73	0.73

Regarding the toxicity tests, the mortality of amphipods and mussels were 14% and 20%, respectively, for the control (not CO_2 injection, RSP sediment). Significant mortality ($p < 0.05$) of amphipods was found at pH 7.0, and for mussels at pH 6.0 when exposed to RSP sediment. However, exposure to TRO sediment control displayed a mortality of 37% for amphipods and 35% for mussels. The absence of acidification treatment with TRO sediments implies that contamination strongly influenced the obtained results [18].

However, for the embryo–larval development test, a percentage of inhibition of less than 20% was observed for both species (sea urchin and mussel) and both sediment and elutriates samples (RSP and TRO) for the control treatment (not CO_2 injection). The mussel–larval development inhibition differed significantly between the control and pH 6.5 and 6.0 for RSP elutriates; though it was statistically significant ($p < 0.05$) for pH 7.5 and below for TRO samples. On the other hand, the sea urchin larval-development inhibition presented significant differences ($p < 0.05$) at pH 7.0 for both sediment samples.

In general, a different bioaccumulation trend for each metal(loid) was shown by *M. galloprovincialis* with acidification. Organisms exposed to RSP samples accumulated greater concentrations of elements than those exposed to TRO samples for all the studied elements, except for Zn and Cr. Results are further detailed in [32].

Results using the ecological integrity showed a decrease in the total abundance, richness, and diversity of benthic fauna for pH 6.0 when compared to the control for both sediment samples tested (Table 2), in agreement with [33].

3.1. Tabular Matrix

Statistically significant differences ($p < 0.05$) between the reference treatment (RSP sediment, pH 8.0—no CO_2 added) and the other treatments and stations are presented as showing a positive response in contamination, toxicity, integrity, or bioaccumulation (Table 3).

Table 3. Matrix results for the different pH treatments used and sediment samples, representing significant differences ($p < 0.05$) were found when compared to the reference case (RSP, pH 8.0). Responses are shown as either positive (+), negative (−), and (+/−) when the result was close to being positive, but not, showed a significant difference.

Cases Site, pH Treatment	Metals		Toxicity		Bioaccumulation	Benthic
	Sediment	Mobility	Acute	Chronic		
RSP, pH 8.0	−	−	−	−	−	−
RSP, pH 7.5	−	−	−	+/−	−	−
RSP, pH 7.0	−	−	+	+	+/−	+/−
RSP, pH 6.5	−	+	+	+	+	+/−
RSP, pH 6.0	−	+	+	+	+	+
TRO, pH 8.0	+	+/−	+/−	+/−	+/−	+
TRO, pH 7.5	+	+	+/−	+	+/−	+
TRO, pH 7.0	+	+	+	+	+/−	+
TRO, pH 6.5	+	+	+	+	+	+
TRO, pH 6.0	+	+	+	+	+	+

In general, the metal concentration in sediment was not significantly affected by acidification for RSP sediment. Nevertheless, significant differences ($p < 0.05$) were found between the metal concentration in the TRO and the control. These results were expected since the metal contamination in the TRO station was higher than in the RSP station. These results agreed with those reported by [16], which characterized the RSP as a low metal-concentration station and, the TRO as a station with moderate metal concentration. On the other hand, the results obtained in the mobility index showed significant differences ($p < 0.05$) between the reference case and pH 6.5 and 6.0 for the RSP and starting at pH 7.5 for the TRO. These findings indicated that acidification could increase the mobility of metals from the sediment samples to the water column, which might cause toxic effects on marine organisms. Previous studies reported the pH as a variable that mostly might influence the availability of elements from sediments [17]. Furthermore, it can be suggested that the significant difference found in the toxicity indexes at pH 7.0, using the RSP sediment, could also be related to the increase in metal mobility. Nevertheless, a significant toxic effect at pH 7.5 was observed, and, according with previous results, showed significant effects on the larval development inhibition of mussels and amphipods, and clam mortalities associated with acidification [7,22,34,35].

The results of the integrity of the benthic community showed significant ($p < 0.05$) differences for all treatments tested using the TRO sediment. These results were expected since the TRO sediment sample presented higher contamination than the RSP sample and the lowest value of total abundance of macrobenthic found in the study. On the other hand, the results of the bioaccumulation of metals in the tissues of mussels showed a significant ($p < 0.05$) increase in the concentration of metals at pH 6.5 for both sediment samples.

3.2. Sector Charts

The obtained results of the present study were also expressed in sector charts that used different levels of the statistical p-value (Figure 4) to assess the influence of ocean acidification promoted by CO_2 enrichment.

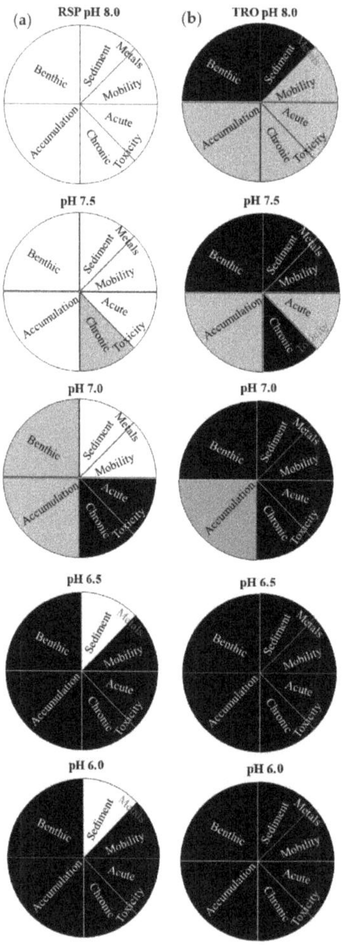

Figure 4. Statistical differences of the LOEs among the pH levels are shown in sector charts: (**a**) RSP treatments and (**b**) TRO treatments. Black colour represents a significant difference ($p < 0.05$) against the control, grey colour represents differences using different p-values ($0.01 \geq p > 0.05$), and white colour represents no significant difference concerning the reference case.

Significant differences were represented with different colours and weights for the LOEs. The RSP site was selected as the reference site at the control pH (not CO_2, pH 8.0 ± 0.1). Then, each case (pH treatment in the sampling site) was compared to this reference case and the significant difference, with respect to the reference case, is represented as black segments when $p < 0.05$, whereas when $0.01 \geq p > 0.05$, they were represented by grey segments, and no significant difference ($p > 0.01$) was represented by white segments (Figure 4).

This approach allows clear visualization of the influence of CO_2 enrichment on marine organisms concerning the pH tested and the contamination levels in each site used. According to [36], pie charts represent the strength of the statistical tool and the simplicity of interpretation, though without quantification. Results showed that both sediment contamination and contamination caused by acidification were associated with the biological responses in this study. Significant results (black area) were observed at pH 7.0 when using noncontaminated sediment. On the other hand, the results obtained using the TRO sediments showed significant effects even in the control treatment (pH 8.0, not CO_2 added) for macrobenthic and contamination by metals in sediment. These results demonstrated that the sediment sampling sites presented the initial differences in their adverse effects associated with the contamination levels being not related to the acidification. Grey areas (moderate effects) indicate the beginning of the adverse effects and are useful for classifying cases in which additional attention should be taken, especially for potential scenarios of acidification.

3.3. Multivariate Analysis

The principal-component analysis (PCA) identified the correlations between sediment contamination, biological responses, and the pH decrease in seawater caused by CO_2 enrichment in the ocean (Table 4). A total of 30 variables were used, including dissolved metals in overlying seawater (As, Cr, Cu, Fe, Ni, and Zn) and elutriate treatment (Cu, Fe, and Zn), toxicity to marine organisms (mortalities, embryo–larval development inhibition, and NRRT time after 24 h, 96 h, and 10 d of exposure), sediment characteristics (concentrations of organic matter, fines, and metals), concentrations of metals bioaccumulated in the whole body of mussels (As, Cr, Cu, Fe, Ni, and Zn) and benthic integrity (total abundance, diversity, and richness). The two main factors explained almost 75% of the original variance in the original data set. The criterion selected to interpret a variable associated with a particular factor was a loading of 0.4 or higher. Each factor is described according to the dominant group of variables.

The first factor (F1, 48.41% of variance) linked all dissolved metals (except Fe) in seawater and elutriates, the metal concentration in mussels (except Zn), mortalities of amphipods and mussels, embryo–larval development inhibition of sea urchins and mussels and the increasing proton (H^+) concentration, in the other words, the CO_2 acidification. Furthermore, the F1 also linked negative values to the results of NRRT at 24 h, 96 h, and 10 d, and the richness of the macrobenthic community. The second factor (24.93% of variance) linked all metals in sediment, dissolved Ni in the seawater and Cu in the elutriate, As and Zn bioaccumulated in the mussels, mortality of mussels, sea urchin larval development inhibition, and the values of total abundance and diversity of the macrobenthic community. Moreover, negative values were linked in factor two for Cr and Fe in seawater.

In general, the multivariate analysis showed that both causes can influence the adverse biological effect measured: acidification by enrichment of CO_2 and the metal concentration analyzed in the sediments.

Table 4. The sorted rotated factor of the original variables on the two principal factors. Just loadings >0.4 are shown.

Variables	Components: Factor 1 (48.41%)	Factor 2 (24.93%)
Metals in sediments		
Cr $_{sed}$		0.98
Cu $_{sed}$		0.97
Fe $_{sed}$		0.97
Zn $_{sed}$		0.95
Metals mobility		
As $_{sw}$	0.48	
Cr $_{sw}$	−0.51	−0.68
Cu $_{sw}$	−0.44	
Fe $_{sw}$		−0.49
Ni $_{sw}$	0.63	0.72
Zn $_{sw}$	0.90	
Cu $_{elu}$	0.42	0.84
Fe $_{elu}$	0.51	
Zn $_{elu}$	0.43	
Protons concentration (H^+)	0.89	
Bioaccumulation		
As $_{bio}$	0.61	0.48
Cr $_{bio}$	0.66	
Cu $_{bio}$	0.64	
Fe $_{bio}$	0.94	
Ni $_{bio}$	0.85	
Zn $_{bio}$		0.89
Acute toxicity		
Amphipod mortality	0.96	
Mussel mortality	0.84	0.38
NRRT at 24 h	−0.94	
NRRT at 96 h	−0.89	
NRRT at 10 d	−0.84	
Chronic toxicity		
Mussels Embryo–larval inhibition	0.78	0.46
Sea urchin Embryo–larval inhibition	0.88	
Ecological integrity		
Total abundance		−0.96
Diversity (H')		−0.87
Richness (d')	−0.74	

The ecological integrity of the macrobenthic community related to the total abundance and diversity showed to be associated with the sediment contamination of the sampling sites, though not with the pH reduction. However, previous studies showed significant changes in the structure and reductions in diversity, abundance, and biomass of the benthic communities exposed to acidification scenarios [8,37,38]. These results confirm the obtained results (Table 3 and Figure 4) that showed a significant ($p < 0.05$) difference in the TRO benthic community at pH 8.0 when compared to the reference case. Therefore, the pH level probably was not correlated with these results, and the macrobenthic community could be affected by the CO_2 enrichment in the ocean only at highly-acidified conditions.

It is known that the forms in which metals are bound in the sediments and the sediment porosity stability and grain size may influence their bioavailability and toxicity [39–43]. Factor one suggests that the majority of metals can increase their mobility in acidified conditions, and might cause adverse effects on marine organisms. These results were also observed in the tabular matrix and pie charts (Table 3 and Figure 4, respectively) that showed a significant difference ($p < 0.05$) in the mobility of metals starting at pH 6.5 for the RSP sediment and at pH 7.5 for the TRO sediment.

Furthermore, acute and chronic toxicity were linked in factor one with the metal's mobility and the increase in the concentrations of H^+ (enrichment of CO_2). These results were indicated in the tabular matrix and pie charts as significantly differenced starting at pH 7.5 and 7.0 for TRO and RSP sediment, respectively, and corroborate with the results found by [9], which showed significant effects starting at pH 7.0 in no contaminated sediment sample. On the other hand, the multivariate analysis identified which metals were correlated with their concentration increase and the toxicity caused to the organisms (As, Ni, and Zn in seawater and Cu, Fe, and Zn in elutriate sediments). These results corroborate previous studies which showed adverse effects on marine organisms associated with increases in the metal concentrations and the pH reduction [10,16,22,34,35] reported with the mortality of amphipod *Hyale youngi* with seawater acidification due to an increase in Ni and As concentrations. In addition, an increase in the mobility of dissolved Zn promoted mortality in bivalves (adults and juveniles) [22,25] and amphipods [16].

Furthermore, it was reported that there was a correlation between the increase in the mobility of Cu, Zn, and As with the mortality of *Hediste diversicolor* polychaete [44]. Furthermore, it has been observed that there is variability in the feeding behaviour of marine amphipods (*O. tuberculata*) in response to ocean acidification [45].

Regarding the results of bioaccumulation of metals, it was correlated that factor one increases in the concentrations of As, Cr, Cu, Fe, and Ni in the tissues of mussels, increases in the mobility of the metals mentioned above, and increases in the concentrations of H^+ protons. These results were described in Table 3 and Figure 4, presenting significant differences from pH 6.5 for both sediment sampling sites. These findings are similar to [44], a study that showed a correlation between the increase of Al, Fe, Mn, Cu, Zn, and As accumulated in the tissues of *H. diversicolor* with the pH reduction. Furthermore, the bioaccumulation of As and Zn was shown in factor two which was correlated with the sediment contamination. These results are expected since mussels exposed to trace elements in seawater can accumulate high amounts in their tissue [6]. In addition, a previous study reported that the bioaccumulation in *Tellina deltoidalis* showed that the bioavailability of metals was likely to be a result of the continued cycling between the pore water and the surface sediments due to the bioturbation process [42].

According to [45], the bioaccumulation of metals may be influenced by some factors, such as metal specificity, environmental influences, exposure routes, and species-specific characterization. Results of metal bioaccumulation in mussels may be associated with different parameters such as CO_2 enrichment in seawater, sediment contamination, metals behaviour, etc.

To confirm the factors' descriptions and to establish the relationship between components associated with contamination, biological effects and acidification at each studied site, it was proposed as a representation of estimated factor scores per case (Figure 5).

The factor scores have demonstrated that the contamination levels of sediment influence the obtained results. Overall, the RSP sediment that is considered a noncontaminated sediment site did not correlate with the contamination levels and biological responses (factor one) for the higher pH values. Moreover, the results of factor scores confirm those previously reported (Table 2) with lower levels of metal contamination in sediments (factor two).

Furthermore, as expected, the results of factor two were correlated even at the control pH values using the TRO which also confirms the results obtained using the tabular matrix and pie charts. Furthermore, it was noted that factor two decreases with the concentration of protons increase (pH reduction) while the opposite results were demonstrated for factor one, which increases its value as the pH values decrease.

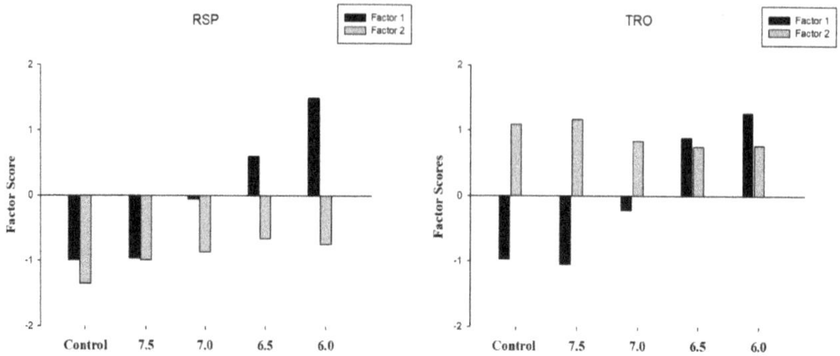

Figure 5. Representation of the estimated factor scores for each sediment sampling site (RSP and TRO) and treatment (control, 7.5, 7.0., 6.5, and 6.0).

In this sense, the pH value of 6.5 was correlated to provoke adverse effects on marine organisms that could also be associated with the increase in the mobility of metals. Moreover, it is important to perform a complete risk assessment in areas affected by CO_2 enrichment considering different LOEs associated with biological responses and sediment characterization for a proper interpretation of the impacts caused by the acidification in a marine environment.

4. Conclusions

The present study applied for the first time a complete and specific design of a weight-of-evidence approach (WOE) to address the potential impacts of CO_2 enrichment in the marine environment using different lines of evidence. It has been demonstrated from the different lines of evidence (LOEs) employed in this study using a WOE that the acidification by enrichment of CO_2 provokes adverse effects on the marine organisms, starting at pH 7.0, which considerably enhances at pH 6.5.

Metal contamination in sediments seems to play a very important role in the environmental risk assessment under the enrichment of CO_2 scenarios. The results showed that dissolved elements such as As, Fe, Cu, Ni, and Zn increased their concentrations and contributed to causing toxic effects on marine organisms under acidified conditions.

The WOE distinguished between contamination sources that are causing biological adverse effects in marine organisms. It differs between the extension of the effects produced by the acidification by enrichment of CO_2, the extension of the effects associated with the levels of contaminants (mainly metals), and the combination of both causes of adverse effects, acidification, and metal contamination at the same time.

Author Contributions: M.C.P. and I.R. contributed to all aspects of the article from the conceptualization to the funding acquisition including methodology software, validation, formal analysis, investigation, resources, data curation, and writing and editing the paper from the start to the final version. E.B. contributes in most of the aspects including conceptualization, investigation, resources, data curation, writing—original draft preparation, writing—review, and editing and visualization. A.C. contributed to most of the aspects including methodology, validation, formal analysis, writing—review and editing, visualization, supervision, and funding acquisition. All authors have read and agreed to the published version of the manuscript.

Funding: This study was funded by the Ministry of Education of Spain (PRX14/00134) and partially by FAPESP (#2017/25936-0). This study was also partially funded by FAPESP through grants #2015/17329-0 and #2018/18456-4.

Institutional Review Board Statement: Not applicable.

Informed Consent Statement: Not applicable.

Acknowledgments: The authors thank Fabio H. Pusceddu for the support in the data analysis and Eric Russell for the English revision. A.C. thanks CNPq (PQ 305417/2019-3) for productivity fellowships, and FAPESP (Process #2017/07353-7). I.R. thanks FAPESP (#2017/25936-0) for her Visiting Researcher Program at Universidade Federal do Sao Paulo. The first author would like to thank the Erasmus Mundus Program for the doctoral fellowship. E.B. thanks to ANID/FONDECYT 11180015.

Conflicts of Interest: The authors declare no conflict of interest.

References

1. United Nations Framework Convention on Climate Change (UNFCCC). Adoptions of the Paris Agreement. 2015. Available online: http://unfccc.int/files/essential_background/convention/application/pdf/english_paris_agreement.pdf (accessed on 2 May 2017).
2. Intergovernmental Panel on Climate Change (IPCC). *Climate Change 2014: Mitigation of Climate Change*; Edenhofer, O., Pichs-Madruga, R., Sokona, Y., Minx, J.C., Farahani, E., Kadner, S., Sevboth, K., Minx, J.C., Eds.; Working Group III Technical Support Unit, Cambridge University Press: Cambridge, UK, 2014; p. 1454.
3. IPCC; Global CCS Institute. The Global Status of CCS 2016: Summary Report. 2016. Available online: https://www.globalccsinstitute.com/publications/global-status-ccs-2016-summary-report (accessed on 2 May 2017).
4. IPCC. Summary for Policymakers. In *Global Warming of 1.5 °C. An IPCC Special Report on the Impacts of Global Warming of 1.5 °C above Pre-Industrial Levels and Related Global Greenhouse Gas Emission Pathways, in the Context of Strengthening the Global Response to the Threat of Climate Change, Sustainable Development, and Efforts to Eradicate Poverty*; Masson-Delmotte, V., Zhai, P., Pörtner, H.-O., Roberts, D., Skea, J., Shukla, P.R., Pirani, A., Moufouma-Okia, W., Péan, C., Pidcock, R., et al., Eds.; World Meteorological Organization: Geneva, Switzerland, 2018; 32p.
5. De Orte, M.R.; Lombardi, A.T.; Sarmiento, A.M.; Basallote, M.; Rodriguez-Romero, A.; Riba, I.; Del Valls, A. Metal mobility and toxicity to microalgae associated with acidification of sediments: CO_2 and acid comparison. *Mar. Environ. Res.* **2014**, *96*, 136–144. [CrossRef] [PubMed]
6. Wang, G.; Cao, R.; Ning, X.; You, L. Effects of ocean acidification on immune responses of the Pacific oyster *Crassostrea gigas*. *Fish Shellfish Immunol.* **2016**, *49*, 24–33. [CrossRef] [PubMed]
7. Conradi, M.; Sánchez-Moyano, J.E.; Galotti, A.; Jiménez-Gómez, F.; Jiménez-Melero, R.; Guerrero, F.; Parra, G.; Bonnail, E.; DelValls, T. CO_2 leakage simulation: Effects of the decreasing pH to the survival and reproduction of two crustacean species. *Mar. Pollut. Bull.* **2019**, *143*, 33–41. [CrossRef] [PubMed]
8. Borrero-Santiago, A.R.; Ribicic, D.; Bonnail, E.; Netzer, R.; Koseto, D.; Ardelan, M. Response of bacterial communities in Barents Sea sediments in case of a potential CO_2 leakage from carbon reservoirs. *Mar. Environ. Res.* **2020**, *160*, 105050. [CrossRef] [PubMed]
9. Passarelli, M.; Bonnail, E.; Cesar, A.; DelValls, T.A.; Riba, I. Integrative assessment of sediments affected by CO_2 enrichment: A case study in the Bay of Santos—SP, Brazil. *Appl. Sci.* **2021**, *11*, 11603. [CrossRef]
10. DelValls, T.A.; Souza, L.S.; Seabra, A.A.; Pereira, C.D.S.; Bonnail, E.; Riba, I. Integrative assessment of sediment quality in acidification scenarios associated with carbon capture and storage operations. *Environ. Rev.* **2019**, *27*, 333–345. [CrossRef]
11. DelValls, T.Á.; Forja, J.M.; Gómez-Parra, A. Integrative assessment of sediment quality in two littoral ecosystems from the Gulf of Cadiz, Spain. *Environ. Toxicol. Chem.* **1998**, *17*, 1073–1084. [CrossRef]
12. Abessa, D.M.S.; Carr, R.S.; Sousa, E.C.P.M.; Rachid, B.R.; Zaroni, L.P.; Pinto, Y.A.; Gasparro, M.R.; Bícego, M.C.; Hortellani, M.A.; Sarkis, J.E.S.; et al. *Integrative Ecotoxicological Assessment of a Complex Tropical Estuarine System*; Hoffer, T.N., Ed.; Nova Science Publishers: New York, NY, USA, 2008; pp. 125–159.
13. Choueri, R.B.; Cesar, D.M.S.; Torres, R.J.; Morais, R.; Riba, I.; Pereira, C.; Nascimento, M.; Mozeto, A.; DelValls, T. Development of site-specific sediment quality guidelines for North and South Atlantic littoral zones: Comparison against national and international sediment quality benchmarks. *J. Hazard. Mat.* **2009**, *170*, 320–331. [CrossRef]
14. Torres, R.J.; Cesar, A.; Pastor, V.A.; Pereira, C.D.S.; Choueri, R.B.; Cortez, F.S.; Morais, R.D.; Abessa, D.M.S.; Nascimento, M.R.L.D.; Morais, C.R.; et al. A critical comparison of different approaches to sediment- Quality Assessments in the Santos Estuarine System in Brazil. *Arch. Environ. Contam. Toxicol.* **2015**, *68*, 132–147. [CrossRef]
15. Bonnail, E.; Riba, I.; Seabra, A.; DelValls, T.A. Sediment quality assessment in the Guadalquivir River (SW, Spain) using caged Asian clams: A biomarker field approach. *Sci. Total Environ.* **2019**, *650*, 1996–2003. [CrossRef]
16. Basallote, M.D.; DelValls, T.Á.; Riba, I. Studying the effect of CO_2-Induced acidification on sediment toxicity using acute amphipod toxicity test. *Environ. Sci. Technol.* **2014**, *48*, 8864–8872. [CrossRef] [PubMed]
17. De Orte, M.R.; Bonnail, E.; Sarmiento, A.M.; Bautista-Chamizo, E.; Basallote, M.D.; Riba, I.; DelValls, Á.; Nieto, J.M. Metal fractionation in marine sediments acidified by enrichment of CO_2: A risk assessment. *Mar. Pollut. Bull.* **2018**, *131*, 611–619. [CrossRef] [PubMed]
18. Passarelli, M.C.; Riba, I.; Cesar, A.; Serrano-Bernando, F.; DelValls, T.A. Assessing the influence of ocean acidification to marine amphipods: A comparative study. *Sci. Total Environ.* **2017**, *595*, 759–768. [CrossRef] [PubMed]
19. Passarelli, M.C.; Cesar, A.; Riba, I.; DelValls, T.A. Comparative evaluation of sea-urchin larval stage sensitivity to ocean acidification. *Chemosphere* **2017**, *184*, 224–234. [CrossRef]

20. CEDEX. Directrices Para la Caracterización del Material Dragado y su Reubicación en Aguas del Dominio Público Marítimo-Terrestre. Comisión Interministerial de Estrategias Marina. 2015. Available online: http://www.mapama.gob.es/es/costas/temas/proteccion-medio-marino/directrices2015_tcm7-325119.pdf (accessed on 2 May 2017).
21. Basallote, M.; Rodríguez-Romero, A.; Blasco, J.; DelValls, A.; Riba, I. Lethal effects on different marine organisms, associated with sediment–seawater acidification deriving from CO_2 leakage. *Environ. Sci. Pollut. Res.* **2012**, *19*, 2550–2560. [CrossRef]
22. Basallote, M.; Rodríguez-Romero, A.; De Orte, M.R.; DelValls, A.; Riba, I. Evaluation of the threat of marine CO_2 leakage-associated acidification on the toxicity of sediment metals to juvenile bivalves. *Aquat. Toxicol.* **2015**, *166*, 63–71. [CrossRef]
23. American Society of Testing and Materials (ASTM). Standard Guide for Conducting Static Toxicity Tests Starting with Embryos of Four Species of Saltwater Bivalve Mollusks. In *Annual Book of ASTM Standards: Water and Environmental Technology*; ASTM: Philadelphia, PA, USA, 1992; Volume 11, pp. 377–394.
24. Cortez, F.S.; Pereira, C.D.S.; Santos, A.R.; Cesar, A.; Choueri, R.; Martini, G.D.A.; Bohrer-Morel, M.B. Biological effects of environmentally relevant concentrations of the pharmaceutical Triclosan in the marine mussel *Perna perna* (Linnaeus, 1758). *Environ. Pollut.* **2012**, *168*, 145–150. [CrossRef]
25. Riba, I.; Kalman, J.; Vale, C.; Blasco, J. Influence of sediment acidification on the bioaccumulation of metals in *Ruditapes philippinarum*. *Environ. Sci. Pollut. Res.* **2010**, *17*, 1519–1528. [CrossRef]
26. Moore, M.N.; Lowe, D.; Allen, I.; McVeigh, A.; Somerfield, P. Predicting health of environment—Lysosomal biomarker responses in mussels. In Proceedings of the Workshop on the Med Pol Biological Effects Programme: Acheivements and Future Orientations, Alessandria, Italy, 20–21 December 2006; p. 245.
27. Dailianis, S.; Domouhtsidou, G.P.; Raftopoulou, E.; Kaloyianni, M.; Dimitriadis, V. Evaluation of neutral red retention assay, micronucleus test, acetylcho-linesterase activity and a signal transduction molecule (cAMP) in tissues of *Mytilus galloprovincialis* (L.), in pollution monitoring. *Mar. Environ. Res.* **2003**, *56*, 443–470. [CrossRef]
28. Lowe, D.M.; Fossato, V.U.; Depledge, M.H. Contaminant-induced lysosomal membrane damage in blood cells of mussels *Mytilus galloprovincialis* from the Venice Lagoon: An in vitro study. *Mar. Ecol. Prog. Ser.* **1995**, *129*, 189–196. [CrossRef]
29. Morales-Caselles, C.; Riba, I.; Sarasquete, C.; DelValls, T.A. Using a classical weight of evidence approach for 4 years monitoring of the impact of an accidental oil spill on sediment quality. *Environ. Int.* **2008**, *34*, 514–523. [CrossRef] [PubMed]
30. DelValls, T.A.; Chapman, P.M. The use of multivariate analysis to link the sediment quality triad components defining site-specific sediment quality values in the Gulf of Cádiz and in San Francísco Bay. In Proceedings of the Second Symposium on the Atlantic Iberian Continental Margin, Cádiz, Spain, 17–20 September 1997.
31. Tabachinic, B.G.; Fidell, L.S. *Using Multivariate Statistics*; Harper Collins, College Publishers: New York, NY, USA, 1996.
32. Passarelli, M.C.; Ray, S.; Cesar, A.; DelValls, T.A.; Riba, I. Effects of CO_2 enrichment on metal bioavailability and bioaccumulation using Mytilus galloprovincialis. *Mar. Pollut. Bull.* **2018**, *133*, 124–136. [CrossRef] [PubMed]
33. Riba, I.; Passarelli, M.C.; Cesar, A. Chapter 3: Risk assessment of CO_2 acidification in aquatic ecosystems: A weight-of evidence approach. In *CO_2 Acidification in Aquatic Ecosystems*; Riba, D.V., Ed.; Elsevier: Amsterdam, The Netherlands, 2022; ISBN 978-0-12-823552-2.
34. Szalaj, D.; De Orte, M.R.; Goulding, T.A.; Medeiros, I.D.; DelValls, T.A.; Cesar, A. The effects of ocean acidification and a carbon dioxide capture and storage leak on the early life stages of the marine mussel *Perna perna* (Linneaus, 1758) and metal bioavailability. *Environ. Sci. Pollut. Res.* **2016**, *24*, 765–781. [CrossRef]
35. Goulding, T.; De Orte, M.R.; Szalaj, D.; Basallote, M.D.; DelValls, T.A.; Cesar, A. Assessment of the environmental impacts of Ocean acidification (OA) and Carbon Capture and Storage (CCS) leaks using the amphipod *Hayle* young. *Ecotoxicology* **2017**, *26*, 521–533. [CrossRef]
36. Riba, I.; Casado-Martínez, C.; Forja, J.M.; DelValls, T.Á. Sediment quality in littoral regions of the Gulf of Cádiz: A triad approach to address the influence of mining activities. *Environ. Pollut.* **2004**, *132*, 341–353. [CrossRef]
37. Christen, N.; Calosi, P.; McNeill, C.L. Structural and functional vulnerability to elevated pCO_2 in marine benthic communities. *Mar. Biol.* **2013**, *160*, 2113–2128. [CrossRef]
38. Almagro-Pastor, V.; Conradi, M.; DelValls, T.A.; Riba, I. Alterations in the macrobenthic fauna from Guadarranque River (Southern Spain) associated with sediment–seawater acidification deriving from CO_2 leakage. *Mar. Pollut. Bull.* **2015**, *96*, 65–75. [CrossRef]
39. Chapman, P.M.; Wang, F.; Janssen, C.R.; Goulet, R.R.; Kamunde, C.N. Conducting ecological risk assessments of inorganic metals and metalloids: Current status. *Hum. Ecol. Risk Assess.* **2003**, *9*, 641–697. [CrossRef]
40. Eggleton, J.; Thomas, K.V. A review of factors affecting the release and bioavailability of contaminants during sediment disturbance events. *Environ. Int.* **2004**, *30*, 973–980. [CrossRef]
41. Atkinson, C.A.; Jolley, D.F.; Simpson, S.L. Effect of overlying water pH, dissolved oxygen, salinity and sediment disturbances on metal release and sequestration from metal contaminated marine sediments. *Chemosphere* **2007**, *69*, 1428–1437. [CrossRef]
42. Maar, M.; Larsen, M.M.; Torring, D.; Petersen, J.K. Bioaccumulation of metals (Cd, Cu, Ni, Pb and Zn) in suspended cultures of blue mussels exposed to different environmental conditions. *Estuar. Coast. Shelf Sci.* **2018**, *201*, 185–197. [CrossRef]
43. Rodríguez-Romero, A.; Basallote, M.D.; Manoela, R.; DelValls, T.Á.; Riba, I.; Blasco, J. Simulation of CO_2 leakages during injection and storage in sub-seabed geological formations: Metal mobilization and biota effects. *Environ. Int.* **2014**, *68*, 105–117. [CrossRef] [PubMed]

44. Benítez, S.; Duarte, C.; López, J.; Manríquez, P.H.; Navarro, J.M.; Bonta, C.C.; Torres, R.; Quijón, P.A. Ontogenetic variability in the feeding behaviour of a marine amphipod in response to ocean acidification. *Mar. Pollut. Bull.* **2016**, *112*, 375–379. [CrossRef] [PubMed]
45. Luoma, S.N.; Rainbow, P.S. Why is metal bioaccumulation so variable? Biodynamics as a unifying concept. *Environ. Sci. Technol.* **2005**, *39*, 1921–1931. [CrossRef] [PubMed]

applied sciences

MDPI

Article

Sediment Contamination and Toxicity in the Guadalquivir River (Southwest, Spain)

Inmaculada Riba [1,*], Angel Luque-Escalona [2] and Maria Helena Costa [3]

[1] Facultad de Ciencias del Mar y Ambientales, Universidad de Cádiz, Polígono Río San Pedro s/n, 11510 Puerto Real, Cádiz, Spain
[2] Departamento de Biología, Universidad de Las Palmas de Gran Canaria, Campus de Tafira, 35017 Las Palmas, Las Palmas de Gran Canaria, Spain
[3] IMAR, Departamento de Ciências e Engenharia do Ambiente, Universidade Nova de Lisboa, 2829-516 Caparica, Portugal
* Correspondence: inmaculada.riba@gmail.com

Abstract: A segment of the Guadalquivir River was assessed between the Alcalá del Río dam and Seville through an integrative sediment quality assessment. Chemical concentrations of metals and toxicity under laboratory conditions were used as lines of evidence. A battery of bioassays with four organisms (the amphipod *Ampelisca brevicornis*, the bacteria *Vibrio fischeri*, the sea urchin *Paracentrotus lividus*, and the oligochaete *Tubifex tubifex*) exposed to sediment made it possible to determine the potential risk associated. The sediments from Seville and Alcalá del Río showed higher values of the concentration of most metals than the Algaba station, with Cu (35–37 μg/g), Zn (70–75 μg/g), Ni (23–26 μg/g), and Pb (27–30 μg/g) being the most abundant metals. An increasing toxicity gradient was shown downstream among the bioassays with the amphipod *A. brevicornis*, the fertilization test using the sea urchin *P. lividus*, and the freshwater worm growth *T. tubifex*. Conversely, an increasing toxicity gradient was shown upstream in the embryo-larval *P. lividus* development. The link between sediment contamination and toxicity makes it possible to obtain a gradient of contaminant concentration comparable with nationally and internationally widely accepted sediment quality guidelines in order to establish the risk associated with this area of study.

Keywords: lines of evidence; sediment quality guidelines; Guadalquivir River; toxicity tests

Citation: Riba, I.; Luque-Escalona, A.; Costa, M.H. Sediment Contamination and Toxicity in the Guadalquivir River (Southwest, Spain). *Appl. Sci.* **2023**, *13*, 3585. https://doi.org/10.3390/app13063585

Academic Editor: Yu Tian

Received: 31 January 2023
Revised: 22 February 2023
Accepted: 8 March 2023
Published: 10 March 2023

1. Introduction

The Guadalquivir River (SW, Spain) is the second longest river in Spain and the only one which allows navigation in the Estuary, being the main water source of a Spanish region with more than 7 million inhabitants. It reaches the Gulf of Cádiz (Atlantic Ocean), bordering Doñana National Park Reserve which is the most important wintering site for many birds in Europe. However, this region has been exposed for centuries to mining activities which have greatly enriched the metal content of its water and soils [1].

Integration of chemical concentration and toxicity responses is the best way to determine the sediment [2,3]. It is known that sediments are able to transport contaminants much more easily than water [4], but also act as a sink and/or source of contaminants [5]. Therefore, biological effects based on laboratory tests can determine toxic responses. Sediment bioassays are simple tests that evaluate the responses of the tested organism to contaminated sediments under controlled conditions [6]. Despite of the fact that some reports have pointed out that biological tests or bioassays are decisive to understand the toxic effects of sediments [7–9], the final decision of the sediment status is based on a comparison between levels of contaminants measured in the sediments against different sediment quality guidelines (SQGs).

A battery of tests was selected using different trophic levels to determine sediment toxicity: (1) a whole sediment survival bioassay using the amphipod *Ampelisca brevicornis*,

(2) the commercial Basic Solid Phase Test (BSPT) of Microtox® based on the bioluminescence of the bacteria *Vibrio fischeri*, (3) the *Paracentrotus lividus* sea urchin fertilization success and (4) embryo-larval development bioassays, and (5) a whole sediment toxicity test with the oligochaete worm *Tubifex tubifex*.

The acute bioassay with the *A. brevicornis* amphipod has shown this species as a sensitive organism valid to assess the toxicity of contaminated sediments [10–14]. The commercial bioassay Microtox® has been used to detect the "hot spots" of field contamination in the screening procedure [15,16], and also to assess oil contaminated sediments [11,17,18]. Sea urchin toxicity tests use fertilization and larval development as endpoints, which are considered to be useful tools for assessing toxicity in marine environments. These acute toxicity tests have been applied to determine the toxicity of sediments [19–21] under laboratory conditions, and have been included and standardized by several national environmental agencies [22,23].

The chronic toxicity test with the oligochaete *T. tubifex* was widely used for sediment bioassessment in the Great Lakes of North America and included in environmental programs [24,25]. In Europe, *T. tubifex* was also used for sediment toxicity assessment [26–28], as well as in bioaccumulation studies [3,29] and organic compounds [30–32]. The followed procedure was developed by [26], and recommended in the American Society for Testing Materials (ASTM) guide [33]. Therefore, they are considered ideal for determining the toxicity of contaminants [34–36] in freshwater sediments. This bioassay was selected to assess toxicity during low tide when salinity falls down, since the tidal influence comes up to the Alcalá del Rio dam (maximum recorded high tide: 1.70 m).

The main aim of this work is to characterize the sediment quality in different areas of the Guadalquivir River affected by different levels of contamination by linking both chemical concentrations and bioassay responses using a weight of evidence approach.

2. Materials and Methods

2.1. Study Area

The present study was performed in the lower course of the Guadalquivir River, between the Alcalá del Río dam and the city of Seville (Figure 1). The study areas are shown in Figure 1: (1) Alcalá del Río town (AR), 1 km downstream the dam; (2) Algaba (AL) municipality, 6 km downstream from the AR station; and (3) Seville (SE) city, 6 km downstream from the AL station, Seville being the main urban and industrial center of this area. According the WFD criteria, the Waterbody Manager of the Guadalquivir River established as "moderate status" the river course from the AR dam to the SE [37]. These sites were selected to test the influence of the historic mining activity [1]. The local water body manager [37] identified different diffuse sources of contaminants from urban, industrial, and agricultural waste waters as the key pressure on Guadalquivir river waters.

Two more sites were chosen as control stations: (1) San Pedro river (RSP), which was selected for the amphipod, the BSPT of Microtox®, and the sea urchin fertilization and embryo-larval bioassays, as well as the fact that it has been previously determined as a clean area [38,39], and (2) Barazar (BAR) for the chronic toxicity test with *T. tubifex* due it's proved quality as a control for this species in particular [40].

2.2. Sediment Sampling

Surface sediment samples (5–10 cm) were collected at the three described sites along the Guadalquivir River (AR, AL and SE) and the control site (RSP) with a 0.025 m^2 Van Veen grab and transferred to a plastic bucket. Sediment was transported in a cooler to the laboratory. There, the contents of the bucket were homogenized, and subsamples were destined for physical-chemical characterization and used for toxicological tests. They were kept at 4 °C in the dark.

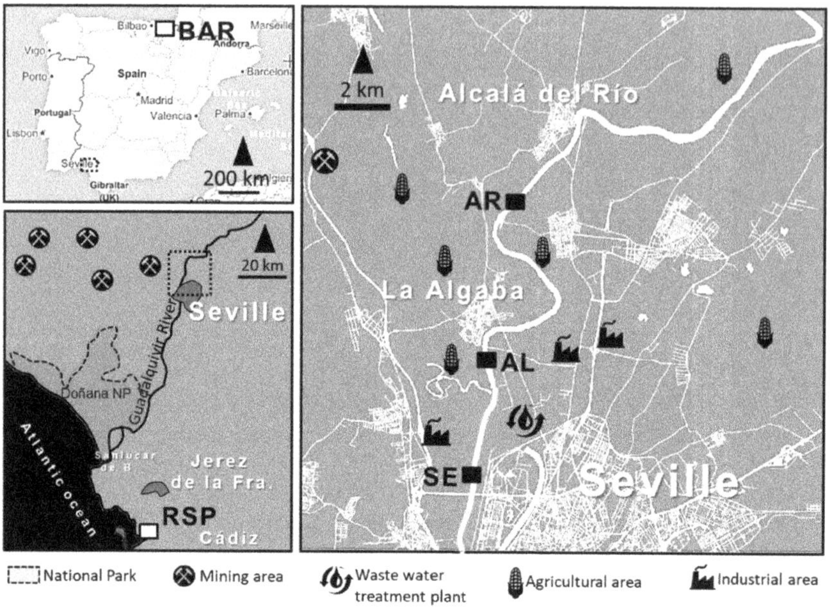

Figure 1. Location of sampling stations in the Guadalquivir River: Alcalá del Río town (AR), Algaba (AL) municipality, and Seville (SE); and controls from Río de San Pedro (RSP) in Cádiz, and Barazar (BAR) in Bilbao.

2.3. Chemical Analyses

Sediment aliquots for chemical analysis were dried at room temperature and then gently homogenized. Organic carbon content was determined by using the method of [41]. Aliquots of sediments were acid-digested (65% H_2NO_3 and 33% H_2O_2) in closed vessels in triplicate. Metals Cu, Zn, Fe, Co, Ni, Cd, and Pb, and the metalloid As were determined using ICP-MS, ICP-AES, and GF-AAS. The concentration was calculated on a dry weight basis. Reactive blanks and reference materials (CRM478 BCR) were used to ensure quality control.

2.4. Toxicity Tests in Laboratory

As mentioned above, sediments from the analyzed points were used to perform a battery of acute toxicity tests in laboratory conditions: (1) a whole sediment survival bioassay using the *A. brevicornis* amphipod, (2) the commercial BSPT of Microtox®, (3) the sea urchin fertilization success and (4) embryo-larval development bioassays, and (5) a whole sediment toxicity test with *T. tubifex*.

2.4.1. Amphipod Toxicity Test

The collection and acclimation of organisms, the collection and preparation of sediment, and the *A. brevicornis* toxicity test procedure were conducted following the methodology pointed out by different authors and standard protocols [42,43]. Water quality and the amphipods' mortality was determined. Results are expressed as survival of organisms after 10 days of exposure.

2.4.2. Bacterial Toxicity Test

The decrease of the bioluminescence of the bacteria *V. fischeri* (Microtox® BSPT) is used as a quality indicator. This bioassay was conducted with the commercial Microtox®(model 500) following the Basic Solid Phase Test (BSPT) according to the standard operating procedure [44] with the modifications reported by [10,45]. Calculations to obtain this concentration were performed with the Microtox Omni® (Version 1.18) software of Strategic Diagnostics Inc. Results were expressed as an IC50 (mg L^{-1}) value, which is the concentration of dry sediment where the luminescence is reduced by half.

2.4.3. Sea Urchin Toxicity Tests

Adults of sea urchins *Paracentrotus lividus* were collected by free diving from an Andalusian beach near to El Estrecho Natural Park. They were placed in a 30 L plastic cooler and transported to the laboratory, where they were acclimated in chambers with aerated clean water at 15 °C for 10 days. Sediment elutriates were obtained using a modification of the [46] method.

Paracentrotus lividus fertilization test: gametes used for fertilization were obtained from a single pair by injection of 1 mL KCl (0.5 M) through the peri-oral membrane. The fertilization procedure followed the one described by [47], adapted from [48]. Each sample, plus the control containing clean water, was run per triplicate.

Paracentrotus lividus embryo development test: The same pools of gametes obtained during the fertilization test were used for the larval development test. The in vitro embryo-larval bioassay was developed using the sea urchin larvae after 48 h of exposition to sediment elutriates. Fecundation and test conditions followed the protocol developed by [49].

The endpoint recorded was the percentage of embryological success, expressed as the percentage of normal pluteus (well-developed four-arm embryos, according to the criteria reported by [47], and calculated by counting 100 larvae per vial with an optical microscope (OLYMPUS CKX41)).

2.4.4. Oligochaete Chronic Toxicity Test

The chronic sediment toxicity test with annelid *Tubifex tubifex* was performed following [33,49,50] protocols and modifications [26,40]. Worms obtained from the Natural Park of Gorbeia (Vizcaya, Spain) were exposed for 28 days to the different sediment samples and a control sediment (called Barazar or BAR) with proved quality [40], which was used for the worm culture.

The endpoints analyzed were growth, reproduction, and mortality after 28 days, calculated respectively as total growth rate (TGR), Total Young Number (TYG), and mortality, following the methodology reported by [40].

2.5. Data Treatment

A one-way analysis of variance (ANOVA) was used to test the significance of the contamination analysis and the toxicity tests compared to control samples, and were followed by multiple comparisons of Dunnett's tests. Statistical analyses of toxicity tests were performed using SPSS 17.0. The acute toxicity for the Microtox BSPT bioassay was analyzed using a gamma model of Microtox Omni software (SDI Europe, Hampshire, UK).

For the analysis of the data from the *T. Tubifex* bioassay, the mortality frequencies in each test sediment were compared with the respective control using Fisher's Exact test. Normality of data and homogeneity of variances were checked through the Shapiro-Wilk and the Levene's F tests, respectively. Data were transformed for normalization when necessary. Single-factor analyses of variance followed by Dunnet's test (if the number of replicates was the same) or a *t*-test with Bonferroni's adjustment (when the number of replicates was different) were used for the comparison of the tests' means and to control for the reproductive and growth variables [43]. When data were not normalized after transformation, the Kruskal-Wallis nonparametric test followed by the Mann-Whitney U test was applied.

The multivariate analysis was performed by means of a factor analysis (Principal Component Analysis (PCA) with varimax normalized rotation). This analysis establishes and quantifies the correlations among the variables from the original data set to facilitate the data interpretation [51,52]. According to [53], only those variables with a loading of 0.30 or higher were associated with a particular factor. The factor analysis procedure was performed using the STATISTICA software tool (Stat Soft, Inc., Tulsa, OK, USA, 2001; version 6.0). The calculation of SQGs followed the methodology presented in [51] slightly modified by [54].

3. Results and Discussion

Different authors have reported the usefulness of the battery of tests used in this work. For instance, [21] showed that Microtox® BSPT, *A. brevicornis*, and *P. lividus* toxicity tests gave supplementary information about particulate and adsorbed contaminants' effects. Additionally, the present report has developed a standardized toxicity test in freshwater sediments with the oligochaete *T. tubifex* to complete the toxicity information during the salinity decreases in low tide events. The suitability of the different bioassays was confirmed through the validity of the controls.

3.1. Sediment Metals Concentration

Table 1 shows the summarized results of the different contaminants analyzed. No increasing or decreasing pattern downstream was observed. In general, sediments from the AR station showed the highest contaminant content and those from AL the lowest, except for Pb, which was higher at SE and lower at AL sediments. The sediments located in the station close to Seville showed significant differences ($p < 0.05$) with the control station (RSP) for every analyzed contaminant and organic matter.

Table 1. Summarized results of chemical analyses in sediments from the studied stations (Adapted from [55]).

Element		AR	AL	SE	RSP
As	µg/g	10.22	9.77	9.84	7.14
Cd	µg/g	0.27	0.26	0.26	0.13
Co	µg/g	8.63	7.73	7.70	5.82
Cu	µg/g	37.42	35.37	35.51	18.00
Fe	µg/g	23,990	21,920	22,140	15,239
Ni	µg/g	25.42	23.39	23.56	14.00
Pb	µg/g	29.65	27.30	31.73	22.30
Zn	µg/g	75.89	66.79	69.65	49.00
OC	%	1.18	1.06	1.16	0.83

3.2. Toxicity Tests in Laboratory

3.2.1. Amphipod Laboratory Bioassay

Mean mortality results after the 10 day amphipod toxicity test are shown in Figure 2a. The highest mortality (60%) was associated with those organisms exposed to sediments at the SE station. An increasing toxicity gradient was shown downstream, while AR showed the lowest mortality values. Furthermore, stations with the highest toxicity values (SE and AL) showed significant differences ($p < 0.05$) compared to the control station (RSP).

3.2.2. Microtox® Bioassay

The highest inhibition of bioluminescence, which corresponds to an $IC_{50} < 1000$ mg L^{-1} dry weight, is shown in the samples collected at SE (Figure 2b). Therefore, this station was only under the limit assumed for sediment toxicity (1000 mg L^{-1} dry weight) according to the Canadian standards [23]. On the other hand, the sediments obtained from the RSP, and used as a control station, showed the highest values of IC_{50} (12,186.67 mg L^{-1} dry weight). No significant differences ($p < 0.05$) were observed between each station and the control.

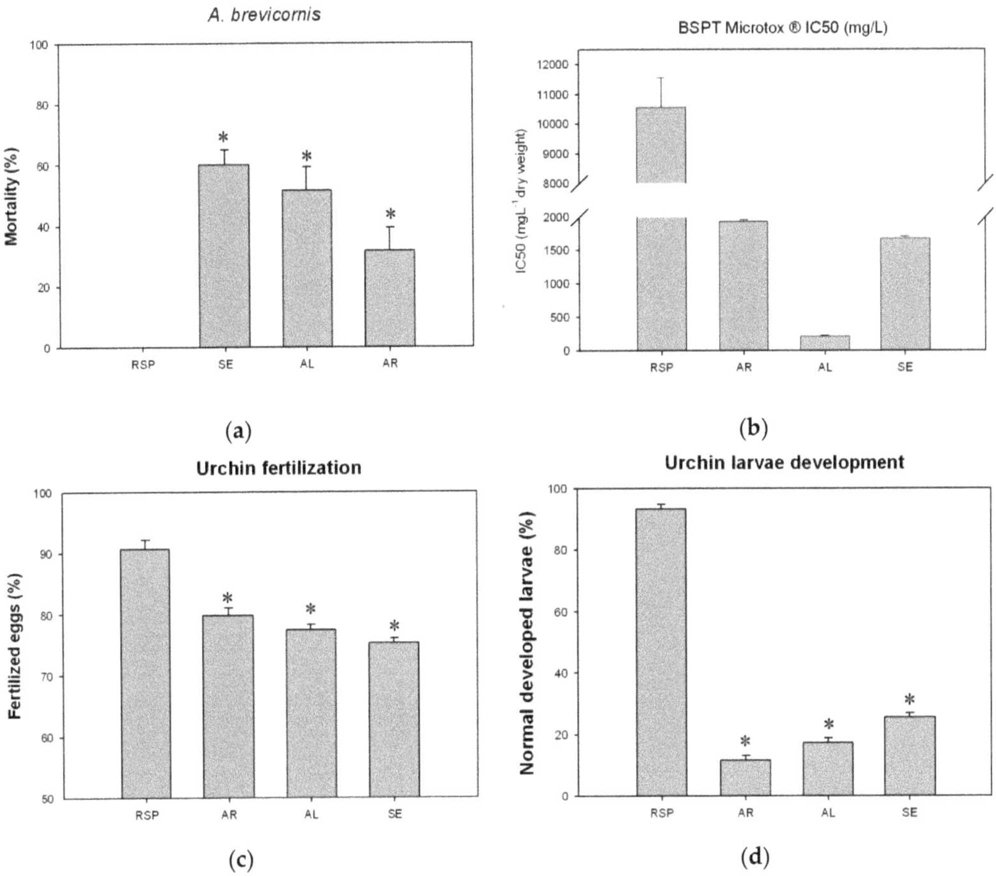

Figure 2. (**a**) *A. brevicornis* mortality (%) after 10 d exposure. (**b**) IC_{50} results (mg L^{-1} dw sediment) obtained from the BSPT Microtox® samples. (**c**) Results of the *P. lividus* fertilization bioassay test: average percentage of fertilized eggs after exposure of the sperm to the sediment samples under laboratory conditions. (**d**) Results of the *P. lividus* embryo-larval bioassay test: average percentage of normal development of larvae after exposure of the fertilized eggs to the sediment samples under laboratory conditions. Asterisk (*) indicates significant differences ($p < 0.05$) between samples and the control (RSP).

3.2.3. Paracentrotus lividus Toxicity Tests

P. lividus Fertilization Bioassay

The percentage of successfully fertilized *P. lividus* eggs in each sampling station is shown in Figure 2c. A gradient of toxicity was observed between the downstream station (SE) with the lowest values of fertilization success and the upstream station AR, which showed the highest values (excluding the control). Significant differences ($p < 0.05$) were observed in all the stations compared with the control (RSP). The control station showed the highest success (more than 90%), confirming the validity of the test.

P. lividus Embryo Development Test

The success percentage of normally-developed *P. lividus* larvae at each station are shown in Figure 2d. An increased number of normal *P. lividus* larvae were observed in the Guadalquivir River stations (AR, AL and SE) and ranged from 12% (AR) to 25% (SE). These results indicated the presence of adverse effects (percentage of normal larvae development < 80%) and showed significant differences ($p < 0.05$) with the control station (RSP), which showed more than 90% normal larvae development. The results confirm the validity of the test with the RSP station as a control.

Tubifex tubifex Chronic Bioassay

In this bioassay the results of the control station (RSP) also confirmed the validity of the bioassay according to the [33] criteria. Dissolved oxygen was maintained above 60% (except twice: once in the control at 34%, and again in AR at 59%), unionized ammonia (NH_3) was maintained above 1 mg L^{-1}, and pH ranged from 6–8.22 (BAR), 7.59–8.59 (AR), 7.98–8.60 (AL), and 7.85–8.62 (SE).

Figure 3 represents the results for growth (TGR) and reproduction (TYG). A decreasing gradient of growth was shown downstream of the Guadalquivir River. TGR was significantly reduced ($p < 0.05$) in sediments from AL and SE (SE being very close to zero). The TYG was significantly reduced ($p < 0.05$) in the sediments from AR and AL, comparing with the control BAR. Nevertheless, no spatial gradient was shown. Regarding mortality, no significant differences in mortality were observed with respect to the control since mortality was not found in any station.

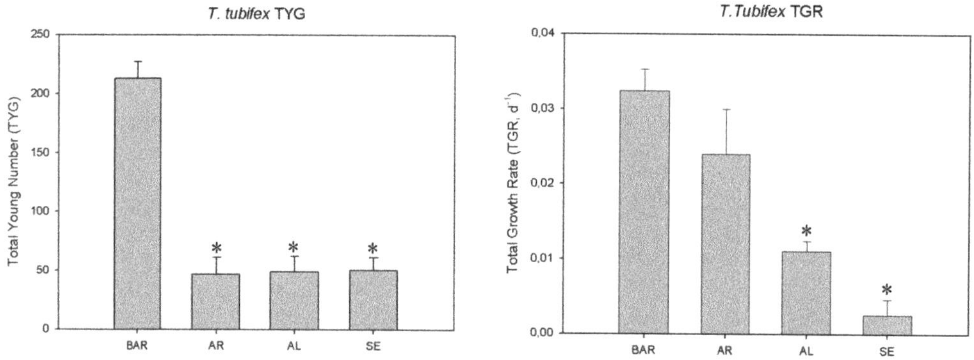

Figure 3. Mean and SD of toxicity endpoints of the *T. tubifex* bioassay: TGR (d^{-1}), TYG, and mortality (%) at each station. Asterisk (*) indicates significant differences ($p < 0.05$) between samples and the control (BAR).

3.3. Linking Contamination and Toxicity

A PCA factor analysis was performed in order to assess the environmental quality of the study area from the Guadalquivir River, employing acute toxicity and chemical analysis. The application of principal component analysis indicated that the original set of variables could be narrowed down to three new factors that explained 100% of the total variance (Table 2).

Factor analysis was performed to link the results of the five bioassays (Microtox® BSPT, *A. brevicornis*, both with *P. lividus*, and *T. tubifex*) with the concentration of the nine contaminants (Cu, Zn, Fe, Co, Ni, Cd, Pb, As, and organic carbon). The first principal factor (F1) explains 78.93% of the total variance and group responses of the sediment toxicity tests (excluding Microtox® and total growth rate measured in *T. tubifex*), metals, and organic carbon analyzed in sediments. However, this low loading of Microtox® in this factor could be due to the fact that this test is a screening bioassay; it is suggested that this biotest by itself may not be representative in certain cases of the full impact of a given pollutant on an ecosystem [17]. The second factor (F2) accounted for 14.31% of the total variance and was correlating the concentration of Cd in sediment with all the responses measured in the bioassays (by presenting positive loadings) except for total young number of *T. tubifex* (which presented a negative value). This correlation between these metals and *T. tubifex* parameters is in agreement with [40], who showed high sensitivity to copper and cadmium. The third factor (F3) accounted for 6.76% of the variance. This factor has associated contamination by Co to *A. brevicornis* and *T. tubifex* mortality and TYG. Only *A. brevicornis* mortality showed negative loadings. Previous results reported by [13] also related the contamination by Co to the *A. brevicornis* mortality indicating a particular sensitivity of this species to Co.

Table 2. Sorted rotated factor loadings of the original variables for the three principal factors results. Variables selected for the interpretation represented a loading of 0.30 or higher.

Variance %	Factor 1 78.93	Factor 2 14.31	Factor 3 6.76
A. brevicornis mortality	0.764	0.555	−0.327
V. fischeri bioluminescence inhibition		0.954	0.301
P. lividus fertilization	0.881	0.436	
P. lividus embriogenesis	0.917	0.322	
T. tubifex mortality	0.410	0.346	0.843
T. tubifex TYG	0.874	−0.413	0.378
T. tubifex TGR		0.982	
Cu	0.948	0.309	
Zn	0.974		
Fe	0.960		
Co	0.939		0.332
Ni	0.956		
Cd	0.948	0.301	
Pb	0.985		
As	0.955	0.344	
OC	0.998		

A graphical representation of the estimated factor values corresponding to each sampling station is presented in order to confirm the descriptions of these factors (Figure 3).

Through the sediment toxicity test under laboratory conditions, it has been demonstrated that the environmental degradation due to metals in the Guadalquivir river water course can be translated into a potential ecological risk. The study here conducted assessments of the synergetic effect of all the contaminants present in the sediment analyzed. The integration of the results from the different lines of evidence makes it possible to address the extension of the metal contamination and the relationship between these metals (and other potential contaminants not measured) on the effects observed in the toxicity tests conducted.

Based on the multivariate analysis results (Table 2 and Figure 4), different contamination and toxicity processes can be defined at each of the sediments studied here: sediments at station AR show the highest score value for Factor 1, Factor 2 being lowest and negative. The toxicity associated with metals is related to factor 1. This fact can be explained by two different sources of metals and the toxicity of them (Cd, and in lower weight Cu and As, from the negative value of Factor 2). It could be related to the presence of a dam 1 km upstream of the AR sample point that could explain potential phenomena of accumulation and thr release of different metals depending on the season.

Sediments at station AL show the highest value score for Factor 2 and a negative value in Factor 1. Factor 2 includes correlations with toxicity test responses lower than F1 except for Microtox. This could be informative regarding the Microtox® responses in these sediments being lower in the rest of the toxicity tests, and as well as regarding a moderate toxic response in this sediment. The screening nature of this test has been reported [17]. In addition, the moderate toxicity observed in this station is related to a moderate correlation of three metals as well (Cu, Cd, and As), but not to the rest of the analyzed metals. Moreover, different sources of these metals could be included in the explanation of these correlations, and Cu, Cd, and As elements could be part of organic compounds, such as pesticides and fertilizers coming from agriculture activities which have a high economic impact in this area. Furthermore, agriculture has economic importance in this area as reported by previous work [56]. In addition, the positive value of Factor 3 could confirm this different origin of the contamination of metals and the moderate toxicity observed in these sediments.

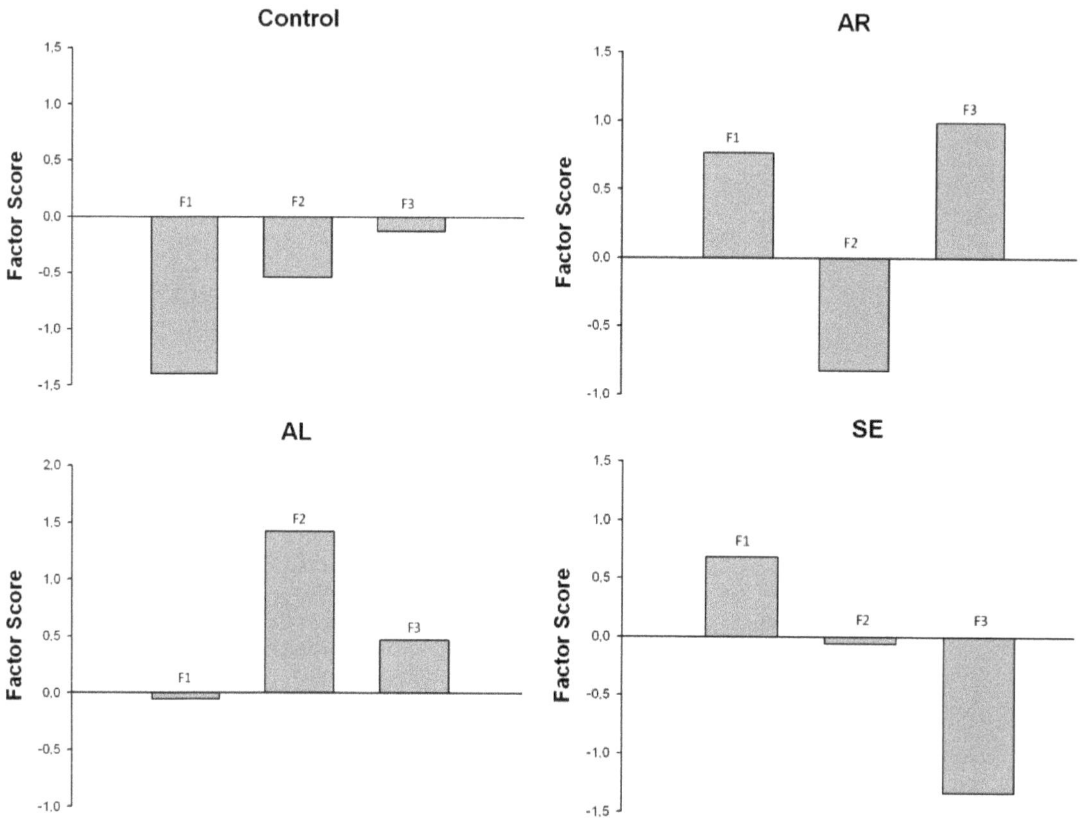

Figure 4. Estimated factor scores from the sampling stations (Control, AR, AL, SE). The factor scores quantify the prevalence of each factor for every station and are established to define each factor.

Sediments at the station located nearby the city of Seville (SE) also show a positive score of Factor 1, informing about the contamination of metals and the toxicity associated with them. However, the values and correlations are lower than station AR and can be considered as intermediate in toxicity in this study. The negative values of Factors 2 and 3 confirm that sediments at this station are less contaminated than those at station AR.

A previous sediment quality assessment was carried out with the biomarker responses of clams in the same stations [55]. AR and AL contamination resulted in the induction of biological stress, and SE contamination (As, Cd, and Cu) caused toxicity as biomarker effects.

From these results, site-specific sediment quality guidelines can be derived for the area studied. These SQGs have been calculated following the procedure reported by [51] and slightly modified by [54]. The SQGs are shown in Table 3.

These values have been compared with nationally and internationally accepted SQGs in order to establish a pollution range in this area, and with the aim of proving the validity of the use of these values to be applied in estuarine environments. In Table 4, this comparison can be observed in detail.

Table 3. Summary of calculated benchmark sediment quality guidelines (SQGs; mg kg^{-1} dw) proposed to evaluate sediment quality in the Guadalquivir River course between the Alcalá del Rio and Seville for the metals Cu, Zn, Fe, Co, Ni, Cd, and Pb, and the metalloid as associated with toxic effects.

Contaminant	Sediment Quality Guidelines (SQGs)		
	Not Polluted	Moderately Polluted	Highly Polluted
Cu	≤35.37	35.37–35.51	≥35.51
Zn	≤66.79	66.79–69.65	≥69.65
Fe	≤21,920	21,920–22,140	≥22,140
Co	≤5.82	5.82–7.70	≥7.70
Ni	≤23.39	23.39–23.56	≥23.56
Cd	≤0.13	0.13–0.26	≥0.26
Pb	≤27.29	27.29–28.40	≥28.40
As	≤9.77	9.77–9.84	≥9.84

Table 4. SQGs comparison for marine sediments.

	CCME [1]		Riba [6]		CEDEX [2]			Present Study
	ISQG	PEL	V1	V2	AL1	AL2	AL3	ALR
As	7.24	41.6	27.4	213	35	70	280	9.77–9.84
Cd	0.7	4.2	0.51	0.96	1.20	2.40	9.60	0.13–0.26
Cu	18.7	108	209	979	70	168	675	35.37–35.51
Ni			-	-	30	63	234	23.39–23.56
Pb	30.2	112	260	270	80	219	600	27.29–28.4
Zn	124	271	513	1310	205	410	1640	66.79–69.95

[1] [55,57]; [2] [58], ISQG: interim marine sediment quality guidelines; PEL: Probable effect levels.

Comparing the results from this study with the different widely accepted SQGs, it is shown that not one of the metal concentrations founded in the Guadalquivir River fit with the action level proposed by these SQGs. However, the action level established in this study was proposed according to several laboratory bioassays, demonstrating sediment toxicity in all the cases with irreversible ecological effects.

The SQGs calculated in the present study are in the interval of the international guidelines for all the metals [57]. Only Zn and Pb showed lower values for the guidelines than the rest of the international proposed guidelines. Ni showed guideline values close to the lower guidelines proposed by [58]. The metals Cu, Cd, and As, showed values between the lowest and highest guideline values proposed by other authors (Table 4). From these values it could be suggested that the toxic effects determined in the area are associated with the lowest SQGs proposed by other authors.

4. Conclusions

Sediments from the Guadalquivir River were assessed through a battery of bioassays with amphipods, sea urchins, bacteria, and worms in laboratory conditions. Based on the results obtained for metal contamination and toxicity assessment, the present study demonstrated alterations associated with the contamination measured in sediments. Sediments from the AL and SE stations showed an important relationship between metals and toxic endpoints which are higher in AL than in the area of SE. Sediments at the AR station showed a complex mixture of sources of metals with a moderate effect compared to the other two stations with a potentially different origin of Cd, Cu, Co, and As at the location of the dam upstream of the estuary.

A sediment quality guideline for metals analyzed was calculated and compared to international and national guidelines showing the action level gap between them, with a contamination range higher than those from this study. For this reason, this set of chemical guidelines derived based on toxic studies could be useful for future riverine/estuarine/coastal management plans. The approach used in this study by means of different lines of evidence (contamination of metals and sediment toxicity) is a first step of sediment quality assessment in the area. Similar studies including other contaminants (organics and an emergent like pharmaceuticals) will improve the integrated analysis for a correct sediment quality assessment in the area.

Author Contributions: I.R. has contributed to all the aspects of the article from the conceptualization to the funding acquisition, including contributions in methodology, software, validation, formal analysis, investigation, resources, data curation and writing, and editing the paper from the start to the final version. A.L.-E. has contributed in most of the aspects including the conceptualization, investigation, resources, data curation, writing, original draft preparation, writing—review and editing, visualization, supervision, and project. M.H.C. has also contributed to most of the aspects including methodology; software; validation; formal analysis; writing—original draft preparation; writing—review and editing; visualization; supervision; project administration; and funding acquisition. All authors have read and agreed to the published version of the manuscript.

Funding: This study was funded by Ministry of Education Spain (PRX14/00134) and partially by FAPESP (#2017/25936-0).

Institutional Review Board Statement: Not applicable.

Informed Consent Statement: Not applicable.

Data Availability Statement: Not applicable.

Acknowledgments: Authors would like to give special thanks to M.J. Salamanca for the assistant in the development of the Microtox® and *P. lividus* bioassays, L. Méndez for her support in the *T. tubifex* bioassay and A. Cesar for his personal clarifications. Riba thanks FAPESP (#2017/25936-0) for her Visiting Reseacher Program at Universidade Federal do Sao Paulo. I. Riba also thanks the Ministry of Education (PRX14/00134) in Spain for funding her placement at University Nova of Lisbon (Portugal) and University of Las Palmas de Gran Canaria (Spain).

Conflicts of Interest: The authors declare no conflict of interest.

References

1. Cabrera, F.; Toca, C.; Diaz, E.; Arambarri, P. Acid mine-water and agricultural pollution in a river skirting the Doñana National Park (Guadiamar river, southwest Spain). *Water Res.* **1987**, *18*, 1469–1482. [CrossRef]
2. DelValls, T.A.; Conradi, M. Advances in marine ecotoxicology: Laboratory tests versus field assessments data on sediment quality studies. *Cienc. Mar.* **2000**, *26*, 39–64. [CrossRef]
3. Chapman, P.M.; Ho, K.T.; Munns, W.R.; Solomon, K.; Weistein, M.P. Issues in sediment toxicity and ecological risk assessment. *Mar. Pollut. Bull.* **2002**, *44*, 271–278. [CrossRef] [PubMed]
4. Ramos-Gómez, J.; Martín-Díaz, M.L.; Rodríguez, A.; Riba, I.; DelValls, T.A. In Situ Evaluation of Sediment Toxicity in Guadalete Estuary (SW Spain) After Exposure of Caged *Arenicola marina*. *Environ. Toxicol.* **2008**, *23*, 643–651. [CrossRef] [PubMed]
5. Förstner, U.; Wittman, G.T.W. (Eds.) *Metal Pollution in Aquatic Environment*; Springer: Berlin/Heidelberg, Germany, 1979; 485p.
6. Riba, I.; Conradi, M.; Forja, J.M.; DelValls, T.A. Sediment quality in the Guadalquivir estuary: Lethal effects associated with the Aznalcóllar mining spill. *Mar. Pollut. Bull.* **2004**, *48*, 144–152. [CrossRef]
7. USEPA/USACE, United States Environmental Protection Agency/United States Army Corps of Engineers. *Evaluation of Dredged Material Proposed for Ocean Disposal-Testing Manual, EPA-503-8-91/001*; Environmental Protection Agency Office of Marine and Estuarine Protection: Washington, DC, USA, 1991.
8. Petrovic, M.; Barceló, D. Seeking harmonisation in assessing sediments and dredged materials, Meeting Report. *Trends Anal. Chem.* **2004**, *23*, 10–12. [CrossRef]
9. Cesar, A.; Choueri, R.B.; Riba, I.; Moralles-Caselles, C.; Pereira, C.D.S.; Santos, A.R.; Abessa, D.M.S.; DelValls, T.A. Comparative sediment quality assessment in different littoral ecosystems from Spain (Gulf of Cadiz) and Brazil (Santos and São Vicente Estuarine System). *Environ. Int.* **2007**, *33*, 429–435. [CrossRef] [PubMed]
10. Casado-Martínez, M.C.; Beiras, R.; Belzunce, M.J.; González-Castromil, M.A.; Marín-Guirao, L.; Postma, J.F.; Riba, I.; DelValls, T.A. Interlaboratory assessment of marine bioassays to evaluate the environmental quality of coastal sediments in Spain. IV. Whole sediment toxicity test using crustacean amphipods. *Cienc. Mar.* **2006**, *32*, 149–157. [CrossRef]
11. Morales-Caselles, C.; Kalman, J.; Riba, I.; DelValls, T.A. Comparing sediment quality in Spanish littoral areas affected by acute (Prestige, 2002) and chronic (Bay of Algeciras) oil spills. *Environ. Pollut.* **2007**, *146*, 233–240. [CrossRef] [PubMed]
12. Morales-Caselles, C.; Riba, I.; Sarasquete, C.; DelValls, T.A. Using a classical weight-of-evidence approach for 4-years monitoring of the impact of an accidental oil spill on sediment quality. *Environ. Int.* **2007**, *34*, 514–523. [CrossRef] [PubMed]

13. Ramos-Gómez, J.; Martín-Díaz, M.L.; DelValls, T.A. Acute toxicity measured in the amphipod *Ampelisca brevicornis* after exposure to contaminated sediments from Spanish littoral. *Ecotoxicology* **2009**, *18*, 1068–1076. [CrossRef] [PubMed]
14. Riba, I.; DelValls, T.A.; Forja, J.M.; Gómez-Parra, A. Comparative toxicity of contaminated sediment from a mining spill using two amphipods species: *Corophium volutator* and *Ampelisca brevicornis*. *Bull. Environ. Contam. Toxicol.* **2003**, *71*, 1061–1068. [CrossRef] [PubMed]
15. Mowat, F.S.; Bundy, K.J. Correlation of field-measured toxicity with chemical concentration and pollutant availability. *Environ. Int.* **2001**, *27*, 479–489. [CrossRef] [PubMed]
16. Stronkhorst, J.; Schipper, C.; Brils, J.; Dubbeldam, M.; Postman, J.; Van De Hoeven, N. Using marine bioassays to classify the toxicity of Dutch Harbor sediments. *Environ. Toxicol. Chem.* **2003**, *22*, 1535–1547. [CrossRef] [PubMed]
17. Brohon, B.; Delolme, C.; Gourdon, R. Complementarity of bioassays and microbial activity measurements for the evaluation of hydrocarbon-contaminated soils quality. *Soil Biol. Biochem.* **2001**, *33*, 883–891. [CrossRef]
18. Pelletier, E.; Delille, D.; Delille, B. Crude oil bioremediation in sub-Antarctic intertidal sediments: Chemistry and toxicity of oiled residues. *Mar. Environ. Res.* **2004**, *57*, 311–327. [CrossRef]
19. Beiras, R.; Fernández, N.; Bellas, J.; Besada, V.; González-Quijano, A.; Nunes, T. Integrative assessment of marine pollution in Galician estuaries using sediment chemistry, mussel bioaccumulation, and embryo-larval toxicity bioassays. *Chemosphere* **2003**, *52*, 1209–1224. [CrossRef]
20. Cesar, A.; Marín, A.; Marín-Guirao, L.; Vita, R. Amphipod and sea urchin tests to assess the toxicity of Mediterranean sediments: The case of Portman Bay. *Sci. Mar.* **2004**, *68*, 205–213. [CrossRef]
21. Silva, C.; Yáñez, E.; Martín-Díaz, M.L.; Riba, I.; DelValls, T.A. Integrated ecotoxicological assessment of marine sediments affected by land-based marine fish farm effluents: Physicochemical, acute toxicity and benthic community analyses. *Ecotoxicology* **2013**, *22*, 996–1011. [CrossRef]
22. USEPA, United States Environmental Protection Agency. *Short-Term Methods for Estimating the Chronic Toxicity of Effluents and Receiving Water to West Coast Marine and Estuarine Organisms*, 3rd ed.; USEPA: Cincinnati, OH, USA, 2002; p. 370.
23. Environment Canada. *Biological Test Method: Reference Method for Determining the Toxicity of Sediment Using Luminescent Bacteria in a Solid-Phase Test*; Report EPS 1/RM/42; Environmental Protection Service: Ottawa, ON, USA, 2002; p. 83.
24. Bailey, R.; Day, K.E.; Norris, R.H.; Reynoldson, T.B. Macroinvertebrate community structure and sediment bioassay results from areas of North American Great Lakes. *J. Great Lakes Res.* **1995**, *21*, 42–52. [CrossRef]
25. Reynoldson, T.B.; Bailey, R.C.; Day, K.E.; Norris, R.H. Biological guidelines for freshwater sediment based on benthic of sediment (THE BEAST) using a multivariate approach for predicting biological state. *Aust. J. Ecol.* **1995**, *20*, 198–219. [CrossRef]
26. Martinez-Madrid, M.; Rodriguez, P.; Pérez-Iglesias, J.I.; Navarro, E. Sediment toxicity bioassays for assessment of contaminated sites in the Nervion River (Northern Spain). 2. *Tubifex tubifex* reproduction sediment bioassay. *Ecotoxicology* **1999**, *8*, 111–124. [CrossRef]
27. Pasteris, A.; Vecchi, M.; Reynoldson, T.; Bonomi, G. Toxicity of copper-spiked sediments to *Tubifex tubifex* (Oligochaeta, Tubificide): A comparison of the 28-day reproductive bioassay with a 6-month cohort experiment. *Aquat. Toxicol.* **2003**, *65*, 253–265. [CrossRef]
28. Bettinetti, R.; Provini, A. Toxicity of 4-nonylphenol to *Tubifex tubifex* and *Chironomous riparius* in 28-days whole sediments tests. *Ecotoxicol. Environ. Saf.* **2002**, *53*, 113–121. [CrossRef]
29. Mosleh, Y.Y.; Paris-Palacios, S.; Biagianti-Risbourg, S. Metallothioneins induction and antioxidative response in aquatic worms *Tubifex tubifex* (Oligochaeta, Tubificidae) exposed to cooper. *Chemosphere* **2006**, *64*, 121–128. [CrossRef] [PubMed]
30. Mosleh, Y.Y.; Paris-Palacios, S.; Couderchet, M.; Biagianti-Risboug, S.; Vernet, G. Metallothionein induction, antioxidative responses, glycogen and growth changes in *Tubifex tubifex* (Oligochaeta, Tubificidae) exposed to the fungicide, fenhexamid. *Environ. Pollut.* **2006**, *135*, 73–82. [CrossRef] [PubMed]
31. Egeler, P.; Rombke, J.; Meller, M.; Knacker, T.; Franke, C.; Studinger, G.; Nagel, R. Bioaccumulation of lindane and hexachlorobenzene by tubificid sludgeworms (Oligochaeta) under standardized laboratory conditions. *Chemosphere* **1997**, *35*, 835–852. [CrossRef]
32. Egeler, P.; Rombke, J.; Meller, M.; Knacker, T.; Nagel, R. Bioaccumulation test with Tubificid Sludgeworms in artificial media—Development of a standardisable method. *Hydrobiologia* **1999**, *406*, 271–280. [CrossRef]
33. American Society for Testing Materials (ASTM). *Standard Test Method for Measuring the Toxicity of Sediments-Associated Contaminants with Freshwater Invertebrates*; American Society for Testing and Materials E1706-05; American Society for Testing and Materials: Philadelphia, PA, USA, 2005.
34. Chapman, P.M.; Farrell, M.O.; Brinkhurst, R.O. Relative Tolerances of Selected Aquatic Oligochaetes to Combinations of Pollutants and Environmental Factors. *Aquat. Toxicol.* **1982**, *2*, 69–78. [CrossRef]
35. Bailey, H.C.; Lui, D.H.W. Lumbriculus variegatus, a Benthic Oligochaete, as a Bioassay Organism. In *Aquatic Toxicology, Third Symposium*; Eaton, J.C., Parrish, P.R., Hendricks, A.C., Eds.; American Society for Testing and Materials: Philadelphia, PA, USA, 1980; pp. 205–215.
36. McMurtry, M.J. Avoidance of sublethal doses of copper and zinc by tubificid oligochaetes. *J. Great Lakes Res.* **1984**, *10*, 267–272. [CrossRef]
37. Confederación Hidrográfica del Guadalquivir. *Memoria Ambiental*; Confederación Hidrográfica del Guadalquivir: Seville, Spain, 2008; p. 41.

38. Blasco, J.; Gomes, T.; García-Barrera, T.; Rodríguez-Romero, A.; Gonzalez-Rey, M.; Morán-Roldán, F.; Tromibini, C.; Miotk, M.; Gómez-Ariza, J.L.; Bebianno, M.J. Trace metal concentrations in sediments from the southwest of the Iberian Peninsula. *Sci. Mar.* **2010**, *74*, 99–106. [CrossRef]
39. Silva, C.; Mattioli, M.; Fabbri, E.; Yáñez, E.; DelValls, T.A.; Martín-Díaz, M.L. Benthic community structure and biomarker responses of the clam *Scrobicularia plana* in a shallow tidal creek affected by fish farm effluents (Rio San Pedro, SW Spain). *Environ. Int.* **2012**, *47*, 86–98. [CrossRef]
40. Maestre, Z.; Martínez-Madrid, M.; Rodriguez, P.; Reynoldson, T. Ecotoxicity assessment of river sediments and a critical evaluation of some of the procedures used in the aquatic oligochaete *Tubifex tubifex* chronic bioassay. *Arch. Environ. Contam. Toxicol.* **2007**, *53*, 559–570. [CrossRef]
41. El Rayis, O.A. Re-assessment of the titration method for the determination of organic carbon in recent sediments. *Rapp. Com. Int. Mer. Medit.* **1985**, *29*, 45–47.
42. Casado-Martinez, C.; Forja, J.M.; DelValls, T.A. Direct comparison of amphipod sensitivities to dredged sediments from Spanish ports. *Chemosphere* **2007**, *68*, 677–685. [CrossRef] [PubMed]
43. ASTM, American Society for Testing and Materials. *Standard Guide for Conducting 10-Day Static Sediment Toxicity Tests with Marine and Estuarine Amphipods*; ASTM: Philadelphia, PA, USA, 1993; p. 26.
44. Azur Environmental. *Microtox Acute Toxicity Basic Test Procedures*; Azur Environmental: Carlsbad, CA, USA, 1998.
45. Campisi, T.; Abbondanzi, F.; Casado-Martínez, C.; DelValls, T.A.; Guerra, R.; Iacondini, A. Effect of sediment turbidity and color on light output measurement or Microtox Basic Solid-Phase Test. *Chemosphere* **2005**, *60*, 9–15. [CrossRef] [PubMed]
46. USEPA/USACE. *Evaluation of Dredged Material Proposed for Discharge in Waters of the U.S. Washington DC-Testing Manual*; EPA-823-B-98-004; USEPA/USACE: Washington, DC, USA, 1998.
47. Carballeira, C.; Martín-Díaz, L.; DelValls, T.A. Influence of salinity on fertilization and larval development toxicity tests with two species of sea urchin. *Mar. Environ. Res.* **2011**, *72*, 196–203. [CrossRef]
48. Ghirardini, A.V.; Novelli, A.A. A sperm cell toxicity test procedure for the Mediterranean species *Paracentrotus lividus* (Echinodermata: Echinoidea). *Environ. Technol.* **2001**, *22*, 439–445. [CrossRef]
49. Reynoldson, T.B.; Thompson, S.P.; Bampsey, J.L. A sediment bioassay using the tubificid oligochaete worm *Tubifex tubifex*. *Environ. Toxicol. Chem.* **1991**, *10*, 1061–1072. [CrossRef]
50. Chapman, P.M.; Benton, M.J.; Brinkhurst, R.O.; Scheuerman, P.R. Use of the aquatic oligochaetes *Lumbriculus variegatus* and *Tubifex tubifex* for assessing the toxicity of copper and cadmium in a spiked-artificial-sediment toxicity test. *Environ. Toxicol.* **1999**, *14*, 271–278. [CrossRef]
51. DelValls, T.A.; Chapman, P.M. Site-specific sediment quality values for the Gulf of Cadiz (Spain) and San Francisco Bay (USA), using the sediment quality triad and multivariate analysis. *Cienc. Mar.* **1998**, *24*, 313–336. [CrossRef]
52. DelValls, T.A.; Forja, J.M.; González-Mazo, E.; Gómez-Parra, A. Determining contamination sources in marine sediments using multivariate analysis. *TrAC Trends Anal. Chem.* **1998**, *17*, 181–192. [CrossRef]
53. Tabachnic, B.G.; Fidell, L.S. *Using Multivariate Statistics*; Harper Collins, College Publishers: New York, NY, USA, 1996.
54. Choueri, R.B.; Cesar, A.; Abessa, D.M.S.; Torres, R.J.; Morais, R.D.; Riba, I.; Pereira, C.D.S.; Nascimento, M.R.L.; Mozeto, A.A.; DelValls, T.A. Development of site-specific sediment quality guidelines for North and South Atlantic littoral zones: Comparison against national and international sediment quality benchmarks. *J. Hazard. Mater.* **2009**, *170*, 320–331. [CrossRef] [PubMed]
55. Bonnail, E.; Riba, I.; de Seabra, A.A.; DelValls, T.Á. Sediment quality assessment in the Guadalquivir River (SW, Spain) using caged Asian clams: A biomarker field approach. *Sci. Total Environ.* **2019**, *650*, 1996–2003. [CrossRef]
56. Linde, A.R.; Arribas, P.; Sánchez-Galán, S.; Marañón, E.; Izquierdo, J.L.; García-Vázquez, F. Preliminary assessment impact of trace metal pollution in Piles river (Spain). In *Water Pollution III: Modelling, Measurement and Prediction*; Wrobel, L.C., Latinopoulos, P., Eds.; Computation Mechanics Publication Ashurst: Southampton, UK, 1995; pp. 407–414.
57. CCME Protocol for the Derivation of Canadian Sediment Quality Guidelines for the Protection of Aquatic Life. CCME, EPC-98E. 1995. Available online: https://www.elaw.org/system/files/sediment_summary_table.pdf (accessed on 12 March 2021).
58. CEDEX. *Directrices Para la Caracterización del Material Dragado y su Reubicación en Aguas del Dominio Público Marítimo-Terrestre*; Comisión Interministerial De Estrategias Marinas: Madrid, Spain, 2014.

applied sciences

MDPI

Article

Integrative Assessment of Sediment Quality in the São Francisco River (Mina Gerais, Brazil)

Inmaculada Riba [1], Estefanía Bonnail [2], María José Salamanca [1], Mercedes Conradi [3,*] and Maria Helena Costa [4]

1 Faculty of Marine and Environmental Sciences, University of Cádiz, Polígono Río San Pedro s/n, 11510 Cádiz, Spain
2 Centro de Investigaciones Costeras, Universidad de Atacama (CIC-UDA), Universidad de Atacama, Copiapó 1530000, Chile
3 Marine Laboratory, Department of Zoology, Faculty of Biology, University of Seville, 41012 Seville, Spain
4 IMAR, Departamento de Ciències e Engenharia do Ambiente, Universidade Nova de Lisboa, 2829-516 Caparica, Portugal
* Correspondence: mconradi@us.es

Abstract: The São Francisco River (one of the most important South American rivers) has many contamination sources, but just a few environmental assessments have been conducted. A weight-of-evidence approach identified the pollution sources (industrial activities, mineral processing, fisheries, and tourism) in the river and the city of Três Marias based on two different lines of evidence: the structure of the benthic community (biological monitoring working party score system, abundance of taxa, number of individuals, Margalef species richness, Pielou evenness, and Shannon–Wiener diversity) and the physicochemical determination of sediments (%fines, TOC, nitrate, ammonium, ammonia, ammoniacal nitrogen, metalloids, and SEM/AVSs). The results show that the wastewater treatment plant was the most important source of pollution. A factory was also detected as a source of contamination, with related adverse effects having been measured downstream. Other sources of contamination and stress were detected in the studied area. The macro-benthic identification study identified three different sentinel species (*Tanytarsus* sp., *Crytochironomus* sp., and *Polypedilum* sp.) for future monitoring assessments of the sediment quality in riverine areas. Thus, an improvement in the management of river effluents and more measures focused on cutting contaminant emissions from the waste treatment plant are recommended.

Keywords: sediment assessment; weight of evidence; SIMPER analysis; biological monitoring working party

Citation: Riba, I.; Bonnail, E.; Salamanca, M.J.; Conradi, M.; Costa, M.H. Integrative Assessment of Sediment Quality in the São Francisco River (Mina Gerais, Brazil). *Appl. Sci.* **2023**, *13*, 3465. https://doi.org/10.3390/app13063465

Academic Editors: Mauro Marini and Angel DelValls

Received: 31 January 2023
Revised: 22 February 2023
Accepted: 7 March 2023
Published: 8 March 2023

1. Introduction

The São Francisco River (SFR) is one of the most important Brazilian and South American rivers; with a 1800 km length, its source is in Serra da Canastra (Mina Gerais), and it drains 641,000 km^2 across different states (Bahía, Pernambuco, Sergipe, and Alagoas) until it reaches the Atlantic Ocean. It also acts as a geographical barrier. Some other important rivers are effluents, such as the Paraopeba River, das Velhas River, or Verde Grande River. This river plays an important economic role due to electricity generation and transportation of goods; nevertheless, some other activities strongly influence it, such as zinc metallurgy, sewage discharges, and agriculture, among others [1]. A few environmental assessments have been performed in this area [1–6].

Sediments are an essential component of aquatic ecosystems and should be considered a resource of high ecological and socio-economic value that requires appropriate protection and management [7]. Sediments in a river could act as a sink of pollutants, releasing them into the waters when the sediment accumulation capacity is saturated [8].

Ecological monitoring using benthic community structure analysis provides information on different changes related to species composition, abundance, and diversity; all these indicators show alterations in the communities in the studied areas [9–11]. This kind of study provides realistic information about the environmental health of aquatic ecosystems. The macro-benthic fauna plays a fundamental role in sediment processes, providing an important measure of the response of a community to environmental perturbation [12–16] and exhibiting the greatest potential for monitoring conditions at a particular site [17]. Biotic indices are useful tools in decision-making processes [18], and those indices synthesize complex scientific data. The structure of the macro-benthic community was analyzed using univariate measures (biological monitoring working party score system, the abundance of taxa, the number of individuals, Margalef species richness, Pielou evenness, and Shannon–Wiener diversity). The biological monitoring working party (BMWP) score system is very useful for water quality management due to its adaptability. However, the use of biotic indices has limitations, since natural variability can be difficult to establish. Therefore, reliable environmental assessment studies must integrate as much information as possible.

Another line of evidence (LOE) was selected in order to establish the cause of potential alteration and/or pollution. This LOE is based on the physicochemical determination of sediments: grain size (fines), organic carbon content (TOC), chemical nitrogen (nitrate, ammonium, ammonia, and ammoniacal nitrogen), metal concentrations in sediments (Al, As, Co, Cr, Cu, Fe, Ni, Pb, Zn, and Hg), and simultaneously extracted metals–acid volatile sulfides (SEM-AVSs). The USEPA [19] establishes some benchmarks for metal(loid)s Cd, Cu, Pb, Ni, Ag, and Zn in freshwater sediments. Ref. [3] classified the toxicity of the sediments according to benthic organisms in total organic carbon fraction (f_{toc}), but also additional elements or factors affecting the area, such as chemical nitrogen or other metals. Therefore, chemical nitrogen was analyzed mainly due to the sewage discharge and agricultural pollution in the Barreiro Grande stream. Furthermore, the concentrations of Al, As, and Fe were measured due to the proximity of the VM-TM refining factory.

The main objective of this study was to establish the sediment quality in this river, addressing the main causes of the potential alteration and/or pollution with an integrative assessment. It was conducted by linking the results of the LOEs of sediment contamination and the descriptive parameters of macro-benthic community. Furthermore, the pollution sources in the upper-middle region of the São Francisco River were located and identified using a weight-of-evidence (WOE) approach based on the integration of both lines of evidence in order to define a management plan approach to determine the best remedial actions for improving the ecological status of the river.

2. Materials and Methods

2.1. Study Area

The area of study (Figure 1) was located between the Três Marias reservoir (SF1), and around 300 m downstream of the effluent discharge point a refining factory (ZM2). Therefore, the region is affected by anoxic water entering from the reservoir, the zinc-refining factory, and human alterations, such as sewage effluents from the Barreiro Grande stream and other diffuse sources. The upper sampling station (SF1) was located on the left bank of the SFR, and upstream of the factory and the confluence with the Barreiro Grande stream. This station was selected as control in the study. Downstream, the BG4 station was located in the Barreiro Grande stream 10 m upstream of the SFR confluence and was selected to determine the sediment characterization before the confluence with the SFR and the effects of the Três Marias wastewater treatment plant (WWTP). In the distance, the SF2 station was located on the right bank of the SFR, just opposite the security dam, in order to determine the sediment quality in the proximity area of the dam. On the right bank of the SFR, the SF3 station was located 100 m downstream of the effluent discharge site of the factory, to assess the direct impact of the factory on the SFR sediments. Upstream, the ZM2 station was located in the mixing zone of water from the SFR, just on the right bank of the SFR, 300 m downstream of the effluent discharge site of the factory and 20 m downstream of the 'Consciência' stream confluence. This station was selected to study the potential contamination associated with factory activity.

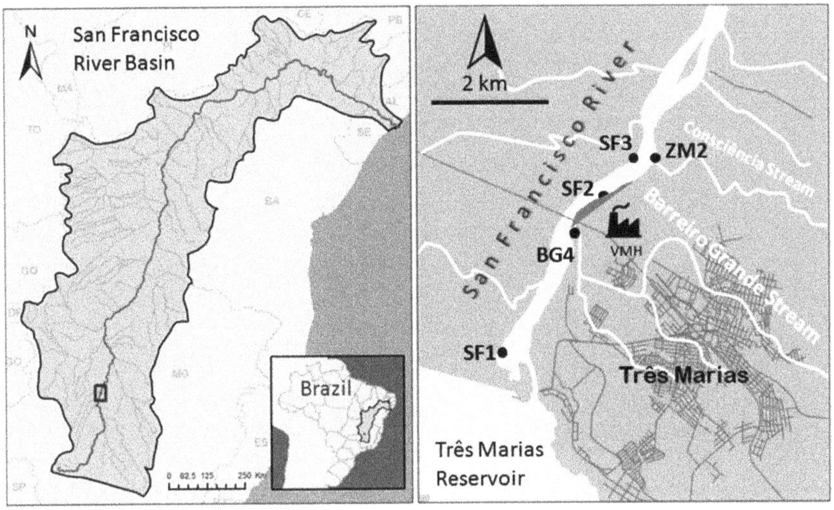

Figure 1. Map showing São Francisco River basin extension (Brazil) and study area with sampling stations (SF1, BG4, SF2, SF3, and ZM2). Some of the sources of contaminants, such as a factory, the city of Tres Marias, and the security dam (in dark grey), are also included.

2.2. Sampling Survey and Characterization

Sediment samples were collected with a Petersen grab of 5 L in capacity at a 7 cm depth at the sampling stations (SF1, BG4, SF2, SF3, and ZM2) (Figure 1). The samples were transported to the laboratory, where they were analyzed for physicochemical characterization. Samples for benthic analysis were separately collected; benthic animals were retrieved with the help of a 0.25 mm sieve and fixed in formalin (4%).

Grain size distribution (% fines) and total organic carbon (TOC) were analyzed using the ABNT NBR 7181/82 procedure [20]. This standard involves the drying, sieving, and sedimentation of the fines. Briefly, 500 g of each sample was dried at 50 °C for 48 h. The dried material was disaggregated in a mortar and sequentially sieved with 0.85 mm, 0.5 mm, 0.25 mm, 0.125 mm, and 0.063 mm meshes. The 0.063 mm fraction was introduced in 1000 mL beakers, and the sedimentation of fines was determined by pipetting/weighing the sediment over time.

Chemical nitrogen forms (nitrate, ammonium, ammonia, and ammoniacal nitrogen) were also measured in sediment samples to analyze diffuse contamination, a common example of which is the leaching of nitrogen compounds from fertilized agricultural lands. Nitrate, ammonium, ammonia, and ammoniacal nitrogen were determined in sediment interstitial waters. Sediment samples were centrifuged at 11,000 rpm and 15 °C for 15 min using an Eppendorf centrifuge. Afterward, the supernatant was filtered with a 0.6 µm pore fiberglass filter (Whatman) and analyzed with ion-exchange chromatography using a DIONEX DX-80 ion analyzer [21].

The methodology used for the analysis of secondary sulfides from sediments followed the procedures of the AVS-SEM technique [22]. In short, a sample of whole sediments (moist) was weighed and introduced into the reaction flask along with 100 mL of Milli-Q water. The flask was sealed, and the suspension was stirred for ultrapure nitrogen flow. Then, 20 mL of previously aeriated HCl (6M) was introduced with nitrogen. The AVSs were purged by the carrier gas (N_2) through the glass tubes that were immersed within the receiving flasks for the time of 30 min. The remaining suspension of the balloon (SEM) was filtered for the determination of metals (Cu, Cd, Pb, Ni, and Zn).

Metal and metalloid concentrations (Al, As, Co, Cr, Cu, Fe, Ni, Pb, Zn, and Hg) were determined using inductively coupled plasma–atomic emission spectroscopy (ICP-AES; Jobin-Ybon Ultima2) using a modification of the USEPA 3050B methodology [23]. The modification was performed to prevent the evaporation loss of the volatile trace elements (Hg, As, and Se). Standards were prepared with SCP SCIENCE standards. Reference certificated material SEM-1640 NIST for freshwater samples and inter-laboratory standard IRMM-N3 were also analyzed. The detection levels were calculated considering half of the signal and the standard deviation of a blank calculated 10 times.

2.3. Benthic Community Analysis

A total of 25 sediment samples were collected with a 0.0928 m^2 Petersen grab. Five replicate samples were taken at each station for benthic faunal analysis. After collection, samples were sieved in situ with a 0.025 mm mesh, and the retained organisms were transported to the laboratory in buffered formalin (4%). There, these samples were washed and transferred to 70% isopropyl alcohol. Then, they were sorted and identified to the lowest taxonomic level.

We used the biological monitoring working party (BMWP) score to study the tolerance of the benthic community. It is based on the principle that different aquatic invertebrates have different tolerance to pollutants, especially organic (i.e., nutrient enrichment that can affect the availability of dissolved oxygen). The presence of mayflies or stoneflies indicated the best water quality, with a score of 10, while worms indicated the worse water quality, with a score of 1 [24].

Moreover, the Shannon–Wiener diversity (H′), species richness (d′ [25,26]), Pielou evenness (J′ [27]), total abundance (A), and abundance of taxa (S) classic indices were calculated using the PRIMER-E (Plymouth Routines in Multivariate Ecological Research, v6) [28,29] suite of computer programs developed at Plymouth Marine Laboratory, UK.

2.4. Statistical Approach

Univariate analyses were based on the community descriptive parameters of the benthic fauna and the physicochemical measurements, which were calculated for each replicate sample and summarized for each site. The results were expressed as arithmetic averages ± standard deviations. In order to determine whether there were significant differences among stations, one-way analysis of variance (ANOVA) of each benthic univariate measure and physicochemical analysis result was used, followed by the multiple comparisons of Dunnett's test. The significance was set at $\alpha < 0.05$. Univariate statistical analyses were performed using SPSS 17.0.

Data were previously checked and/or transformed to ensure normality and heteroscedasticity and conduct the ANOVA test. For multivariate analyses, species abundance was transformed with the fourth root prior to performing non-metric multidimensional scaling (MDS) and similarity percentage analysis (SIMPER). Relative similarities in the abundance or biomass of benthic fauna among treatments (plus the reference samples or D0) were graphically analyzed using MDS with the Bray–Curtis similarity index [29]. The reliability of the MDS representations of the assemblage similarities was assessed with their stress values; stress < 0.01 was considered acceptable [28]. Clustering was performed using a hierarchical, agglomerative method using group average sorting, the results of which are displayed in a dendrogram. SIMPER analysis of abundance (cut-off percentage: 90%) was used to identify the taxa with the greatest contribution to differences among samples [28]. The significance of differences among stations was tested using the randomization/permutation test ANOSIM [30]. All analyses in MDS, SIMPER, and BIOENV were performed using PRIMER 6.1.6.

3. Results

3.1. Physicochemical Analyses

The summarized results of the physicochemical parameters analyzed are shown in Table 1. The BG4 station registered the greatest concentrations of compounds (nitrate, ammonium, ammonia, and ammoniacal nitrogen). However, the greatest element concentrations of Fe and Cr were found at SF1; those of Cu, Pb, and Zn, at SF3; and those of Co and Ni, at SF2. The ZM2 station displayed the highest concentrations of Al and As. The concentrations of metals Pb, Zn, Hg, and Cu showed a smooth gradient that increased from station SF1 as follows: SF1 < SF2 < SF3 < ZM2. Station ZM2 was located in the area of mixing between the SFR and the heavily factory-influenced stream, which could justify the high values of Al and As.

Table 1. Summarized results of physicochemical parameters analyzed in sediments from the San Francisco River (SFR): percentage of fines, total organic content (TOC), nitrogen species (nitrate, ammonium, ammonia, and ammoniacal nitrogen), SEM-AVSs, and metal(loid) concentrations (Al, As, Co, Cr, Cu, Fe, Ni, Pb, Zn, and Hg).

	SF1	BG4	SF2	SF3	ZM2
Fines (%)	7.37	50.16	26.28	37.82	27.81
TOC (μg/g)	20,970	10,590	11,960	12,430	17,590
Nitrate (μg/L)	83.46	617.32	50.87	24.52	35.28
Ammonium (μg/L)	1511	8760	1210	1432	1424
Ammonia (μg/L)	33.99	35.18	4.86	8.65	8.59
Ammoniacal nitrogen (μg/L)	1545	8795	1215	1441	1432
SEM-AVS (μmol/g)	58.7	3687	2720	2947	96.1
Al (μg/g)	2197	3993	2596	4558	32,767
As (μg/g)	5.51	1.36	1.48	2.56	8.83
Co (μg/g)	6.83	3.48	8.36	8.27	5.91
Cr (μg/g)	15.78	10.55	14.63	11.99	12.50
Cu (μg/g)	4.76	14.50	21.19	43.64	16.41
Fe (μg/g)	22,017	13,059	8030	14154	12,307
Ni (μg/g)	5.15	3.56	5.64	4.56	4.71
Pb (μg/g)	4.28	14.5	36.7	132	15.3
Zn (μg/g)	25.1	958	1239	3676	639
Hg (μg/g)	0.02	0.02	0.09	0.17	0.12

Due to the placement of BG4, situated on the Barreiro Grande stream, where the Três Marias water treatment plant discharges, the nitrogen species concentrations (nitrate, ammonium, ammonia, and ammoniacal nitrogen), well known as domestic waste indicators, were higher than the values obtained at the other stations. On the other hand, the sediments from the SF1 sampling point, although not directly influenced by the factory, received many contaminants from all over the drainage basin, since this sampling point was located under the Três Marías reservoir. According to the results, we could establish a decreasing pattern related to domestic waste contamination as follows: BG4 > SF1 > SF3 ≈ ZM2 > SF2.

3.2. Benthic Community Structure

Concerning benthic community structure analysis, a total of 3915 individuals were identified in the benthic fauna in sediments from the five sampling stations and along the river. They were classified into 60 species (1 hirudinean, 12 oligochaetes, 1 crustacean, 43 insects, 3 mollusks, and 1 nematode), and 4 species were not fully identified (17 *hirudinean* annelids, 14 *conchostracan* crustaceans, 2 *dryiopidae* beetles, and 63 individuals classified as *nematode*). Figure 2 shows a relative abundance of different taxonomic groups (hirudinea, oligochaeta, mollusca, and nematoda) observed at each sampling station. Therefore, the benthic community at BG4 clearly differed from the rest of the stations; it was dominated by nematodes (around 68%), unlike the communities from the rest of the stations, where insects were the dominant group (60–77%). Relative nematode abundance varied from 5% (at the SF1 station) to 0% (at the SF3 and ZM2 stations), and relative insect abundance was 2% at the BG4 station.

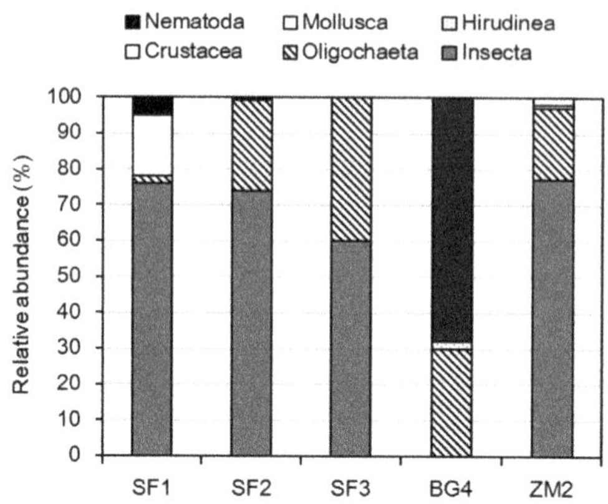

Figure 2. Relative abundance of different taxonomic groups (Hirudinea, Crustacea, Insecta, Oligochaeta, Mollusca, and Nematoda) identified in sediments from each sampling station at the São Francisco River (SF1, SF2, SF3, BG4, and ZM2).

The univariate metrics of the community structure for all the sampling stations are shown in Table 2. According to the BMWP results, all stations showed a bad ecological status. The BMWP score showed significant ($p < 0.05$) differences among sampling sites (BG4 and SF2), close to the affluent, and the control site (SF1). As expected, the SF1 site presented the best BMWP results, which were still considered 'bad' but just at the limit between degraded and regular. Conversely, BG4 showed the lowest BMWP score, defined by the abundance, number of individuals, richness, and diversity values, and the SF2 station showed the lowest Pielou evenness values. The benthic community structure at the SF2 station showed the highest Pielou evenness, while BG4 showed the lowest values.

Table 2. Mean and standard deviation (mean ± SD) values of different biotic indices (abundance of taxa—S′; number of individuals—N′; Margalef species richness—d′; Pielou evenness—J′; Shannon–Wiener diversity—H′; and BMWP score) per square meter at each sampling station.

Index	SF1		BG4		SF2		SF3		ZM2	
S′	3.60	±3.65	0.60	±0.89	6.20	±1.30	6.40	±1.14	4.60	±2.70
N′	185	±245	94.3	±206	896	±667	366	±144	1459	±2588
d′	0.51	±0.62	0.08	±0.12	0.79	±0.20	0.93	±0.21	0.57	±0.28
J′	0.82	±0.33	0.88	±0.00	0.65	±0.14	0.72	±0.12	0.69	±0.25
H′	0.79	±0.84	0.12	±0.27	1.16	±0.24	1.33	±0.32	0.87	±0.28
BMWP	45.7	±2.5	23.3 *	±1.75	25.6 *	±2.14	28.13	±5.02	27.87	±3.29

Asterisks (*) indicate significant differences ($p < 0.05$) compared with the control site (SF1).

Figure 3 shows the results of clustering the different stations using the Bray–Curtis index for abundance similarities. Cluster analysis distinguished three main groups of sites (BG4, SF1, and the rest of the sites), which were considered to represent different communities on the basis of their geometric means. As mentioned above, BG4 was clearly different from the other stations, and the control site (SF1) was different from the other stations but closer than BG4. The SF2, SF3, and ZM2 stations showed similar communities among them.

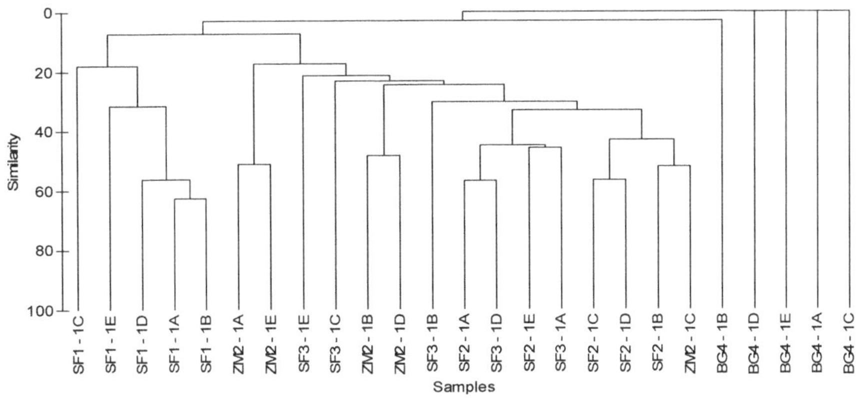

Figure 3. Bray–Curtis similarity dendrogram based on abundance, showing the classification of each station replicates.

The SIMPER analysis showed how the structure of the benthic community contributed to the differences among stations, indicating the main taxa and their contribution to the dissimilarity. The abundance of the macrofauna showed the highest dissimilarity (100%) between the SF3 and BG4 stations, while the least dissimilar (69.38%) stations were SF2 and SF3. Moreover, BG4 showed large differences (or dissimilarities) from the rest of the stations: BG4-ZM2 (99.52%), BG4-SF1 (99.33%), and BG4-SF2 (98.77%). The abundance of species of *Tanytarsus* spp., *Cryptochironomus* spp., and *Polypedilum* spp. was responsible for the dissimilarities between the most different stations (SF3-BG4 and ZM2-BG4), showing dissimilarities of around 8.37–13.33 (*Tanytarsus* spp.), 6.44–11.8 (*Cryptochironomus* spp.), and 5.25–11.63 (*Polypedilum* spp.). Pairwise comparisons derived from the ANOSIM test on species abundance data showed that the stations were significantly different from each other ($p < 0.01$).

3.3. Linking Physicochemical and Benthic Community Results

The relation between benthic fauna abundance and physicochemical variables was studied using the BIOENV procedure (Table 3). This statistical analysis selects environmental variables 'best explaining' community patterns by maximizing the correlations among their respective similarity matrices. This approach is advantageous, since it identifies not only the best correlation but also the best combination of them.

Table 3. BIOENV correlations (Spearman rank correlation coefficients, r_s) between benthic community structure and environmental variables (n).

n	Variables	r_s
5	Nitrate + ammonium + ammonia + Al + Zn	0.915
5	Nitrate + ammonia + ammoniacal nitrogen + Al + Zn	0.915
5	Ammonium + ammonia + ammoniacal nitrogen + Al + Zn	0.915
1	Nitrate	0.879
2	Nitrate + SEM-AVSs	0.879
3	Nitrate + ammonium + SEM-AVSs	0.879
2	Nitrate + ammonium	0.879
2	Nitrate + ammoniacal nitrogen	0.867
5	Nitrate + ammonium + ammonia + ammoniacal nitrogen + SEM-AVSs	0.867

The results of the BIOENV analysis showed that 91.5% of the variance in the different stations was due to the nitrate, ammonium, ammonia, Al, and Zn contents; the nitrate, ammonia, ammoniacal nitrogen, Al, and Zn contents; or the ammonium, ammonia, ammoniacal nitrogen, Al, and Zn contents. This analysis always showed Al and Zn as the contaminant metals of concern in the analyzed macro-benthic community. This variance was lower (87.9%) when one (nitrate), two (nitrate and SEM-AVSs), or three variables were considered (nitrate, ammonium, and SEM-AVS; or nitrate, ammoniacal nitrogen, and SEM-AVS). These results associate the nitrogen species with those measured changes in the macro-benthic community structure.

4. Discussion

This study identified the degradation in the surrounding area of the SFR course using the chemical measurement of sediments and macro-benthic community structure. This area, located between the Três Marias reservoir and around 300 m downstream of the effluent discharge point of the refining factory, showed several sources of contamination. In this kind of situation with multiple stressors, it is necessary to assess the overall sediment quality, and the use of multiple lines of evidence with an integrative approach is recommended [31–36]. The use of these multiple lines of evidence integrated with a weight-of-evidence approach (WOE) links the results obtained in each LOE, in this case, physicochemical parameters and benthic community structure indexes. Metal contamination has been extensively studied, but metal bioavailability and its effects are not commonly studied in this region [1–3,37,38]. The environmental state of the SFR was evaluated in this study using different lines-of-evidence (LOEs), which were integrated using the BIOENV procedure.

Benthic infaunal communities, as living components of sediments, represent the integrated response of the biological effects of pollutant contents on sediment samples. The results obtained using the classical descriptive parameters show the lowest values of Margalef richness (0.08) and Shannon–Wiener diversity (0.12), as well as the BMWP score system (23.34), at the BG4 station. Moreover, this station also showed the highest values of Pielou evenness (0.88). Furthermore, this station showed a significant ($p < 0.05$) difference in the macro-benthic community structure results compared with those obtained for the other stations. Thus, the highest descriptive parameters were observed at different stations, showing, at SF3, the highest values of the abundance of taxa (6.40), Margalef richness (0.93), and Shannon–Wiener diversity (1.33). Significant differences ($p < 0.05$) were only found in the BMWP score results obtained at the BG4 and SF2 stations, which can be observed in the MDS analysis (Figure 3). Furthermore, the SIMPER analysis showed two of three species (*Polypedilum* sp. and *Tanytarsus* sp.) that were classified as sensitive by [39], and the *Tanitarsus formosanus* species, which was classified as sensitive [40–42]. Additionally, the other marked species (*Cryptochironomus* sp.), according to [43,44], typically prefers well-oxygenated sandy substrates and is generally intolerant to polluted conditions. [*Polypedilum trigonous* species showed a positive relationship with most of the organic pollutants, such as TOM, ammonium, and nitrateas well as trace metals Zn, Cu, and Mn [45]. Ref. [41] reported that *Polypedilum* sp. was very tolerant to metal contamination.

Despite the use of biotic indices having some limitations, Ref. [3] used the BMWP score system in this area with good results. Moreover, according to this index, the analyzed area of the SFR showed a 'low' environmental quality classification, except for the station located upstream of the factory, which showed a 'regular' environmental quality classification. Based on these results, it can be considered that the factory exerts negative effects that are reflected in downstream stations. However, the BG4 station, which showed the worst quality classification, was located near the Três Marias WWTP and far away from the influence of the mining factory. The community at this station seemed to be significantly affected ($p < 0.05$) by this plant because of the discharge of urban waste, presenting a very different specific composition (abounding nematode and oligochaete species). Indeed, these species are well-known pollution bioindicators, and WWTPs can deposit large amounts of organic matter and nutrients into receiving waterways, leading to eutrophication [46] and temporary oxygen deficits [47].

The BMWP score system itself, or a modified version of this index, has been previously applied in other countries in order to study the water quality of rivers. The results in [48] show significant differences between quite unpolluted areas and anthropogenically impacted sites along the Nysa Klodzka River (Poland), as well as high correlations with some chemical variables. The modified version of the IBMWP (Iberian biomonitoring working party) score was successfully applied by [49] to 50 Iberian rivers, showing strong correlations between this index and different physicochemical parameters. Both studies show results and relationships similar to those reported in the present work.

Sediment quality assessment has demonstrated the importance of sediment characterization in terms of organic matter, grain size, metals [50–53], and SEM-AVS-SEM [54] in determining the environmental quality conditions, mainly in heavily populated and industrialized regions. The results here obtained related to sediment characterization show three main different groups of findings: (i) station SF1 presented the lowest values of fines, SEM-AVSs, Al, Cu, Pb, Zn, and Hg, but the highest values of TOC, Cr, and Fe; (ii) station BG4 showed the lowest values of TOC, As, Co, Cr, and Ni, but the highest values of fines, nitrate, ammonium, ammonia, ammoniacal nitrogen, SEM-AVSs, and Al; (iii) the other two stations, SF3 and SF2, showed the highest Cu, Pb, Zn, and Hg concentrations, and the highest Co and Ni concentrations, respectively. As reported by [55], the nitrate values found in the Hindon River, which is also polluted by wastewater, were in the range of 96–245 μg/L, i.e., lower values than those found in sediments from the BG4 station. Many rivers around the world are suffering contamination by industrial, agricultural, and domestic activities. The case study on the Axios/Vardar River in Southeastern Europe exposed the results of the water quality in an area surrounding a metal factory that receives domestic wastewater and agricultural runoff. In this case, the maximum values of nitrates and ammonium were 0.22 mg/L and 1.13 mg/L, respectively, which are again lower than those found in this study. The metal concentrations in that river were Pb (13.126 μg/L), Zn (49 μg/L), Cd (0.72 mg/L), and Cr (64 μg/L), indicating that it was fairly less polluted than the river studied here (SFR).

Taking into account the SEM-AVS results, all the sampling stations showed potential alterations in benthic organisms [19], showing SEM-AVS values higher than 0. Station BG4, had the highest value (3687.98), also had the highest alteration values of the calculated quality indices based on the benthic community structure (BMWP, S′, N′, d′, H′, and J′), as mentioned above. Moreover, according to the benchmarks of [3], the toxicity of our sediment samples was classified as unlikely (SF1 and ZM2 stations), uncertain (SF2 and SF3 stations), or likely (BG4 station).

The highest concentrations of nitrogen species were analyzed in sediments from station BG4, which was located downstream of the Três Marias WWTP. These results are in concordance with those reported by [56]. These authors concluded that wastewater contains significant amounts of ammonium, as only a small proportion is oxidized by conventional treatment plants [57]. Furthermore, as exposed by [58], sediments that are rich in organic matter and microbial biomass host the greatest number of denitrification processes, since they require sub-oxic conditions and a source of nitrate (from external inputs or the nitrification of ammonium).

The link between cause (sediment contamination) and effect (macro-benthic community structure) was determined using BIOENV analysis. The obtained results indicate that a large proportion of the among-site variance in the abundance of benthic fauna along the studied river course was strongly associated with the measured environmental variables, particularly, nitrogen chemical species and metals Al and Zn. Changes in benthic fauna abundance and diversity were related to biogeochemical transformations, such as nitrification and denitrification, and thus to the impacts of stress and disturbance on microbial communities. Much research on bioindicators in aquatic systems has been limited to heterotrophic bacteria in relation to the decomposition of dissolved organic matter and as a measure of sewage pollution [59,60]. In the same way, BIOENV analysis indicated changes related to Al and Zn contamination. These metals were at their highest concentrations at the BG4 and SF3 stations. In this sense, these adverse effects may be provoked by other sources of contaminants not measured in this study (e.g., by the influx of oxygen-depleted water or sewer discharge), as exposed by [61].

The biology/ecology status clearly reflects the environmental health of a river. However, apart from contamination [18], there are other problems associated with the basin. Water scarcity in the SFR basin has been detected as a significant base-flow reduction [62]. Further, a recent study on the SFR basin [63] assessed the influence of dams on eco-hydrological conditions, and the socio-economic impact due to the transfer of water from the SFR [64]. The study also highlighted that the intensification of natural resources provokes impacts on land and water resources, as also shown by the current study. Therefore, hydrological management plans should include environmental monitoring programs for river quantity and quality, especially because the Brazilian government is promoting the expansion of irrigated agriculture [63] and due to the increasing population and other associated overexploitation problems. Therefore, the SFR basin presents several problems that are tackled by the United Nations Sustainable Development Goals, such as clean water, poverty, health and wellness, and the distribution of equality [65].

Based on the obtained results, different approaches to remediation and/or mitigation should be addressed in this area. The integration of the two LOEs used in this study suggests that the environmental risk present in the area is associated with different sources of contamination (WWTP, metal factory, and reservoir). The monitoring of the risk identified in the area studied (between the reservoir and the metal factory) is highly recommended.

5. Conclusions

This study characterized the contamination in the middle area of the San Francisco River (SFR) by linking contamination and macro-benthic community structure results of the sediments studied. The main stressor identified and characterized in the area was associated with the activities of the WWTP and, to a lower extent, with the mining factory. ZM2, which was located in the mixed zone between the SFR and the stream produced by the metal factory, showed the highest metal concentrations. BG4, situated on the Barreiro Grande stream where the Três Marias water treatment plant discharges, showed the highest nitrogen species concentrations, but also presented a particular specific community composition (abounding nematode and oligochaete species) related to the discharge of urban waste.

The study identified the macro-benthic community structure as a valuable tool for monitoring the environmental quality of riverine sediments. Thus, we determined three different species (*Tanytarsus* sp., *Crytochironomus* sp., and *Polypedilum* sp.) as useful bio-indicator species for monitoring plans.

The links between cause (sediment contamination) and effect (community structure) showed that a large proportion of the variance in the abundance of benthic fauna along the studied river course was strongly associated with the measured contaminants, especially the nitrogen chemical species, and Al and Zn, related to the WWTP and the mining factory, respectively. The monitoring plan for the area needs to be designed using variables similar to those used in this study. Potential new LOEs such as sediment toxicity could be considered in the WOE approach.

This kind of study (integrative assessment) could allow the identification of the sources of contamination, the localization of the areas affected by pollution, and the characterization of sediment quality to be conducted. Furthermore, integration helps in the identification of the risk associated with different human activities and consequent pollution. Finally, it could be an input to the potential decision to perform remedial and/or mitigating actions to improve the environmental quality of the river.

Author Contributions: I.R. and M.C. contributed to all the aspects of the article from conceptualization to funding acquisition, including contributions to methodology, software, validation, formal analysis, investigation, resources, data curation, and writing and editing the paper from the first version to the final version; E.B. and M.J.S. contributed to most of the aspects, including conceptualization, investigation, resources, data curation, writing—original draft preparation, writing—review and editing, visualization, supervision, and project; M.H.C. also contributed to most of the aspects, including methodology, software, validation, formal analysis, writing—original draft preparation, writing—review and editing, visualization, supervision, project administration, and funding acquisition. All authors have read and agreed to the published version of the manuscript.

Funding: Votorantim Metais co-funded this project (OT2009/113 and OT2010/149 contracts with University of Cádiz). IIEGA provided the equipment and facilities necessary for the performance of sampling and analysis. This study was funded by Ministry of Education Spain (PRX14/00134) and partially by FAPESP (No. 2017/25936-0).

Institutional Review Board Statement: Not applicable.

Informed Consent Statement: Not applicable.

Data Availability Statement: Not possible.

Acknowledgments: Special thanks are given to A.M. Peret, K. Pereira, and M. del Real for their help in this work. I. Riba thanks the Ministry of Education (PRX14/00134) in Spain for funding her placement at University Nova of Lisbon (Portugal) and University of Las Palmas de Gran Canaria (Spain). Riba thanks FAPESP (No. 2017/25936-0) for her Visiting Researcher Program at Universidade Federal do Sao Paulo. E. Bonnail is grateful for ANID/FONDECYT 11180015.

Conflicts of Interest: The authors declare no conflict of interest.

References

1. Almeida, D.F.; Martins, A.H.; Tundisi, J.G. Weight-of-evidence on environmental impact assessment of metal-contaminated sediments in the São Francisco River Três Marias—Minas Gerais—(Brazil)—A case study. *Braz. J. Biol.* **2011**, *71*, 961–973. [CrossRef]
2. Golder Associates Brasil. *Zoneamento da Distribuição da Contaminação de Sedimentos do leito Submerso do rio São Francisco RT-079-515-6014-0002-01-J*; Relatório Interno; GOLDER: Belo Horizonte, Brazil, 2006.
3. Mozeto, A.A.; Nascimento, M.R.L.; Silva, E.F.A.; Fioravanti, M.I.A. *Avaliação Da Contaminação Por Metais E Metalóide Água, Sedimento E Peixe) No Rio São Francisco Em Três Marias MG- Brasil*; Projeto de Pesquisa Participativa com a Comunidade Local; Departamento de Química da UFSCar: São Carlos, Brazil, 2007.
4. Oliveira, M.R. Investigação da Contaminação por Metais Pesados da Água e do Sedimento de Corrente nas Margens do Rio São Francisco e Tributários, a Jusante da Represa da Cemig, no Município de Três Marias, Minas Gerais. Ph.D. Thesis, Universidade Federal de Minas Gerais, Belo Horizonte, Brazil, 2007.
5. Paschoalini, A.L.; Savassi, L.A.; Arantes, F.P.; Rizzo, E.; Bazzoli, N. Heavy metals accumulation and endocrine disruption in *Prochilodus argenteus* from a polluted neotropical river. *Ecotoxicol. Environ. Saf.* **2019**, *169*, 539–550. [CrossRef] [PubMed]
6. Pinto, R.; Patrício, J.; Baeta, A.; Fath, B.D.; Neto, J.M.; Marques, J.C. Review and evaluation of estuarine biotic indices to assess benthic condition. *Ecol. Indic.* **2009**, *9*, 1–25. [CrossRef]
7. Stumm, W.; Morgan, J.J. *Aquatic Chemistry. Chemical Equilibria and Rates in Natural Waters*; Wiley Interscience: Toronto, ON, Canada, 1996.
8. Tundisi, J.G.; Jorgensen, S.E.; Vollenweider, R.A. *Diretrizes para o Gerenciamento de Lagos. VoL1: Princípios para o Gerenciamento de Lagos*; United Nations Environmental Programme—UNEP: São Carlos, Brazil; Instituto Internacional de Ecología—IIE: São Carlos, Brazil; International Lake Environmental Committee—ILEC: São Carlos, Brazil, 2000; 202p.
9. DelValls, T.A.; Forja, J.M.; Gómez-Parra, A. An integrative assessment of sediment quality in littoral ecosystems from the Gulf of Cádiz. *Environ. Toxicol. Chem.* **1998**, *17*, 1073–1084. [CrossRef]
10. Long, E.; Chapman, P.M. A sediment quality triad: Measures of sediment contamination, toxicity, and infaunal community composition in Puget Sound. *Mar. Pollut. Bull.* **1985**, *16*, 405–415. [CrossRef]
11. Feld, C.K.; da Silva, P.M.; Sousa, J.P.; de Bello, F.; Bugter, R.; Grandin, U.; Hering, D.; Lavorel, S.; Mountford, O.; Pardo, I.; et al. Indicators of biodiversity and ecosystem services: A synthesis across ecosystems and spatial scales. *Oikos* **2009**, *118*, 1862–1871. [CrossRef]
12. Baldó, F.; García-Martín, S.F.; Drake, P.; Arias, A.M. Discrimination between disturbed coastal ecosystems by using macro-benthos at different taxonomic levels. *Boletín Inst. Español Oceanogr.* **1999**, *15*, 489–493.
13. Engle, V.D.; Summers, J.K.; Gaston, G.R. A benthic index of environmental condition of Gulf of Mexico estuaries. *Estuaries* **1994**, *17*, 372–384. [CrossRef]
14. Frouin, P. Effects of anthropogenic disturbances of tropical soft-bottom benthic communities. *Mar. Ecol. Prog. Ser.* **2000**, *194*, 39–53. [CrossRef]
15. McGill, B.J.; Etienne, R.; Gray, J.S.; Alonso, D.; Anderson, M.J.; Benecha, H.K.; Dornelas, M.; Enquist, B.; Green, J.L.; He, F.; et al. Species abundance distributions: Moving beyond single prediction theories to integration within an ecological framework. *Ecol. Lett.* **2007**, *10*, 995–1015. [CrossRef]
16. Warwick, R.M. A new method for detecting pollution effects on marine macro-benthic communities. *Mar. Biol.* **1986**, *92*, 557–562. [CrossRef]

17. Pagola-Carte, S.; Urkiaga-Alberdi, J.; Bustamante, M.; Saiz-Salinas, J.I. Concordance degrees in macrozoobenthic monitorng programmes using different sampling methods and taxonomic resolution levels. *Mar. Pollut. Bull.* **2002**, *44*, 63–70. [CrossRef]
18. Pinto, C.C.; Calazans, G.M.; Oliveira, S.C. Assessment of spatial variations in the surface water quality of the Velhas River Basin, Brazil, using multivariate statistical analysis and nonparametric statistics. *Environ. Monit. Assess.* **2019**, *191*, 164. [CrossRef]
19. United States Environmental Protection Agency—EPA. *Procedures for the Derivation of Equilibrium Partioning Sediments Benchmarking for the Protection of Benthic Organisms: Metal Mixtures Cd, Cu, Pb, Ni, Ag, Zn. EPA-600-R-02-011*; Office of Research and Development: Washington, DC, USA, 2005.
20. *NBR 7181/82*; Solo: Análise Granulométrica. Associação Brasileira de Normas Técnicas (ABNT): Rio de Janeiro, Brazil, 1984; 13p.
21. Michalski, R.; Kurzyca, I. Determination of nitrogen species nitrate, nitrite and ammonia ions) in environmental samples by ion chromatography. *Pol. J. Environ. Stud.* **2006**, *15*, 5–18.
22. United States Environmental Protection Agency—USEPA. *Draft Analytical Method for Determination of Acid Volatile Sulfide in Sediment. EPA-821-R-91-100*; Proper Storage of Samples Will Mitigate These Limitations; Office of Water, U.S. Environmental Protection Agency: Washington, DC, USA, 1991.
23. United States Environmental Protection Agency—USEPA. *Method 3050B. Acid Digestion of Sediments Sludge and Soils. Revision 2*; U.S. Environmental Protection Agency: Washington, DC, USA, 1996.
24. Barbour, M.T.; Gerritsen, J.; Snyder, B.D.; Stribling, J.B. *Rapid Bioassessment Protocols for Use in Streams and Wadeable Rivers: Periphyton, Benthic Macroinvertebrates, and Fish*, 2nd ed.; U.S. Environmental Protection Agency: Washington, DC, USA, 1999; Document No. 841-B-99-002.
25. Margalef, R. On certain unifying principles in ecology. *Am. Nat.* **1963**, *97*, 357–374. [CrossRef]
26. Margalef, R. *Perspectives in Ecological Theory*; University of Chicago Press: Chicago, IL, USA, 1968.
27. Pielou, E.C. *Population and Community Ecology: Principles and Methods*; CRC Press: New York, NY, USA, 1974.
28. Clark, K.R.; Warwick, R.M. *Change in Marine Communities: An Approach to Statistical Analysis and Interpretation*, 2nd ed.; Plymouth Marine Laboratory: Plymouth, UK, 2001.
29. Clark, K.R.; Gorley, R.N. *PRIMER v6: User Manual/Tutorial*; PRIMER-E, Plymouth Marine Laboratory: Plymouth, UK, 2006.
30. Clark, K.R.; Green, R.H. Statistical design and analysis for a biological effects' study. *Mar. Ecol. Progress Ser.* **1988**, *91*, 47–56. [CrossRef]
31. Bonnail, E.; Sarmiento, A.M.; DelValls, T.A. The use of the aWeight-of-Evidence approach to address sediment quality in the Odiel River basin SW, Spain. *Ecotoxicol. Environ. Saf.* **2016**, *133*, 243–251. [CrossRef]
32. Bonnail, E.; Riba, I.; de Seabra, A.A.; DelValls, T.A. Sediment quality assessment in the Guadalquivir River (SW, Spain) using caged Asian clams: A biomarker field approach. *Sci. Total Environ.* **2019**, *650*, 1996–2003. [CrossRef]
33. Chapman, P.M.; McDonald, B.G.; Lawrence, G.S. Weight-of-evidence issues and frameworks for sediment quality and other assessments. *Hum. Ecol. Risk Assess.* **2002**, *8*, 1489–1515. [CrossRef]
34. Cesar, A.; Choueri, R.; Riba, I.; Morales-Caselles, C.; Pereira, C.; Santos, A.R.; Abessa, D.; Del Valls, T. Comparative sediment quality assessment in different littoral ecosystems from Spain (Gulf of Cadiz) and Brazil Santos and São Vicente estuarine system. *Environ. Int.* **2007**, *33*, 429–435. [CrossRef]
35. DelValls, A.; Souza, L.S.; de Seabra, A.A.; Dias Seabra, C.; Bonnail, E.; Riba, I. Integrative assessment of sediment quality in acidification scenarios associated with carbon capture and storage operations. *Environ. Rev.* **2019**, *27*, 333–345. [CrossRef]
36. Riba, I.; Forja, J.M.; Gómez-Parra, A.; DelValls, T.A. Sediment quality in littoral regions of the Gulf of Cádiz: A triad approach to address the influence of mining activities. *Environ. Pollut.* **2004**, *132*, 341–353. [CrossRef] [PubMed]
37. Saraiva, K.V. Critérios para Avaliação da Qualidade de Sedimentos—Estudo de Caso: Sub-bacia do Ribeirão Espírito Santo, Afluente do Rio São Francisco. Master's Thesis, Universidade Federal de Minas Gerais, Belo Horizonte, Brazil, 2007.
38. Procópio, M.S.; Ribeiro, H.J.; Pereira, L.A.; Lopes, G.A.O.; Castro, A.C.S.; Rizzo, E.; Sato, Y.; Russo, R.C.; Junior, J.D.C. Sex-response differences of immunological and histopathological biomarkers in gill of *Prochilodus argenteus* from a polluted river in southeast Brazil. *Fish Shellfish. Immunol.* **2014**, *39*, 108–117. [CrossRef]
39. Odume, O.N.; Muller, W.J. Diversity and structure of Chironomidae communities in relation to water quality differences in the Swartkops River. *Phys. Chem. Earth* **2011**, *36*, 929–938. [CrossRef]
40. Helson, J.E.; Williams, D.D.; Turner, D. Larval chironomid community organization in four tropical rivers: Human impacts and longitudinal zonation. *Hydrobiologia* **2006**, *559*, 413–431. [CrossRef]
41. Mousavi, S.K.; Primcerio, R.; Amundsen, P. Diversity and structure of Chironomidae Diptera) communities along a gradient of heavy metal contamination in a subarctic watercourse. *Sci. Total Environ.* **2003**, *307*, 93–110.
42. Carew, M.E.; Pettigrove, V.; Cox, R.L.; Hoffmann, A.A. The response of Chironomidae to sediment pollution and other environmental characteristics in urban wetlands. *Fresh Biol.* **2007**, *52*, 2444–2462. [CrossRef]
43. McMahon, R.F. Mollusca: Bivalvia. In *Ecology and Classification of North American Freshwater Invertebrates*; Thorp, J., Covich, A., Eds.; Academic Press Inc.: São Diego, CA, USA, 1991.
44. Coffman, W.P.; Farrington, L.C., Jr. Chironomidae. In *An Introduction to the Aquatic Insects of North America*, 3rd ed.; Merritt, R., Cummins, K., Eds.; Kendall/Hunt Publishing Company: Dubuque, IA, USA, 1996.
45. Al-Shami, S.A.; Rawi, C.S.M.; HassanAhmad, A.; Nor, S.A.M. Distribution of Chironomidae Insecta: Diptera) in polluted rivers of the Juru River Basin, Penang, Malaysia. *J. Environ. Sci.* **2010**, *22*, 1718–1727. [CrossRef]

46. Gücker, B.; Brauns, M.; Pusch, M.T. Effects of wastewater treatment plant discharge on ecosystem structure and function of lowland streams. *J. N. Am. Benthol. Soc.* **2006**, *25*, 313–329. [CrossRef]
47. Rueda, J.; Camacho, A.; Mezquita, F.; Hernandez, R.; Roca, J.R. Effects of episodic and regular sewage discharges on the water chemistry and macroinvertebrate fauna of a Mediterranean stream. *Water Air Soil Pollut.* **2002**, *140*, 425–444. [CrossRef]
48. Czerniawska-Kusza, I. Comparing modified biological monitoring working party score system and several biological indices based on macroinvertebrates for water-quality assessment. *Limnologica* **2005**, *35*, 169–176. [CrossRef]
49. Alba-Tercedor, J.; Jáimez-Cuellar, P.; Álvarez, M.; Avilés, J.; Bonada i Caparrós, N.; Casas, J.; Mellado, A.; Ortega, M.; Pardo, I.; Prat i Fornells, N.; et al. Caracterización del estado ecológico de ríos mediterráneos ibéricos mediante el índice IBMWP antes BMWP'. *Limnetica* **2002**, *21*, 175–185. [CrossRef]
50. De Biasi, A.M.; Bianchi, C.N.; Morri, C. Analysis of macro-benthic communities at different taxonomic levels: An example from an estuarine environment in the Ligurian Sea NM Mediterranean. *Estuar. Coast. Shelf Sci.* **2003**, *58*, 99–106. [CrossRef]
51. Mucha, A.P.; Vasconcelos, M.T.S.D.; Bordalo, A.A. Vertical distribution of the macro-benthic community and its relationships to trace metals and natural sediment characteristics in the lower Douro estuary, Portugal. *Estuar. Coast. Shelf Sci.* **2004**, *59*, 663–673. [CrossRef]
52. Currie, D.R.; Small, K.J. Macro-benthic community responses to long-term environmental change in an east Australian sub-tropical estuary. *Estuar. Coast. Shelf Sci.* **2005**, *63*, 315–331. [CrossRef]
53. Hatje, V.; Barros, F.; Magalhães, W.; Riatto, V.B.; Amorim, F.N.; Figueiredo, M.B.; Spanó, S.; Cirano, M. Trace metals and benthic macrofauna distributions in Camamu Bay, Brazil: Sediment quality prior oil and gas exploration. *Mar. Pollut. Bull.* **2008**, *56*, 363–370. [CrossRef]
54. Fang, T.; Li, X.; Zang, G. Acid volatile sulfide and simultaneously extracted metals in the sediment cores of the Pearl River Estuary, South China. *Ecotoxicol. Environ. Saf.* **2005**, *61*, 420–431. [CrossRef]
55. Suthar, S.; Sharma, J.; Chabukdhara, M.; Nema, A.K. Water quality assessment of river Hindon at Ghaziabad, India: Impact of industrial and urban wastewater. *Environ. Monit. Assess.* **2009**, *165*, 103–112. [CrossRef]
56. Wakelin, S.A.; Colloff, M.J.; Kookana, R.S. Effect of wastewater treatment plant effluent on microbial function and community structure in the sediment of a freshwater stream with variable seasonal flow. *Appl. Environ. Microbiol.* **2008**, *74*, 2659–2668. [CrossRef]
57. Strom, P.F.; Matulewich, V.A.; Finstein, M.S. Concentrations of nitrifying bacteria in sewages, effluents and a receiving stream and resistance of these organisms to chlorination. *Appl. Environ. Microbiol.* **1976**, *31*, 731–737. [CrossRef]
58. Sandler, B.C.; Kalff, J. Factors controlling bacterial production in marine and freshwater sediments. *Microb. Ecol.* **1993**, *26*, 79–99.
59. Geldreich, E.E. Faecal coliform and faecal streptococcus density relationships in waste discharges and receiving waters. *Crit. Rev. Environ. Control* **1976**, *6*, 349–369. [CrossRef]
60. Miescier, J.J.; Cabelli, V.J. Enterococci and other microbial indicators in municipal wastewater effluents. *J. Water Pollut. Control Fed.* **1982**, *54*, 1599–1606.
61. Khosrovyan, A.; Marin, M.J.S.; Peret, A.M.; De Almeida, D.F.; DelValls, A.; Riba, I. Weight-Of-evidence-based assessment of sediment quality of the São Francisco River Brazil) with the help of software tool SQA. iEMSs—Managing Resources of a Limited Planet. In Proceedings of the 6th Biennial Meeting of the International Environmental Modelling and Software Society, Leipzig, Germany, 1–5 July 2012; pp. 2958–2965.
62. Lucas, M.C.; Kublik, N.; Rodrigues, D.B.B.; Meira Neto, A.A.; Almagro, A.; Melo, D.d.C.D.; Zipper, S.C.; Oliveira, P.T.S. Significant Baseflow Reduction in the Sao Francisco River Basin. *Water* **2021**, *13*, 2. [CrossRef]
63. Nascimento do Vasco, A.; Aguiar Netto, A.O.; da Silva, M.G. The influence of dams on ecohydrological conditions in the São Francisco River Basin, Brazil. *Ecohydrol. Hydrobiol.* **2019**, *4*, 556–565. [CrossRef]
64. Dantas, J.C.; da Silva, R.M.; Santos, C.A.G. Drought impacts, social organization, and public policies in northeastern Brazil: A case study of the upper Paraíba River basin. *Environ. Monit. Assess.* **2020**, *192*, 317. [CrossRef]
65. United Nations. Sustainable Development Goals. Available online: https://sdgs.un.org/es/goals (accessed on 17 February 2023).

applied sciences

MDPI

Article

Understanding the Distributions of Benthic Foraminifera in the Adriatic Sea with Gradient Forest and Structural Equation Models

Masoud A. Rostami [1,2,*], Fabrizio Frontalini [3], Eric Armynot du Châtelet [4], Fabio Francescangeli [5], Maria Virginia Alves Martins [6,7], Rocco De Marco [8], Enrico Dinelli [9], Mario Tramontana [3], Lee A. Dyer [1], Romain Abraham [4], Viviane Bout-Roumazeilles [4], Marion Delattre [4] and Federico Spagnoli [8,10]

1 Department of Biology, University of Nevada, Reno, NV 89557, USA
2 Data Science Program, University of Texas at Arlington, Arlington, TX 76019, USA
3 Department of Pure and Applied Sciences, Università degli Studi di Urbino, 61029 Urbino, Italy
4 Laboratoire d'Océanologie et de Géosciences, UMR 8187 LOG. CNRS, Université de Lille, Université du Littoral–Côte d'Opale, IRD, F-59000 Lille, France
5 Department of Geosciences, University of Fribourg, Chemin du Musée 6, 1700 Fribourg, Switzerland
6 Departament of Stratigraphy and Paleontology, Faculty of Geology, Rio de Janeiro State University (UERJ), R. São Francisco Xavier, 524, LabMicro 4037F, Maracanã, Rio de Janeiro 20550-900, Brazil
7 Department of Geosciences, GeoBioTec, Campus de Santiago, Aveiro University, 3810-197 Aveiro, Portugal
8 Istituto per le Risorse Biologiche e le Biotecnologie Marine, Consiglio Nazionale delle Ricerche, 60125 Ancona, Italy
9 Dipartimento di Scienze Biologiche, Geologiche e Ambientali, Alma Mater Studiorum Università di Bologna, 40126 Bologna, Italy
10 Geology Division, School of Science and Technology, University of Camerino, 62032 Camerino, Italy
* Correspondence: rostami@nevada.unr.edu

Citation: A. Rostami, M.; Frontalini, F.; Armynot du Châtelet, E.; Francescangeli, F.; Alves Martins, M.V.; De Marco, R.; Dinelli, E.; Tramontana, M.; Dyer, L.A.; Abraham, R.; et al. Understanding the Distributions of Benthic Foraminifera in the Adriatic Sea with Gradient Forest and Structural Equation Models. *Appl. Sci.* **2023**, *13*, 794. https://doi.org/10.3390/app13020794

Academic Editors: Angel DelValls and Mauro Marini

Received: 24 November 2022
Revised: 28 December 2022
Accepted: 29 December 2022
Published: 6 January 2023

Abstract: In the last three decades, benthic foraminiferal ecology has been intensively investigated to improve the potential application of these marine organisms as proxies of the effects of climate change and other global change phenomena. It is still challenging to define the most important factors affecting foraminiferal communities and derived faunistic parameters. In this study, we examined the abiotic-biotic relationships of foraminiferal communities in the central-southern area of the Adriatic Sea using modern machine learning techniques. We combined gradient forest (Gf) and structural equation modeling (SEM) to test hypotheses about determinants of benthic foraminiferal assemblages. These approaches helped determine the relative effect of sizes of different environmental variables responsible for shaping living foraminiferal distributions. Four major faunal turnovers (at 13–28 m, 29–58 m, 59–215 m, and >215 m) were identified along a large bathymetric gradient (13–703 m water depth) that reflected the classical bathymetric distribution of benthic communities. Sand and organic matter (OM) contents were identified as the most relevant factors influencing the distribution of foraminifera either along the entire depth gradient or at selected bathymetric ranges. The SEM supported causal hypotheses that focused the factors that shaped assemblages at each bathymetric range, and the most notable causal relationships were direct effects of depth and indirect effects of the Gf-identified environmental parameters (i.e., sand, pollution load Index–PLI, organic matter–OM and total nitrogen–N) on foraminifera infauna and diversity. These results are relevant to understanding the basic ecology and conservation of foraminiferal communities.

Keywords: depth; machine learning; ecology; benthic communities

1. Introduction

Benthic foraminifera are protozoa, widely occurring in both marine and transitional marine ecosystems, including lagoons, coastal lakes, and estuaries. These organisms have been traditionally used by geologists for paleoenvironmental reconstruction because of their fossilizable shells (i.e., tests), and they have been more recently used for defining biotopes [1] and as bioindicators [2–6], particularly for Ecological Quality Status (EcoQS) assessment [7–10]. The paleoenvironment and biomonitoring focus has obscured questions about how foraminifera communities are dependent on specific environmental pressures, their complex interplay (i.e., additive, synergistic, and antagonistic relationships), and the effect of sizes of these different determinants [11,12]. Indeed, the determination of thresholds over which an environmental parameter or a set of them significantly elicits variation in the abundance of a foraminiferal species or an assemblage index is needed for effective EcoQS assessment [13]. Another shortcoming of the existing literature is that most of the benthic foraminiferal investigations have focused on limited areas (e.g., a specific estuary, lagoon or coastal area) and barely cover regional or basinal areas [14–16]. Multiple abiotic factors, including grain-size and type of substrates, nutrients (i.e., quantity and quality of organic matter, OM), physio-chemical and biological disturbance (e.g., sub-areal exposition, pollution, bioturbations) contribute to the distribution of benthic foraminifera [17–21]. For instance, sediment composition was found to be a determinant in different environmental settings [22,23] such that a coarsening in the grain-size corresponds to a decrease of foraminiferal densities until a certain threshold where no more living foraminifera are found [24,25]. Usually, a significant increase in OM is associated with a lowering in foraminiferal diversity, but foraminiferal species do not identically respond to OM enrichment; some foraminiferal species can even benefit from these organic-rich substrates [26,27]. The same observations can be made for other factors that affect marine diversity, such as pollution and salinity. Understanding these assemblages revolves around the concept of niche and the relationship between optimal and limiting factors, and there is a great deal that is still known about these factors for global assemblages of foraminifera [28]. Although we can identify patterns of distribution along broad environmental gradients, in most cases, it is challenging to rank important factors shaping foraminiferal communities and to estimate the effect sizes of these relationships. Therefore, there is a need to develop new approaches to improve our present state of knowledge on foraminiferal ecology, especially given the fact that applied statistics has experienced great advances in handling larger amounts of data and in testing for complex causal relationships.

The development of machine learning approaches, an extension of generalized linear models from classic statistics, have provided important tools to community ecologists. Machine learning methods focus on predictions as opposed to parameter estimates, and the focus is on specific dataset-driven predictions rather than mechanism or previously collected data [29]. Gradient forest (Gf) is a machine learning method that ranks predictors of nonlinear responses, such as the response of biotic communities to environmental variables, and it can include matrices of both predictor and response variables [30]. The advantages of Gf models are that they are assumption free, there is high accuracy with checks on overfitting, pre-processing requirements are trivial, transformations are unnecessary, and clear ranks of variable importance are output [31,32]. Structural equation modeling (SEM) is a robust extension to path analysis that includes latent variables or statistical comparisons of different path models and has been used with increasing frequency in ecological studies [33]. These models allow for tests of a priori causal hypotheses, relationships among latent and measured variables, and for statistical comparisons of various causal hypotheses. Statistical learning approaches are not often combined with traditional statistics focused on mechanisms or causality, such as SEM, but the fact that these approaches are complementary, with one focused on prediction and the other on mechanisms and causal relationships makes them ideal for combining to test hypotheses in community ecology [29]. The alternative to such an approach would be to include latent variables in the SEM models (or to precede the models with factor analysis and include factors), but those approaches yield less interpretable patterns and assume that measured variables are proxies for something that is unmeasurable. In addition, the advantage of SEM is that it is accompanied by path diagrams that allow for clear visualization of multiple variables, causal relationships, and correlations among the variables in a graphic model.

Over the last 40 years, foraminiferal distribution and ecology from the Adriatic Sea have been studied from transitional areas to more open water [34]. For this reason, the area was regarded as optimal for examining the best predictors of community assemblages and testing causal hypotheses using machine learning techniques and SEM with data collected at a regional scale. The main aims of this investigation are: (1) to detect which among the measured environmental parameters contribute the most to determining the composition of benthic foraminiferal communities in the central-southern part of the Adriatic Sea, and (2) to disentangle the direct and indirect effects of multiple environmental variables in forming these communities. These goals were achieved by combining machine learning with classic statistical models to analyse data generated from extensive sampling in the central-southern Adriatic Sea.

2. Study Area

The Adriatic Sea is a mostly shallow semi-enclosed basin connected to the Mediterranean Sea (Figure 1) and is commonly divided into three sectors (i.e., northern, central and southern). The northernmost sector is very shallow (i.e., 100 m). The central part of the Adriatic Sea represents the transition zone between the northern shallower and the southern deeper sub-basins and is limited by the 100 m isobath to the north and the Palagrüza (Pelagosa) sill to the south. In this sector, a depression, the Mid Adriatic Depression (MAD), is present and subdivides the basins into three adjacent sub-basins. The western pit has a depth of ca. 255 m, whereas the central and eastern pits are ca. 270 m and 240 m deep, respectively [35]. The southern sector, between the Palagruža sill and the Otranto Strati, presents a wide depression, the Southern Adriatic Depression (SAD, >1200 m).

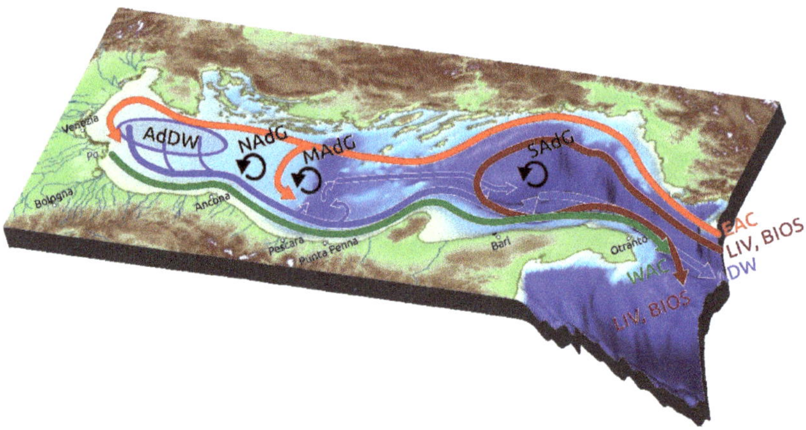

Figure 1. Bathymetric map of the Adriatic Sea showing the general circulation (from Spagnoli et al. [35]). Western Adriatic Current (WAC), Eastern Adriatic Current (EAC), North Adriatic Gyre (NAdG), Mid Adriatic Gyre (MAdG), South Adriatic Gyre (SAdG), Adriatic Dense Water (AdDW) North Adriatic AdW (NAdDW), Levantine Intermediate Water (LIW), Adriatic-Ionian bimodal oscillating system (BIOS).

The circulation of the Northern and Central Adriatic Sea (Figure 1) is represented by anticlockwise currents parallel to the coasts in winter [36], whereas in summer, the hydrodynamic pattern slows down and involves clockwise and counter-clockwise gyres. The circulation is also influenced by dense waters forming in the northern part of the Adriatic Sea that reach the MAD and partly flow further south [37].

The main circulation pattern in the Adriatic Sea is represented by the inflow of Mediterranean Sea waters along the eastern part of the Otranto Strait and the outflow across the western side of it [38]. During summer, the riverine waters and the associated fine sediments predominantly flow in the centre of the basin, which are instead limited near the coast in winter [39]. Strong winter and autumn storms occur in the northern and central Adriatic Sea, causing a marked bottom sediment resuspension up to a sediment thickness of 5–10 cm [40]. Overall, the current and wave actions control the distribution of sediments to the south and, partly, in the open sea [41].

The principal inputs of sediment are placed along the western side of the Adriatic Sea, which is primarily due to the Po River [42–44], and secondarily, by the Italian rivers.

The changes in sea-level in the recent past have markedly shaped the present bottom sea in the northern and central sectors of the Adriatic Sea as well as the shelves in the southern one. The last fast transgression formed an accretionary sedimentary wedge parallel to the Italian coastline over an erosional unconformity [45–48]. The thickness of the pelitic wedge reduces in the centre of the basin between the MAD and the Italian coast [49]. The bottom is marked by a belt of coastal sands and by a finer grain size area and again by coarser sediments offshore in the southern sector [47]. The grain size of sediment is silt and clay in the SAD.

3. Materials and Methods

3.1. Sample Collection

In total, 107 sediment samples were collected between the Ancona Harbor and Marina di Leuca Harbor from 13 to 703 m water depth (Figure 2) during the sampling PERTRE cruise organized by the Institute of Marine Sciences–National Research Council from 16 September to 4 October 2016. Sampling sites were placed along perpendicular transects to the coast at a with about 35—40 km between them, and their position was determined with Global Position System. Sediment was sampled by a box corer (Figure 2). Once retrieved, the box-cores were photographed and described, and the surface sediment (0–2 cm) was sampled for in situ measurement of pH and of the redox potential (Eh) with a Crison 570 multi-parametric probe. Sediment samples were sub-sampled, stored in polyethylene jars, and preserved in cold storage at +4 °C for the grain size. Additionally, an aliquot of sediments was stored in the freezer (−20 °C) for geochemistry, and another one (50 mL) was stored at room temperature for benthic foraminiferal analysis.

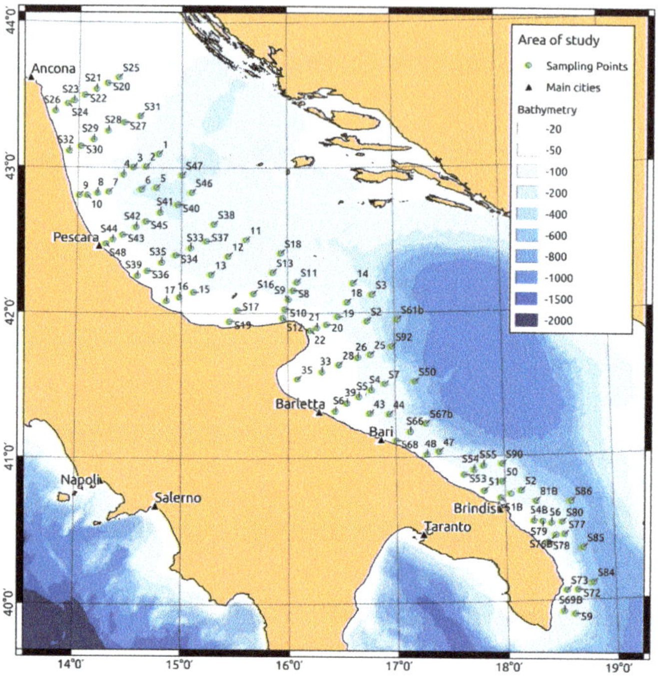

Figure 2. Location of the study area along with sampling stations in the central-southern Adriatic Sea. Supplementary [S], Bis [b].

3.2. *Grain-Size and Geochemical Analyses*

The analyses of grain-size were based on diffraction and diffusion on suspended particles (Malvern Mastersizer 2000, red He-Ne laser, 632 and 466 nm wavelength) [50,51]. Results were organized as three descriptive groups: clay, silt, and sand.

Sediments were analysed for Total organic carbon (OC), total nitrogen (N), and total sulphur (S) contents using a FlashEA 1112 Elemental Analyser (Thermo) equipped with an autosampler. The analyses were carried out on 1.5 to 2 mg of finely dried and crushed sediment to which was added ca. 5 mg of vanadium pentoxide. The OC content was then computed by subtracting mineral carbonate (i.e., carbonate carbon) from the total carbon (TC). Carbon from calcium carbonate content was determined using a Bernard calcimeter. Measurements were performed in triplicate at each station. Trace element analyses were performed by X-ray fluorescence (XRF) spectrometry (Philips PW1480 spectrometer). Major elements were determined on fused glass disks. Trace elements were analysed following the methods of Leoni et al. [52]. Accuracy was tested by analysing international reference standards. The loss on ignition (LOI) content was determined gravimetrically after overnight heating at 950 °C. The quantity of OM was based on OC contents, and the quality was estimated from the OC/N ratio. The OC/S was used to estimate the redox state of the sediments. The Pollution Load Index (PLI) of Tomlinson et al. [53] was calculated and applied for assessing the level of contamination. The PLI was estimated according to the equation: $PLI = \sqrt[n]{CF_1 \times CF_2 \times \ldots . \times CF_n}$, where *CF* is the concentration factor of selected analyzed trace elements, and n is the number of trace elements considered in the study. The concentration factor is determined as follows: $CF = \frac{C_{metal}}{C_{background}}$, where C_{metal} is the element concentration in the analyzed sample and $C_{background}$ is the background concentration of the element. In this study, the background values of Surricchio et al. [54] were used.

3.3. Benthic Foraminifera

The sediment for foraminiferal samples was immediately stained with rose Bengal dye solution (2 g of rose Bengal in 1000 mL of 90% ethanol) to identify living (stained)specimens. A quantitative analysis of the only living benthic foraminifera was performed on >63 μm fraction. The benthic foraminifera were taxonomically identified following Loeblich and Tappan [55], Jorissen [34], Cimerman and Langer [56], and Fiorini and Vaiani [57]. Out of 107 sampled sites, two (i.e., 13 and 28) were not sampled for foraminifera. The identified species were classified according to their microhabitat based on Murray [11]. The percentage of epifaunal and infaunal species was determined, and Hill diversity was calculated as the exponential of Shannon diversity.

3.4. Data Visualization and Analysis

Stations with less than 15 living benthic foraminiferal specimens were removed for the Gf analysis because a minimal number of abundances can create bias in the models. The following environmental variables, namely, sand, mud (silt+ clay), pH, Eh, N, OC, OC/N, and PLI were considered for this study. These are among the most used variables in ecological investigations on foraminifera. We used a correlation heatmap to discover the Pearson correlation coefficient between all the independent variables in our dataset (Figure 3). Here, the heatmap shows that a couple of the independent variables are highly correlated (Pearson coefficient > 0.7) with each other; therefore, we ascertain the presence of multi-collinearity in the data. Multi-collinearity (the existence of a linear relationship between independent variables) might bias the ability of our models and cause problems when we fit the model and interpret the results. For this reason, we carried out a Principal Component Analysis (PCA) to measure the severe multi-collinearity and to remove highly correlated independent variables from further analyses. A PCA takes advantage of multi-collinearity and integrates the positively correlated variables into a set of uncorrelated variables. Therefore, a PCA can effectively annihilate multi-collinearity between independent variables. Based on our PCA result (Figure S1), we removed pH, OC/N, mud and retained sand, Eh, N, OC, and PLI for creating the Gf analysis.

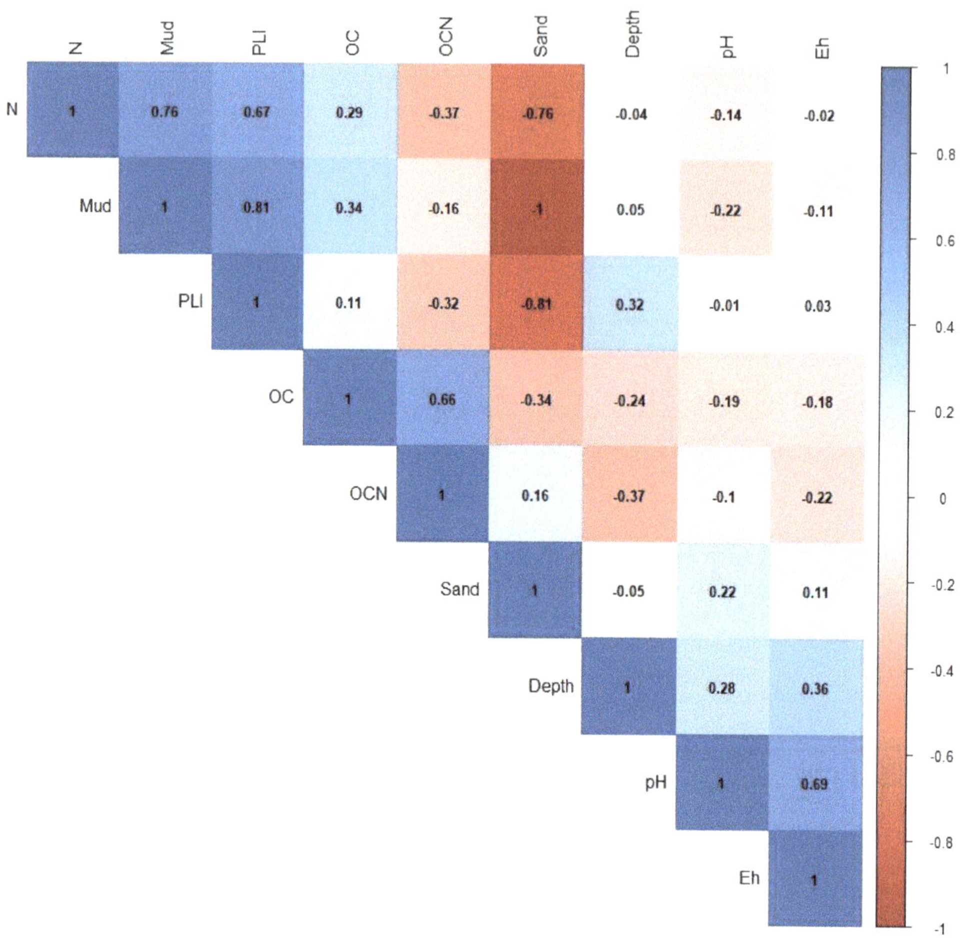

Figure 3. Correlation matrix of environmental variables. Total nitrogen [N], Pollution Load Index [PLI], Total Organic Carbon [OC], Total Organic Carbon/Total Nitrogen [OCN].

3.4.1. Gradient Forest

The Gf, as an extension of the Random Forest [32] and an ensemble machine learning approach, has been applied to unravel the non-linear changes of benthic foraminiferal species in relation to environmental gradients. This algorithm use bootstrap aggregating and random subspace approaches to create a set of decision trees. Based on this approach, the importance of each parameter was evaluated to define the ranking value among the variables for each species [58]. The performance of the Gf model was computed using the proportion of out-of-bag data and weighting that variance on all benthic foraminifera species. The advantage of this approach compared to other models is that it can handle a high number of interactions among different independent variables, and multi-collinearity is not an issue [32,58]. The Gf analyses were completed using R software (4.2.0) with two packages: "extendedForest" and "gradientForest" and 2000 trees. The cumulative split importance in the Gf modelling approach is used to define the pattern and magnitude of cumulative variations in benthic foraminifera assemblages along a depth gradient (Figure 4). In this nonlinear curve, abrupt changes in slope are related to high rates of variations in species composition, whereas superficial slopes suggest low ones. The steep parts of the curve suggest rapid turnover in the benthic foraminifera assemblages and based on these rapid changes, we divide the basin into the four depth intervals (i.e., 13–28 m, 29–58 m, 59–215 m, and >215 m) (Figure 4). A Gf model was built for each interval and the three most relevant variables influencing the benthic foraminiferal distribution were used to create the SEM models for each depth interval.

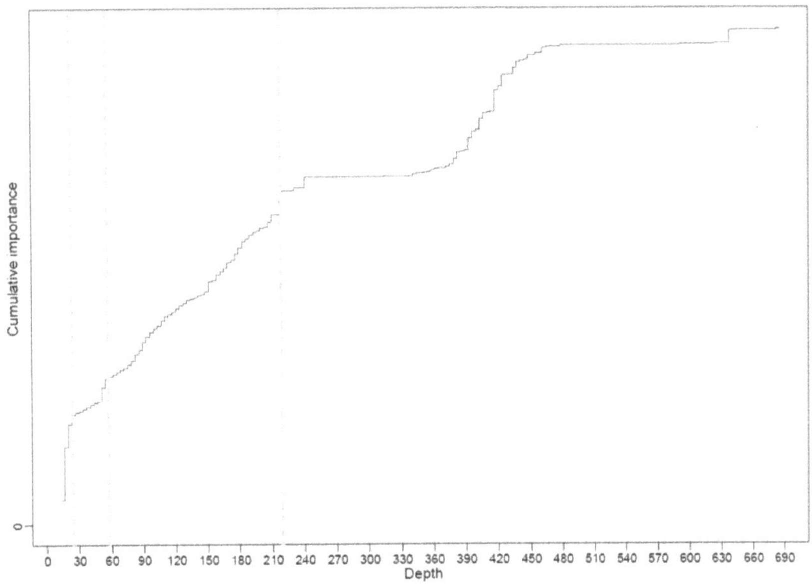

Figure 4. Gradient forest model over benthic foraminiferal assemblages. Dashed lines reflect major foraminiferal turnover.

3.4.2. Structural Equation Models

Structural Equation Models (SEM) were used to test specific causal hypotheses to evaluate the influence of the three most important environmental variables (determined by Gf) on diversity indices (i.e., Hill) and microhabitat (i.e., infauna). SEM are probabilistic models that generate variance-covariance matrices that can be compared with the original data variance-covariance matrices. This approach helps to identify direct and indirect associations between multiple independent and dependent variable in a single model and can compare causal hypotheses with unresolved correlational structures. A strength of SEM over other approaches to multiple regression is that different causal pathways, including indirect effects, can be compared directly, and endogenous variables can be both predictive and dependent on exogenous variables. It is one of the best developed statistical methods for testing causal hypotheses [33,59,60]. We used a Lavaan package in R (version 4.2.0) to test SEM models [61]; model fits were based on the comparative fit index (CFI) and the root mean square error of approximation (RMSEA); models were compared using Akaike information criteria (AIC) and Bayesian information criteria (BIC) (Table S1).

4. Results

4.1. Grain-Size, Organic Matter, and Geochemistry

The pH ranged between 5.74 and 7.80, with a mean value of 7.42 $\pm$ 0.77 (Table S2). The pH clearly shows a North-South and a West-East gradient towards lower values; accordingly, the highest values occurred in the northern part of the study areas and at the shallower stations (Figure 5). The redox potential (Eh) measured in the sediment samples varied between -243 to 273 mV (Table S2). The sediments were mostly composed of silt that accounted for 79.5 $\pm$ 20.5%. Sand content varied between 0.2 to 89.1%, with the highest values nearby the coast in the northern and southernmost parts of the study area and in two deeper areas (Figure 5). The OC (0.61 $\pm$ 0.28%) ranged from 0.004 to 1.31% and showed the highest value in the northern part of the study area at the farthest stations from the coast where the sediments were finer (Figure 5). On the other hand, the lowest OC values were found in the relict sand area and South of the Gargano promontory. The OC/N and OC/S ranged from 0.17 to 74.3 and from 0.04 to 44.4, respectively (Figure 5). The PLI (1.12 $\pm$ 0.25) varied from 0.44 to 1.63 with the highest values associated with finer grained sediments (i.e., silt and clay) (Figure 5).

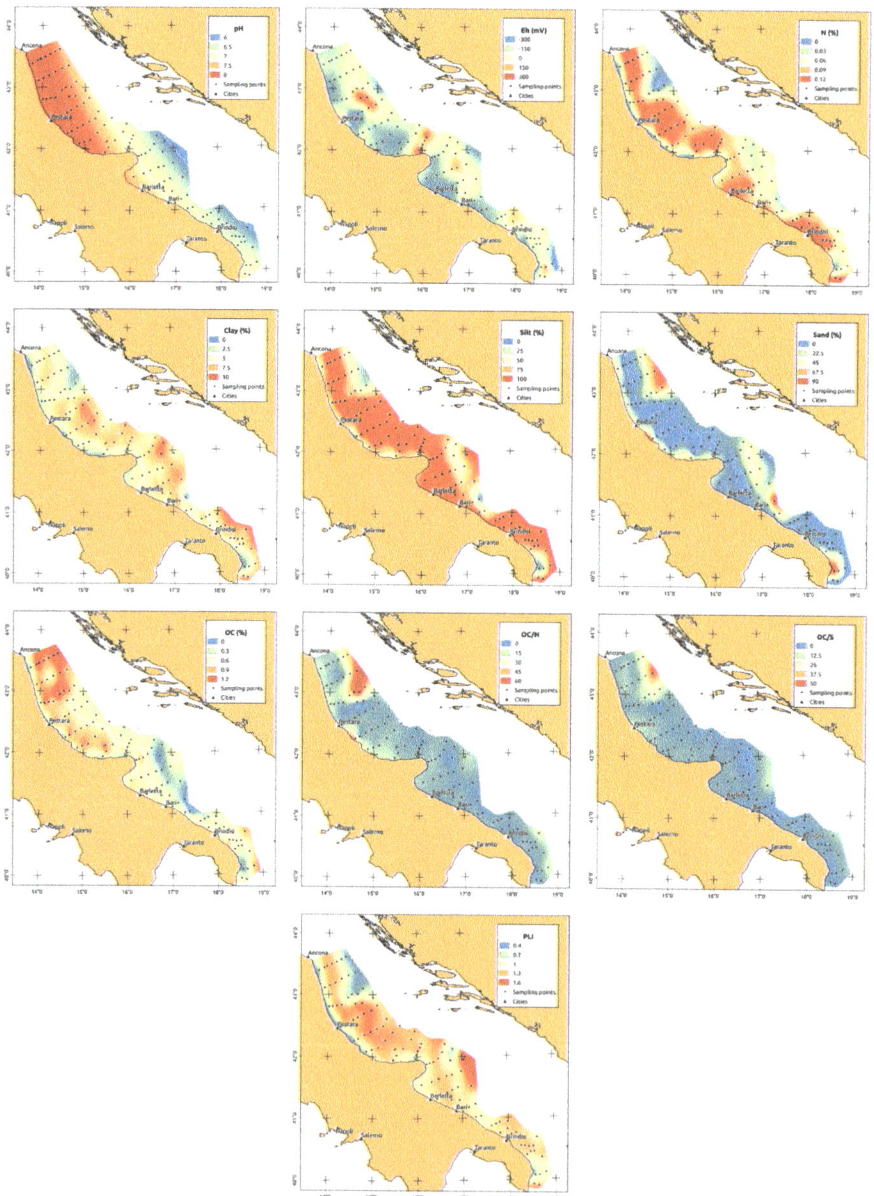

Figure 5. Interpolation map of the environmental variables (i.e., pH, Eh, Total Nitrogen [N], clay, silt, sand, Total Organic Carbon [OC], Total Organic Carbon/Total Nitrogen [OC/N], Total Organic Carbon/total Sulphur [OC/S] and Pollution Load Index [PLI]) in the central-southern Adriatic Sea.

4.2. Benthic Foraminifera

Seven species (33a, S17, S30, S36, S44, S61B and S92) exhibited a low number of specimens and were, therefore, not used for the statistical analysis. Overall, a total of 109 benthic foraminiferal species were recognized. The assemblages were dominated by hyaline (69.1 ± 19.7%) forms, followed by porcelaneous (22.2 ± 20.4%) and agglutinated (8.7 ± 7.1%) wall-type forms (Figure 6) (Table S3). Infaunal groups represented 62.8 ± 23.1% of the assemblages with a clear West-East gradient. Higher percentages of infauna were recorded around the Gargano promontory and at the shallowest stations in the southern part of the investigated area (Figure 6). The Hill varied from 0.83 to 3.48 (2.56 ± 0.6), whereas Fischer α index ranged from 0.91 to 36.13 (12.21 ± 6.12). Both these diversity indices exhibited the lowest values at the shallow stations (i.e., close to the coast) in the northern and central parts of the study area (Figure 6). The most abundant taxa were *Uvigerina mediterranensis* (11.5 ± 13%), *Ammonia parkinsoniana* (7.7 ± 16.4%), *Bulimina elongata* (5.7 ± 11.4%), *Bulimina marginata* (5.2 ± 8.4%), *Cassidulina carinata* (5.1 ± 6.0%), and *Bolivina spathulata* (4.2 ± 5.5%).

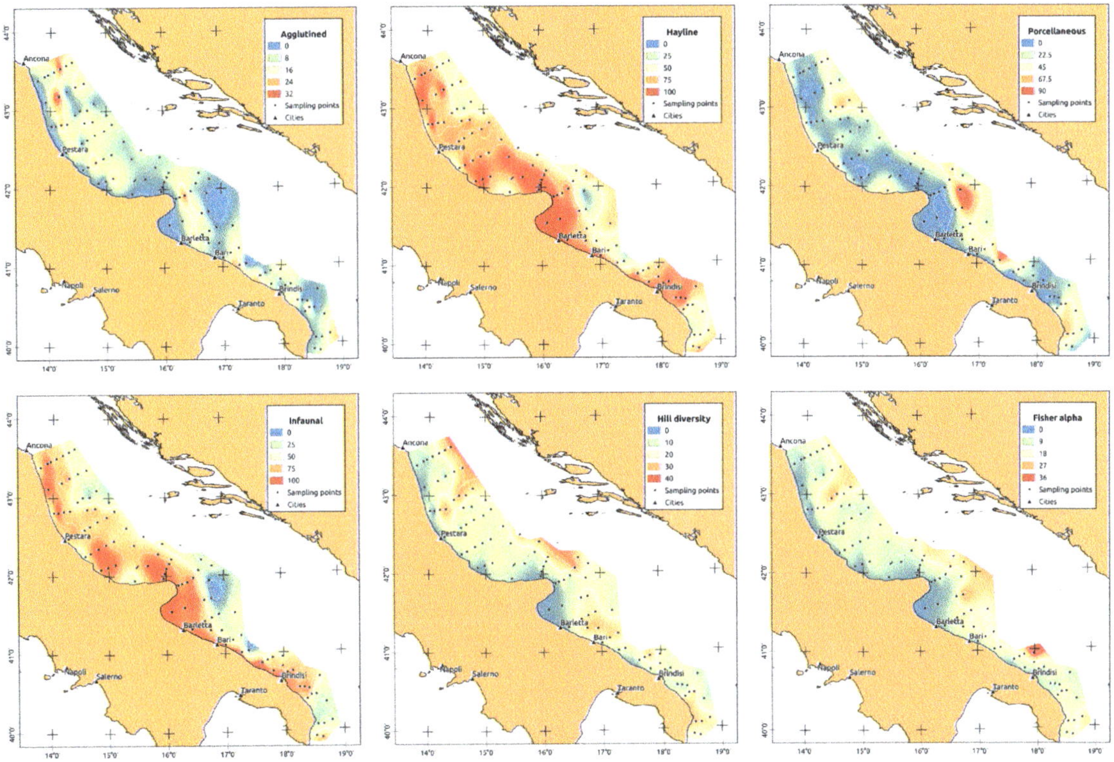

Figure 6. Interpolation map of selected foraminiferal assemblages' parameters and indices in the central-southern Adriatic Sea.

4.3. Models

The Gf modelling revealed cumulative changes in the assemblages of benthic foraminifera, resulting in the identification of four depth intervals (i.e., 13–28 m, 29–58, 59–215, and >215 m) (Figure 4). Unfortunately, the intervals 29–58 m and deeper than 215 m were characterized by very few samples (i.e., 10 and 8, respectively) and were therefore not used for further analyses (i.e., Gf and SEM).

The most important environmental variable in determining compositional changes of the benthic foraminiferal community along the entire bathymetric gradient was OC (R^2 weighted importance >0.035). Other environmental parameters, namely, sand, PLI, Eh, and N had intermediate (R^2 weighted importance range between 0.01 and 0.02) effects on changes in assemblages (Figure 7A). A Gf model was also analysed for two identified depth intervals, namely, 13–28 m and 59–215 m (Figure 7B,C). Sand was the most important variable, followed by PLI, N, and OC at the 13–28 m depth interval (Figure 7B). At deeper intervals (i.e., 59–215 m), OC played a major role in shaping the benthic foraminiferal assemblages, followed by N and sand (Figure 7C).

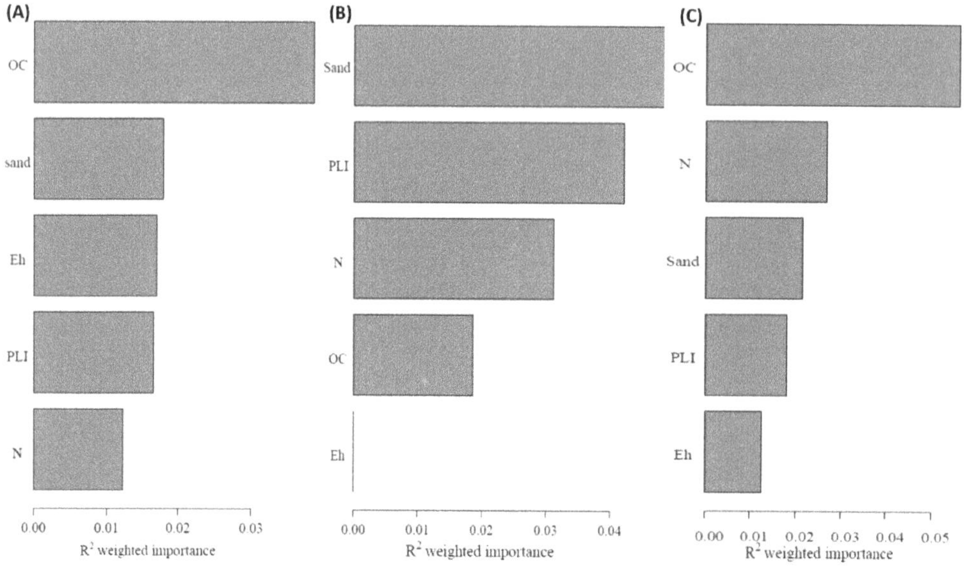

Figure 7. Relative contribution of environmental variables in shaping the benthic foraminiferal assemblages along three depth ranges (**A**) 13–703 m, (**B**) 13–28 m, and (**C**) 59–215 m. Total nitrogen [N], Pollution Load Index [PLI], Total organic carbon [OC].

We also used the SEM approach to investigate the specific a priori causal hypotheses and direct and indirect relationships to estimate the effect of the three most important environmental parameters for the overall bathymetric gradient and for each depth interval resulting from the Gf on different diversity index (i.e., Hill) and lifestyle (i.e., infaunal) in benthic foraminifera. Path coefficients from SEM indicated a minor effect of depth on OC, sand, and PLI when the entire bathymetric range is considered (Figure 8A), with a negative (standardized path coefficient (SPC) = −0.35) influence on infauna and a positive influence (SPC = 0.16) on Hill. Sand had a direct negative effect (SPC = −0.68) on infauna and a positive one (SPC = 0.51) on Hill (Figure 8A). When the depth interval 13–28 m was considered, depth had strong effects on sand, PLI, and N (Figure 8B), with a positive effect on infauna (SPC = 0.14) and a negative one on Hill (SPC = −0.23). In this interval, infauna is negatively related to sand (SPC = −0.96), PLI (SPC = −0.29), and N (SPC = −0.27). Sand had a direct effect on Hill (SPC = 0.81) and N (SPC = 0.41) (Figure 8B). At deeper stations (i.e., 59–215 m), depth did not seem to affect infauna or Hill (Figure 8C) but had a negative influence on OC (SPC = −0.43) and N (SPC = −0.32). Of these two environmental parameters, only N appears to highly influence both infauna (SPC = 0.66) and Hill (SPC = −0.69) (Figure 8C).

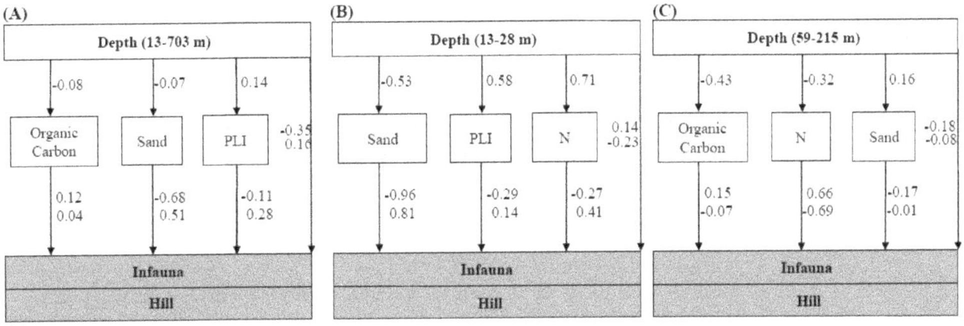

Figure 8. Performance of structural equation modelling along three depth ranges (**A**) 13–703 m, (**B**) 13–28 m and (**C**) 59–215 m. Total Nitrogen [N], Pollution Load Index [PLI].

5. Discussion

A very limited number of foraminiferal studies has included such a high number of samples, and despite intensive studies in the Adriatic Sea, the only research is likely the comprehensive investigation of Jorissen [34], analysing the distribution of benthic foraminifera in 285 samples. A wide range of environmental factors simultaneously shape foraminiferal assemblages, as functions of Adriatic circulation and hydrodynamics. However, determining the effects and magnitude of those factors is challenging, especially when considering such regional distribution. Nevertheless, we identified depth intervals that reflect rapid foraminiferal assemblage turnovers that are recognized at 28–29 m, 58–59 m, and 215 m water depth intervals. It is worth noting that the 28–29 m and 215 m intervals match well the boundary between the inner to middle neritic (i.e., 30 m) and outer neritic to upper bathyal (i.e., 200 m), respectively. These subdivisions have been widely used for depth reconstructions based on foraminiferal zonation in the Cenozoic [62,63]. On the other hand, the foraminiferal turnover identified at 58–59 m cannot be associated with the middle-outer neritic boundary (i.e., 100 m); this change corresponds to the bathymetric subdivision between infralittoral and circalittoral zones of the neritic Mediterranean zones [64]. This has also been applied for benthic foraminiferal distribution in the Gulf of Naples (Tyrrhenian Sea) [65], where the circalittoral zone is defined as between 40–50 m and 150–200 m. In the wide-ranging overview of foraminiferal distribution in the Adriatic Sea, Jorissen [34] inferred that the most important parameters influencing the distribution of the 50 most abundant foraminiferal species were bathymetry, sand fraction, calcium carbonate, and organic matter. Indeed, the author identified several biofacies units (i.e., 7.5–25 m, 20–100 m, and 50–1225 m) that are not only defined on the bathymetric range but also on sand and organic matter contents. Our Gf model does not exactly identify the same "biofacies" units of Jorissen [34] and this might be ascribed to a different sampling strategy: (a) sampling device (i.e., grab and piston core vs. box-core); (b) assemblages (i.e., total vs. living assemblages); (c) sieve-size (i.e., 150 μm vs. 63 μm) and/or investigated areas: (i) geographical extension (i.e., the entire Adriatic Sea vs. the central-southern part of the western side of the Adriatic Sea); (ii) bathymetric range (i.e., 7.5–1152 m vs. 13–703 m) and/or temporal changes in the benthic environment.

In our study, the factors that most contributed to shaping foraminiferal assemblages across the entire depth gradient were organic matter contents (OC), followed by grain-size (sand) and anthropogenic contaminations (PLI). These variables match well those (organic matter and sand) identified by Jorissen [34]. Interestingly, our Gf models for 13–28 m and 59–215 m bathymetric ranges reveal which factors shape the foraminiferal assemblages. While organic matter (i.e., OC and N) is still the most important factor at greater depths (59–215 m), at shallower depths the grain-size (i.e., sand content) and pollution (i.e., PLI) seem to turn out to be the most important variables.

The interpretations of putative causal relationships between abiotic and biotic variables are quite challenging. For the full depth gradient, abiotic and biotic parameters are scarcely-to-moderately correlated with depth, exhibiting only a direct negative influence on infauna (SPC = −0.35). The percentages of infaunal taxa are expected to decrease in shelf/bathyal areas, where the organic matter supplied to the bottom is limited. These results corroborate the TROX model [26], which suggests a decrease in infaunal percentages in oligotrophic environments, where nutrients are a limiting factor; this is represented by deeper stations in our study system. The TROX model also predicts an increase in infaunal percentages in relation to enhanced food supply at the seafloor and to limited oxygen penetration within it. This pattern is also related to the sediment grain-size (i.e., fine) enriched in OM. This is well supported in our Gf model; in fact, when the entire bathymetric range is considered, the two most important components shaping the foraminiferal assemblages are nutrients (i.e., OC) availability and grain-size (i.e., sand). Based on the SEM model, diversity (Hill) is positively, though weakly, related to depth. Variations in species diversity (i.e., species richness) along bathymetric gradients have been deeply documented in both the eastern and western Mediterranean Sea [66]. These authors revealed an optimum (maximum) in species richness between 200 and 1000 m water depth, followed by a steady decline towards depth. In our record, the increase of diversity at greater depths (SPC = 0.19) matches well their finding only when the entire bathymetric range is considered. This implies that diversity might be affected by spatial heterogeneity of various physicochemical and hydrodynamic parameters, caused by geomorphological differences and relict sedimentary processes.

At the 13–28 m depth interval, depth positively affected infauna (SPC = 0.14) and negatively affected Hill (SPC = −0.24). In this bathymetric range, sand, PLI and N are strongly affected by depth, and sand negatively affects infauna but positively affects diversity. Here, sand content, indicative of stronger hydrodynamics, plays a major role in shaping the benthic foraminiferal assemblages as defined by infauna and diversity. Coarse sediments are commonly well-oxygenated and exhibit a lower organic matter availability compared to finer ones; shallow water sandy bottoms are frequently unstable moving substrates subject to remobilization by waves and transport by coastal drift currents; these substrates preferentially host epifaunal species. The increase in diversity with sand content could be due to more heterogeneous microhabitats and increased availability of nutrients and oxygen within the sediments. Total nitrogen (N) also contributes to shape the foraminiferal assemblages (i.e., infauna and diversity) but to a lesser extent than sand content.

At the 59–215 m depth interval, depth caused decreases in organic matter (i.e., OC and N). The main food source for benthic organisms is available via the fraction of OC produced in surface water and can effectively reach the seafloor [67–69]. The organic matter degrades along the water column; therefore, the greater the distance between the surface and the bottom, the less labile nutrients reach the benthic communities. This explains the negative relation between OC and N with depth. At this bathymetric range, the only factor influencing infauna and diversity is nitrogen. Increases in nitrogen in marine sediment are commonly associated with the presence of phytoplankton with a higher protein content, which is preferentially used by benthic foraminifera as labile organic matter [70].

6. Conclusions

The opportunities offered by combining traditional statistics with machine learning, in this case SEM and gradient forest analyses, have allowed us to identify the most important environmental factors controlling benthic foraminiferal assemblages in the central-southern part of the eastern side of the Adriatic Sea. Using this integrative approach, we identified turnovers at 13–28, 29–58, 59–215, and >215 m depth intervals in the benthic foraminiferal compositions, which are good matches to the classical bathymetric distribution of benthic communities.

Among a large set of environmental variables, sand content and organic matter are the most important determinants of foraminiferal distributions either along the entire depth gradient or at selected bathymetric ranges. These variables are commonly influenced by depth that indirectly shapes foraminiferal assemblages with distinct structural characteristics. The potential applications of merging statistical learning with classical statistical modelling, such as path analysis, are just now being realized and should contribute to basic ecology and environmental monitoring.

Supplementary Materials: The following supporting information can be downloaded at: https://www.mdpi.com/article/10.3390/app13020794/s1. Figure S1. Dimensional reduction with Principal Component Analysis method. Table S1. Structural Equation Modeling results including Comparative Fit Index (CFI), Root Mean Square Error of Approximation (RMSEA), Akaike information criteria (AIC), and Bayesian information criteria (BIC). Table S2. Raw environmental data. Table S3. Raw foraminiferal data.

Author Contributions: Methodology, M.A.R., V.B.-R., M.D., F.F. (Fabrizio Frontalini), F.S., E.A.d.C. and M.T.; Conceptualization, F.F. (Fabrizio Frontalini), M.V.A.M., L.A.D. and R.A.; Software, M.A.R.; Validation, M.A.R., L.A.D. and F.F. (Fabio Francescangeli); Formal Analysis, F.F. (Fabrizio Frontalini), E.A.d.C., F.F. (Fabio Francescangeli); Investigation, F.F. (Fabrizio Frontalini), F.F. (Fabio Francescangeli), V.B.-R., F.S.; Resources, M.A.R., F.F. (Fabrizio Frontalini), E.A.d.C., F.F. (Fabio Francescangeli); Data curation, F.F. (Fabrizio Frontalini), E.A.d.C., F.F. (Fabio Francescangeli), F.S.; Writing—original draft preparation, F.F. (Fabrizio Frontalini), F.F. (Fabio Francescangeli), M.A.R., L.A.D.; Writing—review and editing, M.A.R., F.F. (Fabrizio Frontalini), E.A.d.C., F.F. (Fabio Francescangeli), M.V.A.M., R.D.M., E.D., L.A.D., R.A., V.B.-R., M.D. and F.S.; Visualization, R.D.M., M.A.R., F.F. (Fabrizio Frontalini) and F.F. (Fabio Francescangeli); Supervision, F.F. (Fabrizio Frontalini) F.F. (Fabio Francescangeli) and M.A.R.; Project administration, F.S., F.F. (Fabrizio Frontalini), and F.F. (Fabio Francescangeli); Funding acquisition, F.S. and F.F. (Fabrizio Frontalini). All authors have read and agreed to the published version of the manuscript.

Funding: This research was funded by RSE SpA–Ricerca sul Sistema Elettrico, Milano, Italy, and by the PERSEUS European Project (grant number 287600). MVAM would like to thank the Conselho Nacional de Desenvolvimento Científico e Tecnológico of Brazil, CNPq (processes # 443662/2018-5 and # 302676/20198 to MVAM) and Fundação Carlos Chagas Filho de Amparo à Pesquisa do Estado do Rio de Janeiro, FAPERJ, Brazil, (process: E-26/202.927/2019).

Institutional Review Board Statement: Not applicable.

Informed Consent Statement: Not applicable.

Data Availability Statement: Our data are available in the supplementary section.

Acknowledgments: Fabio Francescangeli was supported by the Alexander von Humboldt Foundation (Ref 3.3–ITA -1203503-HFST-P).

Conflicts of Interest: The authors declare no conflict of interest.

References

1. Martins, V.; Isabel, A.; Grangeia, G.; Paula, M.; Nagai, R.; Sousa, S.H.M.; Laut, L.L.M.; Dias, J.M.A.; Ferreira da Silva, E.; Rocha, F. Records of sedimentary dynamics in the continental shelf and upper slope between Aveiro–Espinho (N Portugal). *J. Mar. Syst.* **2012**, *96*, 48–60. [CrossRef]
2. Martins, V.M.; Pinto, A.F.S.; Frontalini, F.; da Fonseca, M.C.M.; Terroso, D.L.; Laut, L.L.M.; Zaaboub, N.; Rodrigues, M.A.; Rocha, F. Can benthic foraminifera be used as bio-indicators of pollution in areas with a wide range of physiochemical variability? *Estuar. Coast. Shelf Sci.* **2016**, *182*, 211–225. [CrossRef]
3. Raposo, D.; Clemente, I.; Figueiredo, M.; Vilar, A.; Lorini, M.L.; Frontalini, F.; Martins, M.V.A.; Belart, P.; Fontana, L.; Habib, R.; et al. Benthic foraminiferal and organic matter compounds as proxies of environmental quality in a tropical coastal lagoon: The Itaipu lagoon (Brazil). *Mar. Pollut. Bull.* **2018**, *129*, 114–125. [CrossRef] [PubMed]
4. Förderer, E.M.; Rödder, D.; Langer, M.R. Global diversity patterns of larger benthic foraminifera under future climate change. *Glob. Chang. Biol.* 2022; 1–13, *Online ahead of print*.
5. Reymond, C.E.; Roff, G.; Chivas, A.R.; Zhao, J.; Pandolfi, J.M. Millennium-scale records of benthic foraminiferal communities from the central Great Barrier Reef reveal spatial differences and temporal consistency. *Palaeogeogr. Palaeoclimatol. Palaeoecol.* **2013**, *374*, 52–61. [CrossRef]
6. Narayan, Y.R.; Lybolt, M.; Zhao, J.; Feng, Y.; Pandolfi, J.M. Holocene benthic foraminiferal assemblages indicate long-term marginality of reef habitats from Moreton Bay, Australia. *Palaeogeogr. Palaeoclimatol. Palaeoecol.* **2015**, *420*, 49–64. [CrossRef]

7. Bouchet, V.M.P.; Goberville, E.; Frontalini, F. Benthic foraminifera to assess the Ecological Quality Status of Italian transitional waters. *Ecol. Indic.* **2018**, *84*, 130–139. [CrossRef]
8. Reymond, C.E.; Bode, M.; Renema, W.; Pandolfi, J.M. Ecological incumbency impedes stochastic community assembly in Holocene foraminifera from the Huon Peninsula, Papua New Guinea. *Paleobiology* **2011**, *37*, 670–685. [CrossRef]
9. Reymond, C.E.; Uthicke, S.; Pandolfi, J.M. Tropical Foraminifera as indicators of water quality and temperature. In Proceedings of the 12th International Coral Reef Symposium, Cairns, Australia, 9–13 July 2012.
10. Narayan, Y.R.; Pandolfi, J.M. Benthic foraminiferal assemblages from Moreton Bay, South-East Queensland, Australia: Applications in monitoring water and substrate quality in subtropical estuarine environments. *Mar. Pollut. Bull.* **2010**, *60*, 2062–2078. [CrossRef]
11. Murray, J.W. *Ecology and Applications of Benthic Foraminifera*; Cambridge University Press: New York, NY, USA, 2006.
12. Frontalini, F.; Greco, M.; Di Bella, L.; Lejzerowicz, F.; Reo, E.; Caruso, A.; Cosentino, C.; Maccotta, A.; Scopelliti, G.; Nardelli, M.P.; et al. Assessing the effect of mercury pollution on cultured benthic foraminifera community using morphological and eDNA meta-barcoding approaches. *Mar. Pollut. Bull.* **2018**, *129*, 512–524. [CrossRef]
13. Rostami, M.A.; Frontalini, F.; Giordano, P.; Francescangeli, F.; Martins, M.V.A.; Dyer, L.; Spagnoli, F. Testing the applicability of random forest modeling to examine benthic foraminiferal responses to multiple environmental parameters. *Mar. Environ. Res.* **2021**, *172*, 105502. [CrossRef]
14. Armynot du Châtelet, E.; Francescangeli, F.; Bouchet, V.M.P.; Frontalini, F. Benthic foraminifera on transitional environments in the English Channel and the southern North Sea: A proxy for regional scale environmental and paleo-environmental characterizations. *Mar. Environ. Res.* **2018**, *137*, 37–48. [CrossRef] [PubMed]
15. Armynot du Châtelet, E.A.; Francescangeli, F.; Frontalini, F. Definition of benthic foraminiferal bioprovinces in transitional environments of the Eastern English Channel and the Southern North Sea. *Rev. Micropaléontologie* **2018**, *61*, 223–234. [CrossRef]
16. Narayan, G.R.; Herrán, N.; Reymond, C.E.; Shaghude, Y.W.; Westphal, H. Local Persistence of Large Benthic Foraminifera (LBF) under Increasing Urban Development: A Case Study from Zanzibar (Unguja), East Africa. *J. Earth Sci.* **2022**, *33*, 1434–1450. [CrossRef]
17. Alve, E. Benthic foraminifera response to estuarine pollution. A review. *J. Foraminifer. Res.* **1995**, *25*, 190–203. [CrossRef]
18. Armynot du Châtelet, E.A.; Frontalini, F.; Francescangeli, F. Significance of replicates: Environmental and paleoenvironmental studies on benthic foraminifera and testate amoebae. *Micropaleontology* **2017**, *63*, 257–274.
19. Francescangeli, F.; Bouchet, V.M.P.; Trentesaux, A.; Armynot du Châtelet, E. Does elevation matter? Living foraminiferal distribution in a hyper tidal salt marsh (Canche Estuary, Northern France). *Estuar. Coast. Shelf Sci.* **2017**, *194*, 192–204. [CrossRef]
20. Reymond, C.E.; Mateu-Vicens, G.; Westphal, H. Foraminiferal assemblages from a transitional tropical upwelling zone in the Golfe d'Arguin, Mauritania. *Estuar. Coast. Shelf Sci.* **2014**, *148*, 70–84. [CrossRef]
21. Humphreys, A.F.; Halfar, J.; Ingle, J.C.; Manzello, D.; Reymond, C.E.; Westphal, H. Effect of seawater temperature, pH, and nutrients on the distribution and character of low abundance shallow water benthic foraminifera in the Galápagos. *PLoS ONE* **2018**, *13*, e0202746. [CrossRef]
22. Francescangeli, F.; Portela, M.; Armynot du Chatelet, E.; Billon, G.; Andersen, T.J.; Bouchet, V.M.P.; Trentesaux, A. Infilling of the Canche Estuary (eastern English Channel, France): Insight from benthic foraminifera and historical pictures. *Mar. Micropaleontol.* **2018**, *142*, 1–12. [CrossRef]
23. Narayan, G.R.; Reymond, C.E.; Stuhr, M.; Doo, S.; Schmidt, C.; Mann, T.; Westphal, H. Response of large benthic foraminifera to climate and local changes: Implications for future carbonate production. *Sedimentology* **2021**, *69*, 121–161. [CrossRef]
24. Mojtahid, M.; Geslin, E.; Coynel, A.; Gorse, L.; Vella, C.; Davranche, A.; Zozzolo, L.; Blanchet, L.; Bénéteau, E.; Maillet, G. Spatial distribution of living (Rose Bengal stained) benthic foraminifera in the Loire estuary (western France). *J. Sea Res.* **2016**, *118*, 1–16. [CrossRef]
25. Francescangeli, F.; Milker, Y.; Bunzel, D.; Thomas, H.; Norbisrath, M.; Schönfeld, J.; Schmiedl, G. Recent benthic foraminiferal distribution in the Elbe Estuary (North Sea, Germany): A response to environmental stressors. *Estuar. Coast. Shelf Sci.* **2021**, *251*, 107198. [CrossRef]
26. Jorissen, F.J.; De Stigter, H.C.; Widmark, J.G.V.A. Conceptual model explaining benthic foraminiferal microhabitats. *Mar. Micropaleontol.* **1995**, *26*, 3–15. [CrossRef]
27. Bouchet, V.M.P.; Frontalini, F.; Francescangeli, F.; Geslin, E.; Martins, V.; Almogi-Labin, A.; Avnaim-Katav, S.; Di Bella, L.; Cearreta, A.; Coccioni, R.; et al. Indicative value of benthic foraminifera for biomonitoring: Assignment to ecological groups of sensitivity to total organic carbon of species from European intertidal areas and transitional waters. *Mar. Pollut. Bull.* **2021**, *164*, 112071. [CrossRef]
28. Murray, J.W. The niche of benthic foraminifera, critical thresholds and proxies. *Mar. Micropaleontol.* **2001**, *41*, 1–7. [CrossRef]
29. Lucas, T.C.D. A translucent box: Interpretable machine learning in ecology. *Ecol. Monogr.* **2020**, *90*, e01422. [CrossRef]
30. Elith, J.; Leathwick, J.R. Species distribution models: Ecological explanation and prediction across space and time. *Annu. Rev. Ecol. Evol. Syst.* **2009**, *40*, 677–697. [CrossRef]
31. Hastie, T.; Tibshirani, R.; Friedman, J. *The Elements of Statistical Learning*; Springer Series in Statistics: Heidelberg, Germany, 2001.
32. Ellis, N.; Smith, S.J.; Pitcher, C.R. Gradient Forests: Calculating importance gradients on physical predictors. *Ecology* **2012**, *93*, 156–168. [CrossRef]
33. Shipley, B. *Cause and Correlation in Biology: A User's Guide to Path Analysis, Structural Equations and Causal Inference*; Cambridge University Press: Cambridge, UK, 2002.
34. Jorissen, F. Benthic foraminifera from the Adriatic Sea; Principles of phenotypic variation. *Utrecht Micropaleontol. Bull.* **1988**, *37*, 174.
35. Spagnoli, F.; De Marco, R.; Dinelli, E.; Frapiccini, E.; Frontalini, F.; Giordano, P. Sources and Metal Pollution of Sediments from a Coastal Area of the Central Western Adriatic Sea (Southern Marche Region, Italy). *Appl. Sci* **2021**, *11*, 1118. [CrossRef]

36. Artegiani, A.; Bregant, D.; Paschini, E.; Pinardi, N.; Raicich, F.; Russo, A. The Adriatic Sea General Circulation. Part I: Air–Sea Interactions and Water Mass Structure. *J. Phys. Oceanogr.* **1997**, *27*, 1492–1514. [CrossRef]
37. Vilibic, I.; Supic, N. Dense water generation on a shelf: The case of the Adriatic Sea. *Ocean Dyn.* **2005**, *55*, 403–415. [CrossRef]
38. Civitarese, G.; Gacic, M.; Lipizer, M.; Eusebi Borzelli, G.L. On the impact of the Bimodal Oscillating System (BiOS) on the biogeochemistry and biology of the Adriatic and Ionian Seas (Eastern Mediterranean). *Biogeosciences* **2010**, *7*, 3987–3997. [CrossRef]
39. Frascari, F.; Spagnoli, F.; Marcaccio, M.; Giordano, P. Anomalous Po river flood event effects on sediments and water column of the Northwestern Adriatic Sea. *Clim. Res.* **2006**, *31*, 151–165. [CrossRef]
40. Cavaleri, L.; Stefanon, A. Bottom features due to extreme meteorological events in the North Adriatic Sea. *Mar. Geol.* **1980**, *79*, 159–170.
41. Spagnoli, F.; Dinelli, E.; Giordano, P.; Marcaccio, M.; Zaffagnini, F.; Frascari, F. Sedimentological biogeochemical mineralogical facies of Northern and Central Western Adriatic Sea. *J. Mar. Syst.* **2014**, *139*, 183–203. [CrossRef]
42. Syvitski, J.P.M.; Kettner, A.J. On the flux of water and sediment into the Northern Adriatic Sea. *Cont. Shelf Res.* **2007**, *27*, 296–308. [CrossRef]
43. Frignani, M.; Langone, L.; Ravaioli, M.; Sorgente, D.; Alvisi, F.; Albertazzi, S. Fine sediment mass balance in the western Adriatic continental shelf over a century time scale. *Mar. Geol.* **2005**, *222–223*, 113–133. [CrossRef]
44. Cattaneo, A.; Correggiari, A.; Langone, L.; Trincardi, F. The Late-Holocene Gargano subaqueous delta, Adriatic shelf: Sediment pathways and supply fluctuations. *Mar. Geol.* **2003**, *193*, 61–91. [CrossRef]
45. Trincardi, F.; Asioli, A.; Cattaneo, A.; Correggiari, A.; Vigliotti, L.; Accorsi, C.A. Transgressive offshore deposits on the Central Adriatic Shelf: Architecture complexity and the record of the younger Dyas short-term event. Il Quaternario. *Ital. J. Quat. Sci.* **1996**, *9*, 753–762.
46. Correggiari, A.; Trincardi, F.; Langone, L.; Roveri, M. Styles of failure in heavily sedimented highstand prodelta wedges on the Adriatic shelf. *J. Sediment. Res.* **2001**, *71*, 218–236. [CrossRef]
47. Colantoni, P.; Asioli, A.; Borsetti, A.M. Evoluzione morfo-sedimentaria recente del Mare Adriatico. Atti del 9° Congresso. *AIOL* **1990**, *LM-75*, 17–24.
48. Colantoni, P.; Gallignani, P.; Lenaz, R. Late Pleistocene and Holocene evolution of the north Adriatic continental shelf Italy. *Mar. Geol.* **1979**, *33*, 41–50. [CrossRef]
49. Cattaneo, A.; Trincardi, F. The late-Quaternary transgressive record in the Adriatic Epicontinental Sea: Basin widening and facies partitioning. In *Isolated Shallow Marine Sand Bodies: Sequence Stratigraphic Analysis and Sedimentologic Interpretation*; Bergman, K.M., Snedden, Eds.; SEPM SpecialPubllication: Tulsa, OK, USA, 1999; Volume 64, pp. 127–146.
50. Loizeau, J.L.; Arbouille, D.; Santiago, S.; Vernet, J.P. Evaluation of a wide range laser diffraction grain size analyser for use with sediments. *Sedimentology* **1994**, *41*, 353–361. [CrossRef]
51. Trentesaux, A.; Recourt, P.; Bout-Roumazeilles, V.; Tribovillard, N. Carbonate grain-size distribution in hemipelagic sediments from a laser particle sizer. *J. Sediment. Res.* **2001**, *71*, 858–862. [CrossRef]
52. Leoni, L.; Menichini, M.; Saitta, M. Determination of S, Cl and F in silicate rocks by X-ray fluorescence analyses. *X-ray Spectrom.* **1982**, *11*, 156–158. [CrossRef]
53. Tomlinson, D.L.; Wilson, J.G.; Harris, C.R.; Jeffrey, D.W. Problems in the assessment of heavy metal level in eastuaries and the formation of a pollution index. *Helol. Meeresunters.* **1980**, *33*, 566–575. [CrossRef]
54. Surricchio, G.; Pompilio, L.; Arizzi Novelli, A.; Scamosci, E.; Marinangeli, L.; Tonucci, L.; d'Alessandro, N.; Tangari, A.C. Evaluation of heavy metals background in the Adriatic Sea sediments of Abruzzo region, Italy. *Sci. Total Environ.* **2019**, *684*, 445–457. [CrossRef]
55. Loeblich, A.R.; Tappan, H. *Foraminiferal Genera and Their Classification*; Van Nostrand, Reinhold Co.: New York, NY, USA, 1987; Volume 2, p. 2047.
56. Cimerman, F.; Langer, M.R. Mediterranean Foraminifera. In *Academia Scientarium et Aritium Slovenica, Dela, Opera 30, Classis IV: Historia Naturalis*; Academia Scientiarum et Artium Slovenica: Ljubljana, Slovenia, 1991.
57. Fiorini, F.; Vaiani, S.C. Benthic foraminifers and transgressive-regressive cycles in the Late Quaternary subsurface sediments of the Po Plain near Ravenna (Northern Italy). *Boll. Soc. Paleontol. Ital.* **2001**, *40*, 357–403.
58. Smith, P.F.; Ganesh, S.; Liu, P.A. Comparison of random forest regression and multiple linear regression for prediction in neuroscience. *J. Neurosci. Methods* **2013**, *220*, 85–91. [CrossRef]
59. Grace, J.B.; Anderson, T.M.; Olff, H.; Scheiner, S.M. On the specification of structural equation models for ecological systems. *Ecol. Monogr.* **2010**, *80*, 67–87. [CrossRef]
60. Pearl, J. The causal foundations of structural equation modeling. In *Handbook of Structural Equation Modeling*; Hoyle, R.H., Ed.; Guilford: New York, NY, USA, 2012; pp. 68–91.
61. Rosseel, Y. lavaan: An R Package for Structural Equation Modeling. *J. Stat. Softw.* **2012**, *48*, 1–36. [CrossRef]
62. van Morkhoven, F.P.C.M.; Berggren, W.A.; Edwards, A.S. Cenozoic cosmopolitan deep-water benthic foraminifera. In *Bull. Cent. Rech. Explor.*; Prod. Elf-Aquitaine, Mem.: Courbevoie, France, 1986; Volume 11, p. 421.
63. Berggren, W.A.; Miller, K.G. Cenozoic bathyal and abyssal calcareous benthic foraminiferal zonations. *Micropaleontology* **1989**, *35*, 308–320. [CrossRef]
64. Pérès, J.M.; Picard, J. Nouveau manuel de bionomie benthique de la Méditerranée. In *Rec. Trav. Stat. Mar.*; Endoume: Marseille, France, 1964; Volume 31, pp. 1–37.
65. Sgarrella, F.; Moncharmont Zei, M. Benthic Foraminifera of the Gulf of Naples (Italy): Systematics and autoecology. *Bollettino Soc. Paleontol. Ital.* **1993**, *32*, 145–264.

66. De Rijk, S.; Hayes, A.; Rohling, E.J. Eastern Mediterranean sapropel S1 interruption: An expression of the onset of climatic deterioration around 7 ka BP. *Mar. Geol.* **1999**, *153*, 337–343. [CrossRef]
67. Gooday, A.J.; Lambshead, P.J.D. Influence of seasonally deposited phytodetritus on benthic foraminferal populations in the bathyal northeast Atlantic the species response. *Mar. Ecol. Prog. Ser.* **1989**, *58*, 53–67. [CrossRef]
68. Gooday, A.J. Benthic foraminifera (Protista) as tools in deep-water palaeoceanography: Environmental influences on faunal characteristics. *Adv. Mar. Biol.* **2003**, *46*, 1–90.
69. De Rijk, S.; Jorissen, F.; Rohling, E.J.; Troelstra, S.R. Organic flux control on bathymetric zonation of Mediterranean benthic foraminifera. *Mar. Micropaleontol.* **2000**, *40*, 151–166. [CrossRef]
70. Koho, K.A.; Langezaal, A.M.; van Lith, Y.A.; Duijnstee, I.A.P.; Van der Zwaan, G.J. The influence of a simulated diatom bloom on deep-sea benthic foraminifera and the activity of bacteria: A mesocosm study. *Deep. Sea Res. Part I Oceanogr. Res. Pap.* **2008**, *55*, 696–719. [CrossRef]

applied sciences

MDPI

Review

The Role of Nitrogen and Phosphorus in Eutrophication of the Northern Adriatic Sea: History and Future Scenarios

Mauro Marini [1,2,*] and Federica Grilli [1,*]

1 Institute for Biological Resources and Marine Biotechnologies, National Research Council (IRBIM, CNR) Largo Fiera della Pesca 2, 60125 Ancona, Italy
2 Fano Marine Center, The Inter-Institute Center for Research on Marine Biodiversity, Resources and Biotechnologies (FMC), Viale Adriatico 1/N, 61032 Fano, Italy
* Correspondence: mauro.marini@cnr.it (M.M.); federica.grilli@cnr.it (F.G.)

Abstract: In the last two decades of the 21st century, a gradual decrease in nitrogen and phosphorus has been observed along the coastal area of the Northern Adriatic Sea. This depletion is attributed to reduced river flows. Studies conducted over the past four decades have indicated that the N/P ratio in the open sea is unlikely to undergo significant change. In fact, it tends to increase due to the unique characteristics of the Northern Adriatic Sea, which experiences slow water turnover and is influenced by strong winds. Additionally, the Northern Adriatic Sea receives a substantial amount of freshwater from rivers, accounting for about one-third of the total freshwater flow into the Mediterranean. These rivers carry nutrient loads that contribute to the high productivity and abundance of fish in this sea, making it one of the most productive areas in the Mediterranean. It has been observed that the cessation of anthropogenic phosphorus input, which has been regulated since the late 1980s with legislation limiting its use in detergents, has significantly affected the trophic chain. The aim of this review is to provide an overview of the eutrophication trend in the Northern Adriatic Sea, highlighting the importance long-term data series.

Keywords: Northern Adriatic Sea; eutrophication; nitrogen; phosphorus; freshwater; plankton; fishing resource; long term research

Citation: Marini, M.; Grilli, F. The Role of Nitrogen and Phosphorus in Eutrophication of the Northern Adriatic Sea: History and Future Scenarios. *Appl. Sci.* **2023**, *13*, 9267. https://doi.org/10.3390/app13169267

Academic Editor: Leonel Pereira

Received: 31 July 2023
Revised: 9 August 2023
Accepted: 13 August 2023
Published: 15 August 2023

1. Peculiarity of the Northern Adriatic Sea

The Northern Adriatic Sea is particularly vulnerable to nutrient pollution due to its unique characteristics, including shallow depths, limited water exchange, and the influence of river discharges. The Northern Adriatic Sea (Figure 1), covering an area of 19,000 km^2 with an average depth of 35 m, receives freshwater input from numerous rivers and streams, including the Po River (1500 m^3 s^{-1}, [1]), which contributes more than 50% of the external nutrient input [2–4]. The Adriatic's salinity is lower than that of the Mediterranean because the Adriatic collects one-third of the freshwater flowing into the Mediterranean Sea [5].

The Po watershed is very large (75,000 km^2, [1]) and includes highly developed Italian regions with a resident population of 16 million people [6,7]. This area is intensively cultivated with heavy use of natural and artificial fertilizers [8]. Being one of the most densely populated and agriculturally productive areas in Europe [9], the Po Valley significantly influences the nitrogen (N) and phosphorus (P) input to the Northern Adriatic Sea, affecting the productivity and trophic dynamics of the whole Mediterranean basin [10–13]. When these N and P inputs enter the sea, they fuel the growth of phytoplankton and macroalgae. However, when present in excessive amounts, they can lead to eutrophication. This process in the Adriatic Sea is significantly influenced by seasonal circulation patterns [14–16]. In winter, the anticlockwise superficial and deeper circulation along the western coasts (Figure 2) brings in nutrient-rich waters from the Po River [5,17–19] and other rivers originating in the Apennines [20–23]. These colder and less saline waters flow close to the Italian coast throughout the water column.

Figure 1. Map of the Mediterranean Sea with details of the Northern Adriatic and the hydrographic basin of the Po Valley (bottom left corner: Lat. 43.324050 N, Lon. 6.547654 E; top right corner: Lat. 46.537197 N, Lon. 16.027382 E; Source: Google Maps).

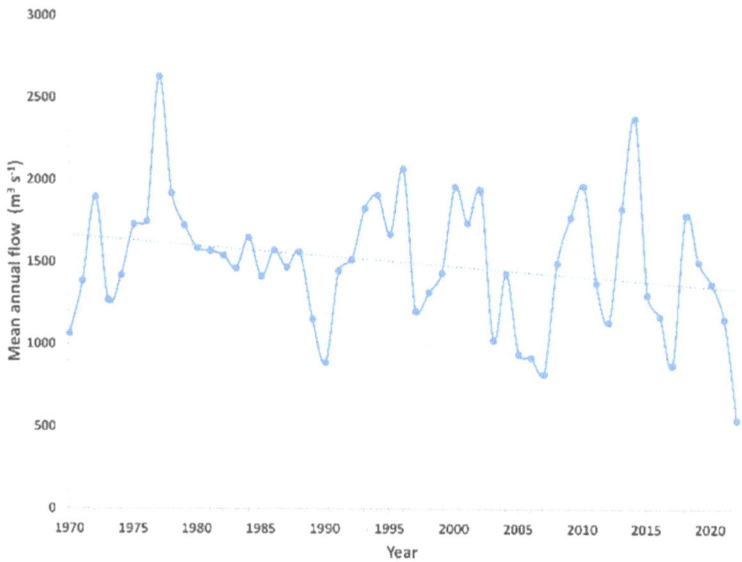

Figure 2. Trends of the daily Po River discharge (m^3 s^{-1}) from 1970 to 2022, with the significant long-term linear regression. Data from the Pontelagoscuro station available through the Regional Open Data Portal https://simc.arpae.it/dext3r/ (accessed on 1 July 2023), provided by the Agenzia Regionale per la Protezione dell'Ambiente of the Emilia Romagna Region (ARPAE-ER).

In contrast, during the summer season, the hydrodynamic conditions change, resulting in weaker currents and a series of clockwise and anticlockwise gyres, as well as strong water column stratification, with colder waters from the previous winter near the bottom [12,24–29]. As a result, nutrient concentrations in the surface waters increase, creating favorable conditions for the growth of phytoplankton and macroalgae. Following the cyclonic circulation, the waters of the Mediterranean Sea flow into the Adriatic Sea from the eastern side of the Strait of Otranto while on the western side, they flow from the Adriatic to the Mediterranean Sea [30–34].

We used long-term data from the Northern Adriatic Sea, in conjunction with a comprehensive literature review, to illustrate the impact of anthropic factors during this post-industrial century on the future state of the continental shelf. Moreover, our investigation aims to assess the responses of aquatic ecosystems to the persistent impacts of climate change.

2. Influence of Climate Change in the Northern Mediterranean Area

Several authors have demonstrated the Mediterranean's particular sensitivity to the North Atlantic Oscillation (NAO). However, the influence of the NAO on the Mediterranean exhibits heterogeneity and distinct trends in two regions: the Northern Mediterranean (southern Europe) and Central/Northern Europe. These regions demonstrate contrasting patterns to the NAO trend [35–39]. An analysis of rainfall data collected in the Mediterranean spanning two centuries, namely the XIX and XX centuries, was conducted, and the trends over the Italian territory starting from the 1970s were analyzed. The findings revealed a negative trend in rainfall, particularly pronounced during the winter season. This decrease in rainfall was more prominent in northern Italy compared to the southern regions. Moreover, the study documented an increase in the occurrence of more intense rainfall events during the analyzed period [36,38].

In recent decades, the fluvial loads of rivers in the Mediterranean Sea have undergone significant changes, particularly in the last two to three decades (see the Po River discharge in Figure 3). These changes have had more pronounced and evident effects in the Mediterranean Sea due to two main factors: firstly, the constrained availability of water resources in the Mediterranean region, and secondly, the increased anthropic pressure, especially in coastal areas and the Po Valley. The construction and utilization of river dams have evolved rapidly since the 1950s [40], substantially altering the natural outflow of Mediterranean rivers. The widespread use of these dams for irrigation and various other purposes has led to profound modifications in the fluvial dynamics of the region's rivers. The second crucial factor contributing to these changes is the ongoing climatic shifts in the Mediterranean region. These changes have been extensively documented through the monitoring programs that have recorded variations since the latter part of the 20th century. Additionally, modeling processes have forecasted an increase in drought occurrences, primarily attributed to the rise in temperatures [41–45]. As we progress through this century, we can anticipate significant effects, including substantial reductions in river water discharges [6]. These climatic changes are expected to further compound the challenges posed to the fluvial loads of Mediterranean rivers and have far-reaching implications for the region's hydrological systems.

Grilli et al. [46] confirmed that during the first two decades of the 21st century, the river flows of the Po River have exhibited a continuous decrease, along with variable trends in flood events. Specifically, the average daily flow of the Po River was observed to be 12% lower in the period from 2006 to 2015 compared to the earlier period from 1971 to 2005. Notably, the annual mean flow of the Po River experienced significant changes due to the occurrence of persistent drought periods. Moreover, the frequency of flow rates higher than 3000 $m^3 s^{-1}$ declined between 2006 and 2015 (the latter period considered in the analysis), and such higher flow events were more concentrated in certain months, namely February, April, November, and December. In contrast, previous decades experienced a greater number of these events that were spread over a larger period throughout the year.

Different authors have reported significant changes in various aspects of marine ecosystems [47,48]. These changes include alterations in the composition of plankton communities, shifts in the distribution of macrobenthos, the introduction and spreading of non-indigenous species, changes in the total biomass of target demersal fishes, and fluctuations in the catches of small pelagic fish [49–57].

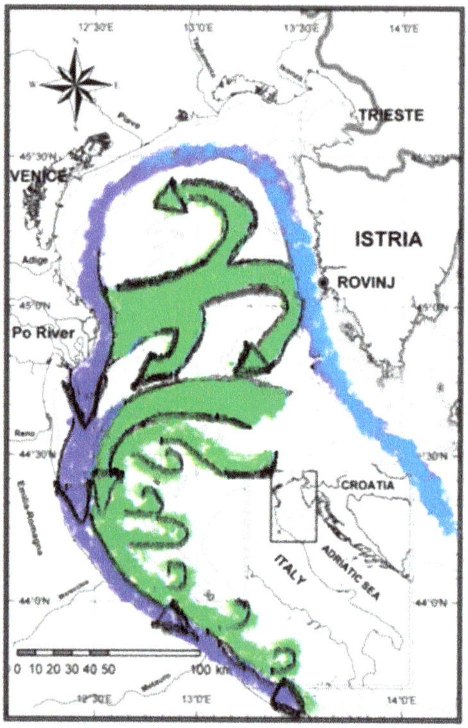

Figure 3. Simplified description of the seasonal surface circulation in the Northern Adriatic Sea, image adapted from [15] in agreement with [57,58]; blue (WAC, Western Adriatic current) and light blue (EAC, Eastern Adriatic current) for winter conditions; green for summer conditions, with eddies along the western coast.

3. Nitrogen and Phosphorus Conditioning in the Sea

The period of significant social changes, robust post-World War II industrial development, and the subsequent globalization of trade from the 1950s to the 1990s led to a substantial increase in the consumption of N and P for various purposes. Notably, N was extensively used as a fertilizer in agriculture, and its subsequent mineralization in surface and ground water, rather than industrial discharges into the atmosphere and urban wastewater, resulted in a substantial input of N watercourses eventually reaching the sea [8]. This pattern was also observed in the Northern Adriatic basin. As for P, its primary uses were prevalent in domestic environments and industry [8]. N and P are two elements that exhibit different behaviors once emitted into terrestrial and aquatic environments. The fate of P in the soil is primarily determined by chemical processes, such as adsorption/desorption and dissolution/precipitation. In contrast, the fate of N is primarily governed by biological processes including mineralization, nitrification, and denitrification [59].

Indeed, the extensive use of P and N during the latter half of the 20th century, along with the effects of climate change, has had significant implications for the Northern Adriatic Sea. The impact of climate change has become particularly evident in this region, starting from the end of the 20th century [36,60]. Notably, these changes are characterized by intense precipitation events, especially in spring and autumn, which significantly influence river outflows, leading to the transport of freshwater and suspended sediment into the sea [18,61,62]. As a consequence, these freshwater and sediment inputs carry pollutants, including N and P [19,23,63,64]. The combination of nutrient inputs from anthropogenic sources and climate-induced changes in hydrological patterns has contributed to alterations in nutrient concentrations and dynamics within the Northern Adriatic Sea.

In Italy, the total use of P in domestic detergents was prevented through Ministerial Decree n. 413 of 13 September 1988, which took effect from January 1989. Previously, from 1985 to 1988, there were other gradual measures implemented to reduce the use of P. Subsequently, in 2012, Europe adopted restrictive measures through Regulation 259/2012/EC, setting limitations for P of 0.5 g/dose in domestic laundry detergents and 0.3 g/dose in domestic dishwasher detergents. These regulatory measures have significantly contributed to the reduction of P inflow in surface watercourses up to the sea. As a result of this reduction, observable effects were recorded in the coastal strip with particular evidence in the Western North Adriatic Sea.

The predictions made by Caggiati & Ferrari [9] anticipated a decrease in P levels by the year 2020. Subsequent observations from the late 1990s to the 2000s showed a leveling off of P inputs into the sea, followed by a downward trend in the first two decades of the 21st century. However, the same trend was not observed for N. Despite the study by Caggiati & Ferrari [9] and subsequent works (e.g., [46]), N levels did not exhibit evident decreases, except in the strictly coastal strip around 3 km from the coast, where observations by Ricci et al. [23] demonstrated lower river flows and consequently reduced N and P loads. Over the years, the annual runoff of the Po River has exhibited significant fluctuations on a multi decadal time scale, with an overall decrease (−33% [65]) observed in all rivers of the Northern Adriatic when comparing recent discharges to those before the 1980s. Notably, during the dry years 2005–2007 (the last period of data analysis considered), the Northern Adriatic ecosystem experienced a considerable reduction in river water flows and nutrient loads compared to previous years characterized by medium-high regimes, as confirmed by Cozzi & Giani [65].

The decreasing temporal trend of N and P levels observed in the Adriatic Sea is consistent with trends seen in other European seas that are characterized by shallow waters (continental shelf) and are strongly influenced by freshwater inputs. For example, similar trends have been observed in the coastal zone of Baltic Sea as well [66].

The excess accumulation of N and P has significantly accelerated the expansion of eutrophication in these seas. In the Northern Adriatic Sea, the eutrophication phenomenon has been perceived as a pollutant due to the excessive proliferation of algae, which have degraded water quality, benthic habitats, and community structures [67–69].

In particular, offshore of the Po Delta and along the western coast, the process of oxygen consumption (hypoxia/anoxia) was accentuated during the summer when waters are warm and calm, and stratification of the water column occurs. This leads to widespread and persistent oxygen deficiencies in bottom waters, resulting in the suffering of benthic communities [3,70–73]. Another important phenomenon that occurred concomitant with eutrophication was the excessive proliferation of mucilage, which was documented especially in the late 1980s and in the early 2000s [3,7,25,74–79].

4. Final Considerations, and Future Directions

The Northern Adriatic Sea has experienced significant changes in eutrophication pressure over the past decades. From the 1970s to the mid-1980s, eutrophication increased, but it reversed course in the 2000s. This shift can be attributed to the combination of factors. Firstly, there was a reduction in anthropogenic impact, mainly due to a substantial decrease in phosphorus loads. Additionally, climatic changes played a role, leading to a decline in atmospheric precipitation and, subsequently, a decrease in runoff into the Northern Adriatic Sea. During the period 2000–2009, there was a significant decrease in the Po River outflow, which caused altered circulation patterns. As a consequence, there was a more frequent inflow of high salinity and oligotrophic water from the central Adriatic, resulting in a change in salinity and nutrient budget of the entire Northern Adriatic [15,33]. A four-decadal data set was analyzed to gain a better perspective of the driving pressures of phytoplankton in the Northern Adriatic. It was found that spring is the more critical season for the overproduction of organic matter [16,80,81]. Consequently, during the period 2000–2009, the intensity of phytoplankton blooms diminished, as evidenced by the decrease in chlorophyll *a* concentration [82]. Kužat et al. [83] analyzed spatio-temporal distribution data extracted from a long-term phytoplankton in situ monitoring data set conducted by monthly to quarterly sampling between the years 1978 and 2020 offshore in the Northern Adriatic Sea. The analysis revealed hydrodynamic regime shifts in the system which had altered the phytoplankton abundance, community composition, and seasonal cycle. Notably, appearance and frequency alterations of some prominent taxa (e.g., *Pseudo-nitzschia delicatissima* group and *Pseudo-nitzschia seriata* group) were detected.

Casabianca et al. [84] described the changes in phytoplankton time series along the western coast of the Northern Adriatic Sea, showing a regular annual period in the composition structure of the coastal strip. The research indicates that the phytoplankton composition is influenced by various environmental factors, including physical variables and nutrient availability. The persistence of species in phytoplankton communities over the long term is normally affected by factors such as nutrient loads, climate change, and human pressures, while still maintaining a stabile structure. In the Northern Adriatic Sea, this may impact the succession of functional groups of species and overall productivity of the ecosystem. However, due to stochastic environmental and climate changes, as well as enhanced environmental pressures, the time series have lost their regularity, causing changes in the succession and species composition of phytoplankton communities. Neri et al. [85] conducted a comparison between one coastal strip station and one offshore station along a transect along the western coast. The Western Adriatic Current (WAC, [14,20,86]), combined with river inflows, exerts a direct impact on coastal stations, resulting in heightened variability, as illustrated in the comparison with an offshore station. This disparity is particularly evident during winter and autumn, whereas similarity increases during the summer months when river waters spread outwards into the open sea. Owing to its oligotrophic nature, the offshore phytoplankton community demonstrated a greater level of biodiversity in contrast to observations along the coastline, where more prominent algal blooms tend to take place.

Over the past four decades, a gradual depletion of N and P has been observed along the coastal strip of the Northern Adriatic [23,87], compared to the previous period (1969–1992) [80]. This trend is expected to continue in the coming decades [19,79], due to lower flows from rivers with minor anthropic impact, resulting in reduced N and P loads. The study conducted by Ivančić et al. [88] on alkaline phosphatase activity during the decade 2004–2013 emphasized the significance of anthropogenic P and the effects of its continuous depletion along the coast. This reduction is attributed to lower river inputs and a delayed mixing of the water column in autumn, which is caused by the rise in sea temperatures. However, according to [46], the N/P ratio in the open sea of the Northern Adriatic Sea is not likely to undergo substantial changes. This is attributed to the unique circulation patterns of the Northern Adriatic Sea, characterized by slow turnover, and influenced by specific factors such as the Bora [72,89–93] and the Scirocco [94,95]. Additionally, the presence of freshwater promotes the reproduction of plankton and supports the entire trophic chain up to nekton, making the Northern Adriatic one of the most abundant fishing grounds in the Mediterranean.

One example of coastal phytoplankton adaption is showed in [96], which demonstrates how the *Ostreopsis* cf. *ovata* becomes highly competitive in P-limited water. Indeed, *Ostreopsis* has the ability to filter and store allochthonous organics, preventing the loss of autochthonous organics, and to rapidly process them. This process could explain why *Ostreopsis* is able to thrive in low-P environments as noted recently in the Northern Adriatic, particularly on the Conero Riviera [20,97]. Xing et al. [98] have also verified that other species of microalgae could take advantage of organic matter, for example during strong storm events involving wastewater flow along the coastal strip as described in [99] and [100]. Kužat et al. [83] also highlight how even micro phytoplankton communities respond with morphological and physiological adaptations to the reduction in P.

Moreover, changes in the zooplankton community have also been observed because zooplankton act as a link between primary producers of organic matter and the higher-order consumers, and they provide grazing control on phytoplankton blooms [54,101]. Mozetič et al. [102] show that while in a regime of reduced resources and biomass, plankton community control can occur seasonally, either through climatic or anthropogenic forcing.

The decline in ground fish, large predators, and small pelagics during 1980–1990 has also been attributed to both overfishing and the reduction in eutrophication [49,103,104]. In fact, the excessive abundance of fish in this sea attributed to the strong eutrophication has led to excessive fishing. As a result, the biological resource has now been reduced, and this will likely lead to a leveling of the resource equivalent to other seas. However, the North Adriatic Sea will remain particularly abundant in fish [103].

Table 1 presents the datasets available in the literature and analysed in this study.

Table 1. Datasets in the Northern Adriatic Sea.

Dataset	Time Period (Years)	References
River load	1970–2022	https://simc.arpae.it/dext3r/ (accessed on 1 July 2023) [5,45,64]
Nitrogen and phosphorus in the sea	1972–2019	[2–4,23,70]
Meteorological data	1850–2003	[36,38,39]
Plankton	1978–2020	[13,15,83–85]
Oceanographic data	1911–2019	[14,23,28,46,73,79,87,94]

5. Conclusions

It is of paramount importance for future studies to continue investigating the consequences of climate change and human impacts on the marine food chain, particularly in the Western Northern Adriatic coastal strip which serves as the main nursery area for many species of fish (e.g., *Solea solea*, *Mullus barbatus*, *Pagellus erythrinus*, *Squilla mantis*, *Engraulis encrasicolus*, and *Eledone moschata* [105–108]).

The recent basin evolution with a significant positive trend in nitrate concentrations in offshore seawater in almost all seasons during the decade 2006–2015 occurred despite a decreasing trend in N concentration in the Po River water. The rise in N/P ratios in river water and the excess of N in the marine environment, which cannot be utilized by phytoplankton due to P limitation, can explain this phenomenon. The change in the hydrodynamic regime was one of the main factors that altered phytoplankton abundance, community composition, and seasonal cycles.

The changes in nutrient dynamics indicate that the productivity of the Northern Adriatic Sea has been largely affected in the last 40 years by the combination of river water discharges and human activities. Further changes can be expected in the future, particularly in the case of a further reductions in anthropogenic phosphorus loads. Furthermore, another UE directive (91/676/EEC) aims to regulate the use of N in agriculture, but specific measures have not yet been adopted by many EU countries, including Italy.

Studies in this direction will be indispensable in the management of fisheries and the development of aquaculture, which are important policy concerns, such as in the EU, where the development of aquaculture is currently supported at the expense of industrial sea fishing. Additionally, the findings from these studies can inform policy decisions and help in designing targeted measures to mitigate the impacts of eutrophication and other environmental stressors. Implementing effective nutrient management strategies, reducing anthropogenic inputs, and promoting sustainable fishing practices are key steps towards safeguarding the Adriatic Sea's ecological integrity and ensuring the continued availability of its marine resources for future generations.

Author Contributions: Conceptualization, M.M. and F.G.; writing—original draft preparation, M.M. and F.G.; writing—review and editing, M.M. and F.G. All authors have read and agreed to the published version of the manuscript.

Funding: This research received no external funding.

Institutional Review Board Statement: Not applicable.

Informed Consent Statement: Not applicable.

Data Availability Statement: Not applicable.

Conflicts of Interest: The authors declare no conflict of interest.

References

1. Cati, L. *Idrografia e Idrologia Del Po*; Istituto poligrafico e Zecca dello Stato: Rome, Italy, 1981.
2. Degobbis, D.; Gilmartin, M.; Revelante, N. An annotated nitrogen budget calculation for the northern Adriatic Sea. *Mar. Chem.* **1986**, *20*, 159–177. [CrossRef]
3. Degobbis, D.; Gilmartin, M. Nitrogen, Phosphorus, and Biogenic Silicon Budgets for the Northern Adriatic Sea. *Oceanol. Acta* **1990**, *13*, 31–45.
4. Djakovac, T.; Supić, N.; Aubry, F.B.; Degobbis, D.; Giani, M. Mechanisms of hypoxia frequency changes in the northern Adriatic Sea during the period 1972–2012. *J. Mar. Syst.* **2015**, *141*, 179–189. [CrossRef]
5. Raicich, F. On the fresh water balance of the Adriatic Sea. *J. Mar. Syst.* **1996**, *9*, 305–319. [CrossRef]
6. Ludwig, W.; Dumont, E.; Meybeck, M.; Heussner, S. River discharges of water and nutrients to the Mediterranean and Black Sea: Major drivers for ecosystem changes during past and future decades? *Prog. Oceanogr.* **2009**, *80*, 199–217. [CrossRef]
7. Degobbis, D. Increased eutrophication of the northern Adriatic sea: Second act. *Mar. Pollut. Bull.* **1989**, *20*, 452–457. [CrossRef]
8. Viaroli, P.; Soana, E.; Pecora, S.; Laini, A.; Naldi, M.; Fano, E.A.; Nizzoli, D. Space and time variations of watershed N and P budgets and their relationships with reactive N and P loadings in a heavily impacted river basin (Po river, Northern Italy). *Sci. Total Environ.* **2018**, *639*, 1574–1587. [CrossRef]
9. Caggiati, G.; Ferrari, F.; Piazza, D. Classificazione Qualitativa Dei Principali Corpi Idrici Superficiali Del Bacino Del Fiume Po. In *I Quaderni del Piano di Bacino*; Autorità di bacino del fiume Po: Parma, Italy, 1997.
10. Gilmartin MRevelante, N. The Phytoplankton of the Adriatic Sea: Standing Crop and Primary Production. *Thalassia Jugosl.* **1983**, *19*, 173–188.
11. Pettine, M.; Barra, I.; Campanella, L.; Millero, F.J. Effect of metals on the reduction of chromium (VI) with hydrogen sulfide. *Water Res.* **1998**, *32*, 2807–2813. [CrossRef]
12. Ravaioli, M.; Bergami, C.; Riminucci, F.; Langone, L.; Cardin, V.; Di Sarra, A.; Aracri, S.; Bastianini, M.; Bensi, M.; Bergamasco, A.; et al. The RITMARE Italian Fixed-Point Observatory Network (IFON) for marine environmental monitoring: A case study. *J. Oper. Oceanogr.* **2016**, *9*, s202–s214. [CrossRef]
13. Morabito, G.; Mazzocchi, M.G.; Salmaso, N.; Zingone, A.; Bergami, C.; Flaim, G.; Accoroni, S.; Basset, A.; Bastianini, M.; Belmonte, G.; et al. Plankton dynamics across the freshwater, transitional and marine research sites of the LTER-Italy Network. Patterns, fluctuations, drivers. *Sci. Total Environ.* **2018**, *627*, 373–387. [CrossRef]
14. Artegiani, A.; Paschini, E.; Russo, A.; Bregant, D.; Raicich, F.; Pinardi, N. The Adriatic Sea General Circulation. Part II: Baroclinic Circulation Structure. *J. Phys. Oceanogr.* **1997**, *27*, 1515–1532. [CrossRef]
15. Marić, D.; Kraus, R.; Godrijan, J.; Supić, N.; Djakovac, T.; Precali, R. Phytoplankton response to climatic and anthropogenic influences in the north-eastern Adriatic during the last four decades. *Estuar. Coast. Shelf Sci.* **2012**, *115*, 98–112. [CrossRef]
16. Marini, M.; Campanelli, A.; Sanxhaku, M.; Zoran, K.; Mattia, B. Grilli Federica Late Spring Characterization of Different Coastal Areas of the Adriatic Sea. *Acta Adriat.* **2015**, *56*, 27–46.
17. Supić, N.; Grbec, B.; Vilibić, I.; Ivančić, I. Long-term changes in hydrographic conditions in northern Adriatic and its relationship to hydrological and atmospheric processes. *Ann. Geophys.* **2004**, *22*, 733–745. [CrossRef]
18. Campanelli, A.; Grilli, F.; Paschini, E.; Marini, M. The influence of an exceptional Po River flood on the physical and chemical oceanographic properties of the Adriatic Sea. *Dyn. Atmos. Oceans* **2011**, *52*, 284–297. [CrossRef]
19. Dunić, N.; Supić, N.; Sevault, F.; Vilibić, I. The northern Adriatic circulation regimes in the future winter climate. *Clim. Dyn.* **2023**, *60*, 3471–3484. [CrossRef]
20. Marini, M.; Fornasiero, P.; Artegiani, A. Variations of Hydrochemical Features in the Coastal Waters of Monte Conero: 1982–1990. *Mar. Ecol.* **2002**, *23*, 258–271. [CrossRef]
21. Campanelli, A.; Fornasiero, P.; Marini, M. Physical and Chemical Characterization of the Water Column in the Piceno Coastal Area (Adriatic Sea). *Fresenius Environ. Bull.* **2004**, *13*, 430–435.
22. Spagnoli, F.; De Marco, R.; Dinelli, E.; Frapiccini, E.; Frontalini, F.; Giordano, P. Sources and Metal Pollution of Sediments from a Coastal Area of the Central Western Adriatic Sea (Southern Marche Region, Italy). *Appl. Sci.* **2021**, *11*, 1118. [CrossRef]

23. Ricci, F.; Capellacci, S.; Campanelli, A.; Grilli, F.; Marini, M.; Penna, A. Estuarine, Coastal and Shelf Science Long-term dynamics of annual and seasonal physical and biogeochemical properties: Role of minor river discharges in the North-western Adriatic coast. *Estuar. Coast. Shelf Sci.* **2022**, *272*, 107902. [CrossRef]
24. Grilli, F.; Paschini, E.; Precali, R.; Russo, A.; Supić, N. Circulation and horizontal fluxes in the northern Adriatic Sea in the period June 1999–July 2002. Part I: Geostrophic circulation and current measurement. *Sci. Total Environ.* **2005**, *353*, 57–67. [CrossRef] [PubMed]
25. Grilli, F.; Marini, M.; Degobbis, D.; Ferrari, C.R.; Fornasiero, P.; Russo, A.; Gismondi, M.; Djakovac, T.; Precali, R.; Simonetti, R. Circulation and horizontal fluxes in the northern Adriatic Sea in the period June 1999–July 2002. Part II: Nutrients transport. *Sci. Total Environ.* **2005**, *353*, 115–125. [CrossRef] [PubMed]
26. Marini, M.; Russo, A.; Paschini, E.; Grilli, F.; Campanelli, A. Short-term physical and chemical variations in the bottom water of middle Adriatic depressions. *Clim. Res.* **2006**, *31*, 227–237. [CrossRef]
27. Marini, M.; Maselli, V.; Campanelli, A.; Foglini, F.; Grilli, F. Role of the Mid-Adriatic deep in dense water interception and modification. *Mar. Geol.* **2016**, *375*, 5–14. [CrossRef]
28. Vilibić, I.; Pranić, P.; Denamiel, C. North Adriatic Dense Water: Lessons learned since the pioneering work of Mira Zore-Armanda 60 years ago. *Acta Adriat.* **2023**, *64*, 53–78. [CrossRef]
29. de Mendoza, F.P.; Schroeder, K.; Langone, L.; Chiggiato, J.; Borghini, M.; Giordano, P.; Verazzo, G.; Miserocchi, S. Deep-water hydrodynamic observations of two moorings sites on the continental slope of the southern Adriatic Sea (Mediterranean Sea). *Earth Syst. Sci. Data* **2022**, *14*, 5617–5635. [CrossRef]
30. Zore-Armanda, M. Water exchange between the Adriatic and the Eastern Mediterranean. *Deep. Sea Res. Oceanogr. Abstr.* **1969**, *16*, 171–178. [CrossRef]
31. Artegiani, A.; Gacic, M.; Michelato, A.; Kovacevic, V.; Russo, A.; Paschini, E.; Scarazzato, P.; Smircic, A. The Adriatic sea hydrography and circulation in spring and autumn (1985–1987). *Deep. Sea Res. Part II Top. Stud. Oceanogr.* **1993**, *40*, 1143–1180. [CrossRef]
32. Civitarese, G.; Gačić, M.; Lipizer, M.; Borzelli, G.L.E. On the impact of the Bimodal Oscillating System (BiOS) on the biogeochemistry and biology of the Adriatic and Ionian Seas (Eastern Mediterranean). *Biogeosciences* **2010**, *7*, 3987–3997. [CrossRef]
33. Grilli, F.; Marini, M.; Book, J.W.; Campanelli, A.; Paschini, E.; Russo, A. Flux of nutrients between the middle and southern Adriatic Sea (Gargano-Split section). *Mar. Chem.* **2013**, *153*, 1–14. [CrossRef]
34. Gacic, M.; Civitarese, G.; Ursella, L. Spatial and Seasonal Variability of Water and Biogeochemical Fluxes in the Adriatic Sea. In *The Eastern Mediterranean as a Laboratory Basin for the Assessment of Contrasting Ecosystems*; Malanotte-Rizzoli, P., Eremeev, V.N., Eds.; NATO Science Series; Springer: Dordrecht, The Netherlands, 1999; Volume 51, pp. 1–23.
35. Hurrell, J.W.; Van Loon, H. Decadal Variations in Climate Associated with the North Atlantic Oscillation. *Clim. Chang.* **1997**, *36*, 301–326. [CrossRef]
36. Brunetti, M.; Maugeri, M.; Nanni, T. Changes in total precipitation, rainy days and extreme events in northeastern Italy. *Int. J. Clim.* **2001**, *21*, 861–871. [CrossRef]
37. Türkeş, M.; Erlat, E. Influences of the North Atlantic Oscillation on Precipitation Variability and Changes in Turkey. *Nuovo C Della Soc. Ital. Di Fis. C* **2006**, *29*, 117–135. [CrossRef]
38. Brunetti, M.; Maugeri, M.; Monti, F.; Nanni, T. Temperature and precipitation variability in Italy in the last two centuries from homogenised instrumental time series. *Int. J. Clim.* **2006**, *26*, 345–381. [CrossRef]
39. Hurrell, J.W. Decadal Trends in the North Atlantic Oscillation: Regional Temperatures and Precipitation. *Science* **1995**, *269*, 676–679. [CrossRef] [PubMed]
40. Margat, J.; Treyer, S. L'eau Des Méditerranéens: Situation et Perspectives. In *Water for the People of the Mediterranean: Present and Future*; Plan Bleu PNUE/PAM, MAP Technical Reports Series N° 158; PAM: Athènes, Greece, 2004.
41. Gibelin, A.-L.; Déqué, M. Anthropogenic climate change over the Mediterranean region simulated by a global variable resolution model. *Clim. Dyn.* **2003**, *20*, 327–339. [CrossRef]
42. Milly, P.C.D.; Dunne, K.A.; Vecchia, A.V. Global pattern of trends in streamflow and water availability in a changing climate. *Nature* **2005**, *438*, 347–350. [CrossRef] [PubMed]
43. Norrant, C.; Douguédroit, A. Monthly and daily precipitation trends in the Mediterranean (1950–2000). *Theor. Appl. Clim.* **2005**, *83*, 89–106. [CrossRef]
44. Christensen, J.H.; Hewitson, B.; Busuioc, A.; Chen, A.; Gao, X.; Held, I.; Jones, R.; Kolli, R.K.; Kwon, W.T.; Laprise, R.; et al. Regional Climate Projections. In *Climate Change 2007: The Physical Science Basis. Contribution of Working Group I to the Fourth Assessment Report of the Intergovernmental Panel on Climate Change*; Solomon, S., Qin, D., Manning, M., Chen, Z., Marquis, M., Averyt, K.B., Tignor, M.M.B., Miller, H.L., Eds.; UNEP: Nairobi, Kenya, 2007; ISBN 0521705967.
45. Giorgi, F.; Lionello, P. Climate change projections for the Mediterranean region. *Glob. Planet. Chang.* **2008**, *63*, 90–104. [CrossRef]
46. Grilli, F.; Accoroni, S.; Acri, F.; Aubry, F.B.; Bergami, C.; Cabrini, M.; Campanelli, A.; Giani, M.; Guicciardi, S.; Marini, M.; et al. Seasonal and Interannual Trends of Oceanographic Parameters over 40 Years in the Northern Adriatic Sea in Relation to Nutrient Loadings Using the EMODnet Chemistry Data Portal. *Water* **2020**, *12*, 2280. [CrossRef]
47. Aubry, F.B.; Berton, A.; Bastianini, M.; Bertaggia, R.; Baroni, A.; Socal, G. Seasonal Dynamics of Dinophysis in Coastal Waters of the NW Adriatic Sea (1990–1996). *Bot. Mar.* **2000**, *43*, 423–430. [CrossRef]

48. Acri, F.; Bastianini, M.; Aubry, F.B.; Camatti, E.; Boldrin, A.; Bergami, C.; Cassin, D.; De Lazzari, A.; Finotto, S.; Minelli, A.; et al. A long-term (1965–2015) ecological marine database from the LTER-Italy Northern Adriatic Sea site: Plankton and oceanographic observations. *Earth Syst. Sci. Data* **2020**, *12*, 215–230. [CrossRef]
49. Giani, M.; Djakovac, T.; Degobbis, D.; Cozzi, S.; Solidoro, C.; Umani, S.F. Recent changes in the marine ecosystems of the northern Adriatic Sea. *Estuar. Coast. Shelf Sci.* **2012**, *115*, 1–13. [CrossRef]
50. Grbec, B.; Dulcic, J.; Morovic, M. Long-term changes in landings of small pelagic fish in the eastern Adriatic-possible influence of climate oscillations over the Northern Hemisphere. *Clim. Res.* **2002**, *20*, 241–252. [CrossRef]
51. Coll, M.; Santojanni, A.; Palomera, I.; Arneri, E. Ecosystem assessment of the North-Central Adriatic Sea: Towards a multivariate reference framework. *Mar. Ecol. Prog. Ser.* **2010**, *417*, 193–210. [CrossRef]
52. Cabrini, M.; Cerino, F.; de Olazabal, A.; Di Poi, E.; Fabbro, C.; Fornasaro, D.; Goruppi, A.; Flander-Putrle, V.; Francé, J.; Gollasch, S.; et al. Potential transfer of aquatic organisms via ballast water with a particular focus on harmful and non-indigenous species: A survey from Adriatic ports. *Mar. Pollut. Bull.* **2018**, *147*, 16–35. [CrossRef]
53. Petrocelli, A.; Antolić, B.; Bolognini, L.; Cecere, E.; Cvitković, I.; Despalatović, M.; Falace, A.; Finotto, S.; Iveša, L.; Mačić, V.; et al. Port Baseline Biological Surveys and seaweed bioinvasions in port areas: What's the matter in the Adriatic Sea? *Mar. Pollut. Bull.* **2019**, *147*, 98–116. [CrossRef]
54. Mozetič, P.; Cangini, M.; Francé, J.; Bastianini, M.; Aubry, F.B.; Bužančić, M.; Cabrini, M.; Cerino, F.; Čalić, M.; D'Adamo, R.; et al. Phytoplankton diversity in Adriatic ports: Lessons from the port baseline survey for the management of harmful algal species. *Mar. Pollut. Bull.* **2019**, *147*, 117–132. [CrossRef]
55. Azzurro, E.; Sbragaglia, V.; Cerri, J.; Bariche, M.; Bolognini, L.; Ben Souissi, J.; Busoni, G.; Coco, S.; Chryssanthi, A.; Fanelli, E.; et al. Climate change, biological invasions, and the shifting distribution of Mediterranean fishes: A large-scale survey based on local ecological knowledge. *Glob. Chang. Biol.* **2019**, *25*, 2779–2792. [CrossRef]
56. Zenetos, A.; Gofas, S.; Morri, C.; Rosso, A.; Violanti, D.; García Raso, J.E.; Cinar, M.E.; Almogi-Labin, A.; Ates, A.S.; Azzurro, E.; et al. Alien species in the Mediterranean Sea by 2012. A contribution to the application of European Union's Marine Strategy Framework Directive (MSFD). Part 2. Introduction trends and pathways. *Mediterr. Mar. Sci.* **2012**, *13*, 328–352. [CrossRef]
57. Flander-Putrle, V.; Francé, J.; Mozetič, P. Phytoplankton Pigments Reveal Size Structure and Interannual Variability of the Coastal Phytoplankton Community (Adriatic Sea). *Water* **2021**, *14*, 23. [CrossRef]
58. Poulain, P.M.; Cushman-Roisin, B. Circulation. In *Physical Oceanography of the Adriatic Sea*; Cushman-Roisin, B., Gačić, M., Poulain, P.-M., Artegiani, A., Eds.; Kluwer Academic: Dordrecht, The Netherlands, 2001; pp. 67–109, ISBN 1-4020-0225-4.
59. Edwards, A.C.; Withers, P.J.A. Soil phosphorus management and water quality: A UK perspective. *Soil Use Manag.* **1998**, *14*, 124–130. [CrossRef]
60. Basterretxea, G.; Font-Muñoz, J.S.; Salgado-Hernanz, P.M.; Arrieta, J.; Hernández-Carrasco, I. Patterns of chlorophyll interannual variability in Mediterranean biogeographical regions. *Remote Sens. Environ.* **2018**, *215*, 7–17. [CrossRef]
61. Zoppini, A.; Ademollo, N.; Bensi, M.; Berto, D.; Bongiorni, L.; Campanelli, A.; Casentini, B.; Patrolecco, L.; Amalfitano, S. Impact of a river flood on marine water quality and planktonic microbial communities. *Estuar. Coast. Shelf Sci.* **2019**, *224*, 62–72. [CrossRef]
62. Toffolo, M.M.; Grilli, F.; Prandi, C.; Goffredo, S.; Marini, M. Extreme Flooding Events in Coastal Lagoons: Seawater Parameters and Rainfall over A Six-Year Period in the Mar Menor (SE Spain). *J. Mar. Sci. Eng.* **2022**, *10*, 1521. [CrossRef]
63. Guzzella, L.; Paolis, A. Polycyclic Aromatic Hydrocarbons in Sediments of the Adriatic Sea. *Mar. Pollut. Bull.* **1994**, *28*, 159–165. [CrossRef]
64. Viganò, L.; Guzzella, L.; Marziali, L.; Mascolo, G.; Bagnuolo, G.; Ciannarella, R.; Roscioli, C. The last 50 years of organic contamination of a highly anthropized tributary of the Po River (Italy). *J. Environ. Manag.* **2023**, *326*, 116665. [CrossRef]
65. Cozzi, S.; Giani, M. River water and nutrient discharges in the Northern Adriatic Sea: Current importance and long term changes. *Cont. Shelf Res.* **2011**, *31*, 1881–1893. [CrossRef]
66. Lønborg, C.; Markager, S. Nitrogen in the Baltic Sea: Long-term trends, a budget and decadal time lags in responses to declining inputs. *Estuar. Coast. Shelf Sci.* **2021**, *261*, 107529. [CrossRef]
67. Provini, A.; Crosa, G.; Marchetti, R. Nutrient Export from the Po and Adige River Basins over the Last 20 Years. In *Proceedings of the International Conference on Marine Coastal Eutrophication, Bologna, Italy, 21–24 March 1992*; Vollenweider, R.A., Marchetti, R., Viviani, R., Eds.; Elsevier: Amsterdam, The Netherlands, 1992; Volume 2, pp. 291–313.
68. Rabalais, N.N.; Turner, R.E.; Díaz, R.J.; Justić, D. Global change and eutrophication of coastal waters. *ICES J. Mar. Sci.* **2009**, *66*, 1528–1537. [CrossRef]
69. Howarth, R.; Chan, F.; Conley, D.J.; Garnier, J.; Doney, S.C.; Marino, R.; Billen, G. Coupled biogeochemical cycles: Eutrophication and hypoxia in temperate estuaries and coastal marine ecosystems. *Front. Ecol. Environ.* **2011**, *9*, 18–26. [CrossRef] [PubMed]
70. Degobbis, D.; Smodlaka, N.; Pojed, I.; Škrivanić, A.; Precali, R. Increased eutrophication of the Northern Adriatic sea. *Mar. Pollut. Bull.* **1979**, *10*, 298–301. [CrossRef]
71. Justić, D.; Legović, T.; Rottini-Sandrini, L. Trends in oxygen content 1911–1984 and occurrence of benthic mortality in the northern Adriatic Sea. *Estuar. Coast. Shelf Sci.* **1987**, *25*, 435–445. [CrossRef]
72. Boldrin, A.; Carniel, S.; Giani, M.; Marini, M.; Aubry, F.B.; Campanelli, A.; Grilli, F.; Russo, A. Effects of bora wind on physical and biogeochemical properties of stratified waters in the northern Adriatic. *J. Geophys. Res. Atmos.* **2009**, *114*, C08S92. [CrossRef]

73. Kralj, M.; Lipizer, M.; Čermelj, B.; Celio, M.; Fabbro, C.; Brunetti, F.; Francé, J.; Mozetič, P.; Giani, M. Hypoxia and dissolved oxygen trends in the northeastern Adriatic Sea (Gulf of Trieste). *Deep. Sea Res. Part II Top. Stud. Oceanogr.* **2019**, *164*, 74–88. [CrossRef]
74. Smodlaka, N. Primary production of the organic matter as an indicator of the eutrophication in the northern Adriatic sea. *Sci. Total Environ.* **1986**, *56*, 211–220. [CrossRef]
75. Degobbis, D.; Fonda-Umani, S.; Franco, P.; Malej, A.; Precali, R.; Smodlaka, N. Changes in the northern Adriatic ecosystem and the hypertrophic appearance of gelatinous aggregates. *Sci. Total Environ.* **1995**, *165*, 43–58. [CrossRef]
76. Precali, R.; Giani, M.; Marini, M.; Grilli, F.; Ferrari, C.R.; Pečar, O.; Paschini, E. Mucilaginous aggregates in the northern Adriatic in the period 1999–2002: Typology and distribution. *Sci. Total Environ.* **2005**, *353*, 10–23. [CrossRef]
77. Giani, M.; Berto, D.; Zangrando, V.; Castelli, S.; Sist, P.; Urbani, R. Chemical characterization of different typologies of mucilaginous aggregates in the Northern Adriatic Sea. *Sci. Total Environ.* **2005**, *353*, 232–246. [CrossRef]
78. Faganeli, J.; Kovač, N.; Leskovšek, H.; Pezdič, J. Sources and fluxes of particulate organic matter in shallow coastal waters characterized by summer macroaggregate formation. *Biogeochemistry* **1995**, *29*, 71–88. [CrossRef]
79. Solidoro, C.; Bastianini, M.; Bandelj, V.; Codermatz, R.; Cossarini, G.; Canu, D.M.; Ravagnan, E.; Salon, S.; Trevisani, S. Current state, scales of variability, and trends of biogeochemical properties in the northern Adriatic Sea. *J. Geophys. Res. Atmos.* **2009**, *114*, C07S91. [CrossRef]
80. Degobbis, D.; Precali, R.; Ivancic, I.; Smodlaka, N.; Fuks, D.; Kveder, S. Long-term changes in the northern Adriatic ecosystem related to anthropogenic eutrophication. *Int. J. Environ. Pollut.* **2000**, *13*, 495–533. [CrossRef]
81. Bernardi Aubry, F.; Berton, A.; Bastianini, M.; Socal, G.; Acri, F. Phytoplankton succession in a coastal area of the NW Adriatic, over a 10-year sampling period (1990–1999). *Cont. Shelf Res.* **2004**, *24*, 97–115. [CrossRef]
82. Djakovac, T.; Degobbis, D.; Supić, N.; Precali, R. Marked reduction of eutrophication pressure in the northeastern Adriatic in the period 2000–2009. *Estuar. Coast. Shelf Sci.* **2012**, *115*, 25–32. [CrossRef]
83. Kužat, N.; Pfannkuchen, D.M.; Tanković, M.S.; Baričević, A.; Ivančić, I.; Vrana, I.; Gašparović, B.; Pfannkuchen, M. Morpho-physiological adaptations of Leptocylindrus aporus and L. hargravesii to phosphate limitation in the northern Adriatic. *Sci. Rep.* **2022**, *12*, 2687. [CrossRef]
84. Casabianca, S.; Capellacci, S.; Ricci, F.; Scardi, M.; Penna, A. A phytoplankton time series in the Northwestern Adriatic Sea: Structure and dynamics of the assemblages in a coastal ecosystem. *Estuar. Coast. Shelf Sci.* **2022**, *278*, 108109. [CrossRef]
85. Neri, F.; Romagnoli, T.; Accoroni, S.; Ubaldi, M.; Garzia, A.; Pizzuti, A.; Campanelli, A.; Grilli, F.; Marini, M.; Totti, C. Phytoplankton communities in a coastal and offshore stations of the northern Adriatic Sea approached by network analysis and different statistical descriptors. *Estuar. Coast. Shelf Sci.* **2023**, *282*, 108224. [CrossRef]
86. Marini, M.; Jones, B.H.; Campanelli, A.; Grilli, F.; Lee, C.M. Seasonal variability and Po River plume influence on biochemical properties along western Adriatic coast. *J. Geophys. Res. Atmos.* **2008**, *113*, 1–18. [CrossRef]
87. Mozetič, P.; Solidoro, C.; Cossarini, G.; Socal, G.; Precali, R.; Francé, J.; Bianchi, F.; De Vittor, C.; Smodlaka, N.; Umani, S.F. Recent Trends Towards Oligotrophication of the Northern Adriatic: Evidence from Chlorophyll a Time Series. *Estuaries Coasts* **2010**, *33*, 362–375. [CrossRef]
88. Ivančić, I.; Kraus, R.; Najdek, M.; Cozzi, S. Ecological Importance of Alkaline Phosphatase Activity in Changing Marine Environmental Conditions. *Water* **2021**, *13*, 2750. [CrossRef]
89. Lee, C.M.; Askari, F.; Book, J.; Carniel, S.; Cushman-Roisin, B.; Dorman, C.; Doyle, J.; Flament, P.; Harris, C.K.; Jones, B.H.; et al. Northern Adriatic response to a wintertime bora wind event. *Eos* **2005**, *86*, 157–165. [CrossRef]
90. Pullen, J.; Doyle, J.D.; Haack, T.; Dorman, C.; Signell, R.P.; Lee, C.M. Bora event variability and the role of air-sea feedback. *J. Geophys. Res. Atmos.* **2007**, *112*. [CrossRef]
91. Raicich, F.; Malačič, V.; Celio, M.; Giaiotti, D.; Cantoni, C.; Colucci, R.R.; Čermelj, B.; Pucillo, A. Extreme air-sea interactions in the Gulf of Trieste (North Adriatic) during the strong Bora event in winter 2012. *J. Geophys. Res. Ocean.* **2013**, *118*, 5238–5250. [CrossRef]
92. Zore-Armanda, M. Natural Characteristics and Climatic Changes of the Adriatic Sea. *Acta Adriat.* **1991**, *32*, 567–586.
93. Zore-Armanda, M.; Gačić, M. Effects of Bora on the Circulation in the North Adriatic. *Ann. Geophys.* **1987**, *5B*, 93–102.
94. Artegiani, A.; Bregant, D.; Paschini, E.; Pinardi, N.; Raicich, F.; Russo, A. The Adriatic Sea General Circulation. Part I: Air–Sea Interactions and Water Mass Structure. *J. Phys. Oceanogr.* **1997**, *27*, 1492–1514. [CrossRef]
95. Franco, P.; Michelato, A. Northern Adriatic Sea: Oceanography of the basin proper and of the western coastal zone. In *Marine Coastal Eutrophication*; Elsevier: Amsterdam, The Netherlands, 1992; pp. 35–62. [CrossRef]
96. Accoroni, S.; Tartaglione, L.; Iacovo, E.D.; Pichierri, S.; Marini, M.; Campanelli, A.; Dell'Aversano, C.; Totti, C. Influence of environmental factors on the toxin production of Ostreopsis cf. ovata during bloom events. *Mar. Pollut. Bull.* **2017**, *123*, 261–268. [CrossRef]
97. Accoroni, S.; Glibert, P.M.; Pichierri, S.; Romagnoli, T.; Marini, M.; Totti, C. A conceptual model of annual Ostreopsis cf. ovata blooms in the northern Adriatic Sea based on the synergic effects of hydrodynamics, temperature, and the N:P ratio of water column nutrients. *Harmful Algae* **2015**, *45*, 14–25. [CrossRef]
98. Xing, D.; Li, X.; Wang, Y.; Deng, S.; Jin, C.; Zhao, Y.; Guo, L. The comprehensive impact of phosphorus sources on microalgae biochemical metabolism and phosphorus transformation. *J. Water Process. Eng.* **2023**, *51*, 103477. [CrossRef]

99. Romei, M.; Lucertini, M.; Renzoni, E.E.; Baldrighi, E.; Grilli, F.; Manini, E.; Marini, M.; Iagnemma, L. A Detention Reservoir Reduced Combined Sewer Overflows and Bathing Water Contamination Due to Intense Rainfall. *Water* **2021**, *13*, 3425. [CrossRef]
100. Manini, E.; Baldrighi, E.; Ricci, F.; Grilli, F.; Giovannelli, D.; Intoccia, M.; Casabianca, S.; Capellacci, S.; Marinchel, N.; Penna, P.; et al. Assessment of Spatio-Temporal Variability of Faecal Pollution along Coastal Waters during and after Rainfall Events. *Water* **2022**, *14*, 502. [CrossRef]
101. Kiørboe, T. Turbulence, Phytoplankton Cell Size, and the Structure of Pelagic Food Webs. *Adv. Mar. Biol.* **1993**, *29*, 1–72. [CrossRef]
102. Mozetič, P.; Francé, J.; Kogovšek, T.; Talaber, I.; Malej, A. Plankton trends and community changes in a coastal sea (northern Adriatic): Bottom-up vs. top-down control in relation to environmental drivers. *Estuar. Coast. Shelf Sci.* **2013**, *115*, 138–148. [CrossRef]
103. Coll, M.; Santojanni, A.; Palomera, I.; Arneri, E. Food-web changes in the Adriatic Sea over the last three decades. *Mar. Ecol. Prog. Ser.* **2009**, *381*, 17–37. [CrossRef]
104. Colloca, F.; Cardinale, M.; Maynou, F.; Giannoulaki, M.; Scarcella, G.; Jenko, K.; Bellido, J.M.; Fiorentino, F. Rebuilding Mediterranean fisheries: A new paradigm for ecological sustainability. *Fish Fish.* **2011**, *14*, 89–109. [CrossRef]
105. Scarcella, G.; Grati, F.; Raicevich, S.; Russo, T.; Gramolini, R.; Scott, R.D.; Polidori, P.; Domenichetti, F.; Bolognini, L.; Giovanardi, O.; et al. Common sole in the northern and central Adriatic Sea: Spatial management scenarios to rebuild the stock. *J. Sea Res.* **2014**, *89*, 12–22. [CrossRef]
106. Colloca, F.; Garofalo, G.; Bitetto, I.; Facchini, M.T.; Grati, F.; Martiradonna, A.; Mastrantonio, G.; Nikolioudakis, N.; Ordinas, F.; Scarcella, G.; et al. The Seascape of Demersal Fish Nursery Areas in the North Mediterranean Sea, a First Step Towards the Implementation of Spatial Planning for Trawl Fisheries. *PLoS ONE* **2015**, *10*, e0119590. [CrossRef]
107. Canu, D.M.; Laurent, C.; Morello, E.B.; Querin, S.; Scarcella, G.; Vrgoc, N.; Froglia, C.; Angelini, S.; Solidoro, C. *Nephrops norvegicus* in the Adriatic Sea: Connectivity modeling, essential fish habitats, and management area network. *Fish. Oceanogr.* **2021**, *30*, 349–365. [CrossRef]
108. Zorica, B.; Pešić, A.; Marčeta, B.; Mandić, M.; Piccinetti, C.; Isajlović, I.; Vrgoč, N.; Keč, V. A review of reproduction biology and spawning/ nursery grounds of the most important Adriatic commercial fish species in the last two decades. *Acta Adriat.* **2020**, *61*, 89–100. [CrossRef]

applied sciences

MDPI

Review

The Role of Sediment Records in Environmental Forensic Studies: Two Examples from Italy of Research Approaches Developed to Address Responsibilities and Management Options

Luca Giorgio Bellucci * and Silvia Giuliani

Institute of Marine Sciences, National Research Council of Italy, 40129 Bologna, Italy; silvia.giuliani@bo.ismar.cnr.it
* Correspondence: luca.bellucci@bo.ismar.cnr.it; Tel.: +39-0516398851

Abstract: The coupling of scientific evidence from sediment cores with historical information represents an effective way to reconstruct and quantify recent anthropogenic impacts in transitional and marine-coastal areas. These are both key points pertaining to studies that aim at establishing the responsibility for envi ronmental pollution. Good practices for the selection of sampling sites and specific survey techniques are fundamental to understand pollution histories and dynamics, together with reliable dating methods and analytical procedures. In addition, a certain degree of flexibility and willingness to explore different research pathways is necessary, particularly when unexpected questions arise from scientific data or from requests posed by authorities in charge of preliminary investigations or court debates. In this paper, two different study cases are reviewed, and the approaches developed to tackle with specific issues are presented. Its main purpose is both to explain study paths undertaken to answer challenging scientific-legal questions and to provide examples for developing countries that present similar risks of uncontrolled industrialization. Results were used in preliminary investigations or court debates for the attribution of responsibility for environmental pollution to past or present industrial managements. In addition, they were fundamental for other studies aiming at implementing models that simulate the fate and distribution of contaminants and human exposure. In the Augusta Harbor, an integrated approach merged archive information, bathymetry, and high-resolution seismic profiles with the results of an independent tracer (hexachlorobenzene, HCB). This approach helped attribute the presence of high surficial Hg concentrations to resuspension and redistribution of deep sediments caused by dredging and maritime traffic and not to active outfalls. In the Venice Lagoon, an extensive literature search supported analytical results for the correct identification of industrial processes responsible for the contamination by polychlorinated dibenzo-p-dioxins and furans (PCDD/Fs) in the canals of the Porto Marghera Industrial Area. In addition, sedimentary profiles of PCDD/Fs in local salt marshes ("barene") recorded well the events relative to the industrial development and management of the area reported by historical documents, confirming their potential for this kind of investigation.

Keywords: sediment records; management; environmental forensic; industrial pollution; Venice Lagoon; Augusta Bay

Citation: Bellucci, L.G.; Giuliani, S. The Role of Sediment Records in Environmental Forensic Studies: Two Examples from Italy of Research Approaches Developed to Address Responsibilities and Management Options. *Appl. Sci.* **2023**, *13*, 6999. https://doi.org/10.3390/app13126999

Academic Editors: Angel DelValls and Mauro Marini

Received: 21 April 2023
Revised: 24 May 2023
Accepted: 7 June 2023
Published: 9 June 2023

1. Introduction

Industrial activities and urban settlements (mostly located close to seas and freshwater bodies) have the potential to discharge contaminants in great quantities through their wastes, especially when major environmental laws or control procedures are missing or insufficient [1–5]. Scientists can then be interested in assessing both contamination levels and input chronologies of such endangered areas. This interest might result not only from curiosity-driven motivations, but as a specific request to provide expert contribution to policy and decision makers who are in charge of health security measures and/or need to assess the responsibility of past and present industrial managements [6–8].

Accreting aquatic sediments can be extremely important sources of information, mostly because they have the potential to behave as natural archives through the retention of significant information about the surrounding environment at the time of their deposition and accumulation [6,9–14]. Specifically, their role as recorders of past chemical signatures makes them ideal matrices for the reconstruction of relatively recent anthropogenic impacts [15–21]. Indeed, the comparison with historical information allows the attribution of observed changes along the core profile to specific activities/management choices. This is possible when geochronological studies supply reliable analytical results [22,23]. It is evident that this kind of information is highly relevant in environmental forensic studies, in particular during court debates that aim at recognizing when environmental pollution started, how it changed, and who has to be held responsible for it. However, sampling strategies should be carefully designed and carried out, in order to avoid the use of inappropriate techniques that can partially or totally obscure the potential information preserved in sediments. For example, sampling devices need to be carefully chosen, since the collection of undisturbed sedimentary profiles is a mandatory request for studies that deal with historic reconstructions of events and changes in the dynamic of the environment. When used in forensic studies that deal with recent events, gravity coring systems are to be favoured with respect to vibro-corers: the first enable the preservation of the sediment water interface, thus ensuring the collection of the recent-most depositions and the least disturbance to subsurface sediment layers. However, gravity coring systems are generally not able to collect sandy sediments that then act as an impenetrable barrier for the sampling of deeper layers [6]. Identifying suitable sampling sites is then an important critical step and is generally obtained through bathymetry and high-resolution seismic surveys prior to sampling campaigns [6,24–26]. These surveys are able to reveal areas where sediment accumulation is more regular or where bottom morphology is affected by slumping, mass accumulation, presence of gas, etc. [6,24–26].

Coupled to the collection of sediments, the acquisition of historical information retained in private and public archives is fundamental. It can help identify: (1) where specific point sources were located in the past; (2) if and how productions have changed through time; (3) which areas sustained substantial human-induced reworking of bottom sediments [27,28].

The number of sampling sites should be chosen in accordance with the purpose of the research: higher for a general site characterization of surficial sediments to check sources, distribution, and level of pollution; lower for more specific questions, such as sediment accumulation rates and dates, chronologies, fluxes, processes, and trends.

In this paper, results of the scientific approaches designed and applied for the study of environmental pollution issues in two Italian industrial areas are summarized. The locations are the Augusta Harbor in Southern Italy and the Venice Lagoon with its nearby industrial area of Porto Marghera in northeastern Italy. In both cases, the coupling of scientific evidence from sediment cores with historical information managed to provide a clear view of polluting sources, timings, and mechanisms.. Results were either debated in courts (Venice Lagoon) or used in the framework of preliminary investigations (Augusta Harbor). They represent perfect examples of how sediment records can be used as tools for environmental forensic studies. Scientific data have been published in specific papers on international journals [27,28], and this article aims to underline the study paths undertaken to answer the scientific-legal questions that have arisen during the investigations, sometimes not foreseen and often quite challenging. The acquired experience can be put at the service of developing countries that present similar risks of uncontrolled industrialization.

2. Study Areas and Historical Information

2.1. The Augusta Harbor

The Augusta Bay (Figure 1a) is a semi-enclosed natural basin occupying around 30 km of the eastern coast of Sicily (Ionian Sea, southern Italy) and covering an area of approximately 4000 ha [29,30]. The Augusta Harbor is the northern sector of the bay and was sheltered in the early 1960s with the construction of artificial breakwaters. Since then, the harbor is connected to the bay and the open sea only through two artificial inlets, called Scirocco (in the south, 300 m wide and 13 m deep) and Levante (in the east, 400 m wide and 40 m deep), respectively [30]. The Augusta Harbor is about 24 km^2 and the average water depth is approximately 15 m, with the maximum depth (40 m) in correspondence of the Levante inlet [31]. Exchanges with the open sea are mainly driven by tidal fluctuations and are consequently correlated with the entry/exit of tidal flows and relative amplitudes. Open sea waters enter predominantly from the Levante inlet (mean speed of 18 cm s^{-1}), then flow northward, parallel to the dam, slow and constantly with their pace and exit at the Scirocco inlet at a speed of 5–6 cm s^{-1} [32]. The north-western sector of the harbor is locally influenced by the poor seasonal outflows of the Mulinello, Marcellino, and Cantera streams but their contribution is generally poor, with localized hydrodynamic effects [31]. Due to its confinement, the hydrodynamic regime of the Augusta Harbor is scarce, favoring the accumulation of nutrients and pollutants in sediments and the occurrence of eutrophication episodes [31,33].

The Augusta Harbor hosts one of the most important ports of the Mediterranean Sea, and many industries are located along its western part, including petroleum and petrochemical plants. The industrial development started in the 1950s with the construction of oil refineries and proceeded in the 1960s with the onset of a Mercury Cell Chlor-Alkali Plant (MCCAP), almost contemporary to the construction of the outer breakwaters [34]. Sediment dredging, disposal, and partial nourishment of the western coast have further modified the pristine environment of the bay. Indeed, 3/5 of the port surface (approximately 14.4 km^2) have been dredged so far while the dredged material might have been released over a large area of the bay on the way to offshore dumping sites (yellow areas; Figure 1a) [27].

Since the 1970s the area has become internationally recognized as a polluted environment from industrial and petrochemical plants, but also from agriculture and urban wastes, with consequent risks to the ecosystem and human health [30,34,35]. Specifically, mercury contamination from the MCCAP in the southern sector of the Augusta Harbor (Figure 1a) has produced sediment concentrations beyond international levels in parts of the area and is by far the most important environmental issue [27,34,36]. It is estimated that ~500 t of Hg have been discharged to the Augusta Harbor [30,34]. In the 1980s the demercurization and waste treatment plants started working in response to the Italian anti-pollution law 319 of 1976 (aka "Legge Merli", in Italian), and discharged treated wastewater outside the harbor. Owing to the high state of environmental degradation, this area was included in the 2002 National Remediation Plan by the Italian Environmental Ministry, then the MCCAP closed in 2005 with the removal of the Hg cells [32,34].

In the early 2000s, scientific evidence became necessary to support legal investigations against the industries that caused the pollution and who could potentially be expected to contribute to remediation costs. In 2003, researchers from CNR-ISMAR Bologna started work to provide such evidence.

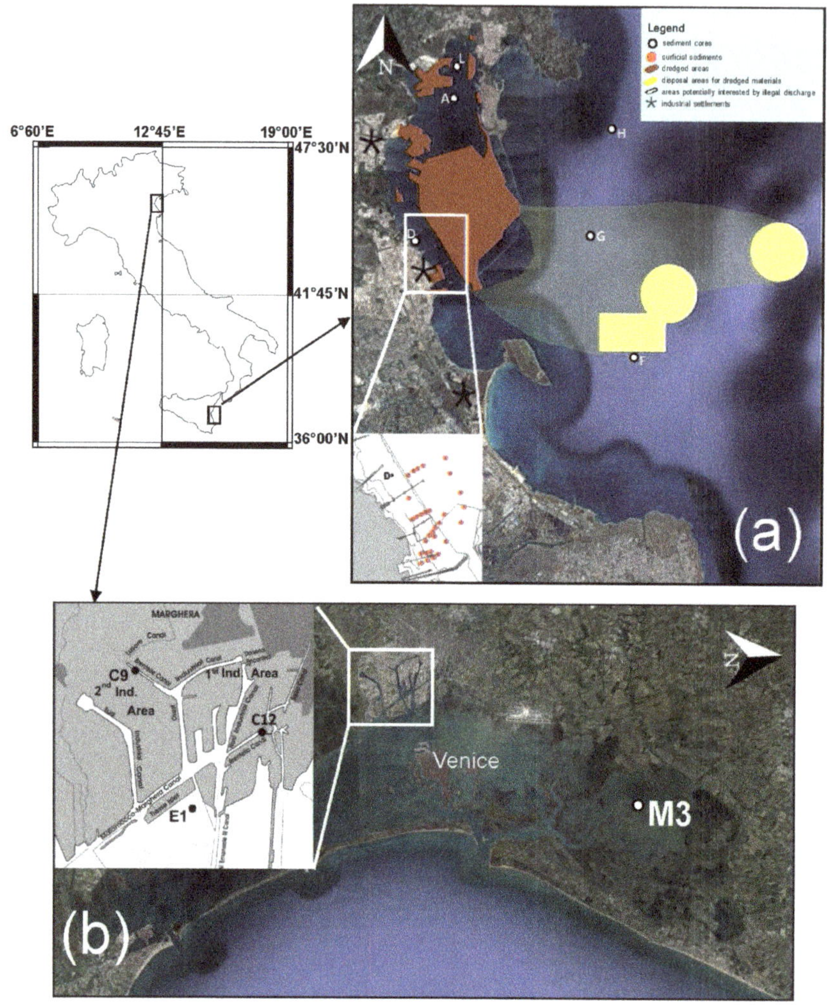

Figure 1. Location of study areas and sampling sites: (**a**) Augusta Bay with sampling sites, dredged and disposal areas, and locations of industrial settlements; (**b**) Venice Lagoon, with a snapshot over the Industrial Area of Porto Marghera. Satellite images are taken from Google maps.

2.2. The Venice Lagoon

The Venice Lagoon (Figure 1b) is part of the Adriatic Sea in northeastern Italy. It has a surface area of 550 km^2, an average depth of 0.6 m, and a salinity of about 25–36‰ [37]. The water exchange with the Adriatic Sea (1.6–5.2 $\times$ 10^8 m^3 day^{-1}) occurs through tidal cycles at three entrance channels that divide the lagoon into three basins (Lido, Malamocco, and Chioggia), while freshwater (30.7 m^3 s^{-1}) enters the lagoon through about 24 tributaries [37,38]. The morphology is characterized by a network of canals of various depths, mud flats, tidal marshes, and islets that were first created during the maximum Holocene transgression 6000 years BP [39]. The sediment deposition rate in the lagoon is quite variable, and is higher in the northern sector, due to the supply from the Lido inlet [40,41]. Bottom sediments are composed almost exclusively of very fine sand and silt. Clay contribution is higher in the inner part of the lagoon and reflects the water dynamic: maximum at the inlets and within the main canals, and lower close to the mainland [38].

The Venice Lagoon is a unique shallow aquatic ecosystem whose complex morphology has been profoundly modified over time by human interventions. In particular, during the last century it has experienced a general degradation with the deepening of tidal flats (present rates of sediment loss are estimated to be 0.8 $Mm^3\ y^{-1}$) and the reduction of salt marsh areas (from 168 km^2 in 1930 to 60 km^2 in 2002) [39,42]. In addition, the development of economic activities on the mainland have led to further changes: (1) the construction of the 1st Industrial Area of Porto Marghera at the beginning of the 20th century; (2) the excavation of the navigation canal Vittorio Emanuele III in 1930; (3) the building of the trans-lagoon bridge in 1934; (4) the development of the 2nd Industrial Area after World War 2; (5) the excavation of the navigation canal Malamocco Marghera, in 1969 [28]. More recently, the construction of the controversial *Modulo Sperimentale Elettromeccanico* (MOSE, Italian for Experimental Electromechanical Module) began in 2003 at the lagoon's inlets, with the aim of protecting the city of Venice from high tides (*Acqua alta* in Italian) through isolation from the adjacent Adriatic Sea [43] and is now fully operative.

Until the 1970s, Porto Marghera was one of the most important industrial areas in Italy, with plants for the production of chloro-alkali, sulfuric acid, ammonia, fibers, fertilizers from phosphorite, glass, coke, paints, detergents, firebricks, Al from bauxite, and the recovery of Zn and Cd from sphalerites [6,28,38]. The contamination of lagoon waters and sediments began soon after the first economic development, when industrial, urban, and agricultural discharges were disposed freely into the lagoon [41]. The situation became so serious that, at the end of the 1990s, a trial against managers of the petrochemical plants began. The defendants were charged with pollution and environmental disaster by polychlorodiphenyl dioxins and furans (PCDD/Fs) in the industrial area of Porto Marghera and in the whole Venice Lagoon. The main subject of the lawsuit was the production of vinylchloride monomer (VCM) that, at that time, was considered the principal industrial source of PCDD/Fs in this region. Researchers from CNR-ISMAR Bologna were involved for their expertise in order to provide scientific evidence in support or against this hypothesis, and the results of their studies served in the court hearing as a technical report.

3. Materials and Methods

3.1. Research Activity in the Augusta Bay Performed by ISMAR-CNR Bologna

A seismic survey in the Augusta Bay was carried out in October 2003 using a CHIRP II (Datasonic) sub-bottom profiler with the following operational characteristics: frequency sweep between 2 and 7 kHz; sampling interval of 125 ms; high resolution of a few centimeters; SeisPrho processing software [27,44]. Internal structures of the sediment layers were recognized and permitted to discriminate between undisturbed, laminated sediment layers, and chaotic sequences (Figure 2a). Bottom reflectivities were calculated in order to obtain a first approximation of grain size structures [44].

The sampling strategy for the definition of Hg contamination in sediment cores was designed according to seismic data, bottom reflectivities, and historical informationvi retrieved from local public (Port Authority, Municipality of Siracusa) and private (industries) archives relative to the dredging and dumping areas. As a result, much attention was paid during sampling to avoid the dredged and dumping areas or zones close to piers and/or influenced by docking and manoeuvring of ships. Core G (Figure 1a) was sampled on purpose in an area of supposed illegal dumping of dredged material from the harbor, in order to check the existence and effects of this practice. Sediment cores were collected with the gravity corer SW-104, patented by CNR and especially designed to preserve undisturbed sediment–water interfaces.

Newly arisen legal questions relative to the Hg sources in surficial sediments asked for a second sampling in 2005. Surficial sediments were sampled in the southern sector of the Augusta Harbor (Figure 1a) in order to be analysed for Hg and hexachlorobenzene (HCB) [27].

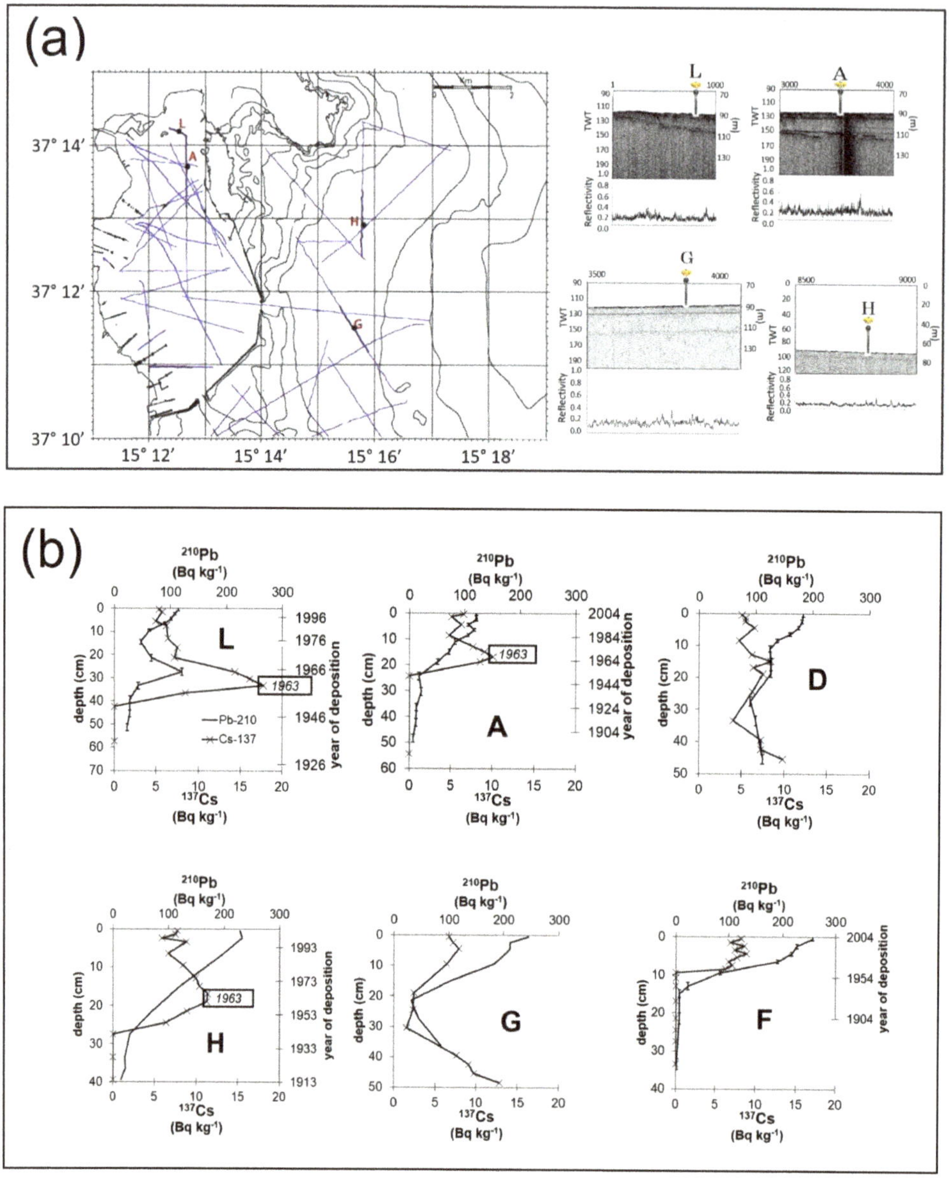

Figure 2. (**a**) Map of the seismic survey realized in the Augusta Bay and Harbor, together with CHIRP profiles and seismic surface reflectivities relative to L, A, H, and G core locations; (**b**) ^{210}Pb and ^{137}Cs activity–depth distributions for cores L, A, D, H, G, and F. Profiles are presented also vs. years of deposition, calculated from the S values. When detected, the ^{137}Cs 1963 peak is indicated. Modified from [27].

After collection, sediment cores were kept vertical (in order to avoid intra-layer mixing and intestitial water migration) while scanning for magnetic susceptibility (using a Bartington meter, on each centimeter every 10 s) and X-radiographies (with a radiogenic industrial tube Gilardoni CPX-91 160 equipped with a Be window and industrial camera films). These two parameters are useful for a first-rate approximation of internal structures and as a support for chronologies. Indeed, if highly magnetic tephra layers are observed, then a correlation with historical eruptions can be made. Cores were then extruded and carefully sectioned in 1–3 cm intervals with higher resolution at the top. Sediments were analysed for bulk dry density (assuming a particle density of 2.5 g cm^{-3}), organic matter content (% of C and N, $\delta^{13}C$ and δ^{15} N on a Carlo Erba CHNS Elemental Analyser), mineral composition (from 2° to 60° on a Philips PW 1710 CuK X-ray diffractometer, operating at 30 mA and 40 kV), grain size (through wet sieving and analyses with an X-ray Micromeritic SediGraph set on a 4–12 ϕ size range) and chronology. This latter was obtained through: (1) acid extraction and alpha spectrometry of ^{210}Po for ^{210}Pb determinations (assuming secular equilibrium), and (2) non-destructive gamma spectrometry of dry samples for ^{137}Cs. For ^{210}Pb, accuracy was estimated by repeated measurements of the standard reference material (SRM) IAEA 300 (Baltic Sea sediment) and the results were within uncertainties. Precision, calculated from independent analyses of the same sample, was 4.6%. For ^{137}Cs, accuracy was checked by counting the international SRM S7 IAEA Baltic Sea sediment, and the results were within uncertainties. Precision, estimated by repeated analyses of the same sample, ranged between 2.05 and 3.07%. Radionuclide profiles for cores L, A, D, H, G, and F are reported in Figure 2b.

Hg was extracted following the procedure outlined in the EPA 3051A REV.1 (https://www.epa.gov/sites/default/files/2015-12/documents/3051a.pdf, accessed on 6 June 2023), including drying at 40 °C, acid microwave extraction, centrifugation, and evaporation. Concentrations were determined following the method EPA 6010 C, by use of an inductively coupled plasma-atomic emission spectrometer (ICP-AES) Spectro Ciros. Reproducibility, obtained with repeated measurements of the same sample, was 8.2%. Recovery, defined with the SRM MESS-1, was 81%. The analytical detection limit was 0.1 mg kg^{-1}.

For HCB analyses, dried samples were spiked with $^{13}C_{12}$-HCB (CLM351; Cambridge Isotope Laboratories, Woburn, MA, USA) as internal standard, extracted by ASE 200 (DIONEX, Sunnyvale, CA, USA; with an n-hexane/dichloromethane solution), and cleaned up using the automatic system, Dioxin Prep (Fluid Management System Inc., Lexington, KY, USA), as defined by method EPA 8270 (https://19january2017snapshot.epa.gov/sites/production/files/2015-07/documents/epa-8270d.pdf, accessed on 6 June 2023). High resolution gas chromatography mass spectrometry (HRGC-MS) analyses were performed on an HP 6890 plus gas chromatograph coupled to a Micromass Autospec Ultima mass spectrometer. The SRM solution for quantification was EC1668 (Cambridge Isotope Laboratories, Woburn, MA, USA). Recovery (calculated with three $^{13}C_{12}$-labeled PCBs from SRM EC4979) was 86%, while reproducibility was 2.1%. The analytical detection limit was 1 mg kg^{-1}.

Hg and HCB concentrations were calculated with respect to dry weight. For the purpose of this paper, only results of radionuclide dating and chemical measurements (Hg and HCB) are summarized. Details of the other measurements can be found in [27].

3.2. Research Activity in the Venice Lagoon and the Porto Marghera Industrial Area Performed by ISMAR-CNR Bologna

A series of published papers and documents issued by public authorities were reviewed in order to define the state of the art of scientific studies. References are reported in detail in [28,38,45–48].

Sediment cores considered in this paper provide a comprehensive view of the potential that can be expressed by sedimentary records involved in environmental forensic studies: (1) cores E1 and M3 represent a lagoon site near the Tresse Islet sampled in 1996 and a salt marsh of the Northern lagoon collected in 1998, respectively; (2) core C9 was collected in 1996 in the 2nd Industrial Area, and (3) core C12 was collected in the same year in the 1st Industrial Area (Figure 1b).

Sediment cores were analysed for chronology and ancillary parameters, with the same methods reported in Section 3.2. Analysis performances were similar to those reported for chronologies.

The analysis of PCDD/Fs on freeze dried sediments included the spiking with a series of 15 ^{13}C-labelled PCDD/F congeners as internal standards, then a Soxhlet-extraction with toluene, as determined by the EPA 3540 method (https://www.epa.gov/sites/production/files/2015-12/documents/3540c.pdf, accessed on 6 June 2023). Then the analysis continued through clean-up and detection procedures defined by the HRGC-MS EPA 1613/94/revision b method (https://www.epa.gov/sites/default/files/2015-08/documents/method_1613b_1994.pdf, accessed on 6 June 2023), for the determination of 17 2,3,7,8 substituted congeners with an HP 6890 Plus gas-chromatograph coupled to a FINNIGAN MAT 9SS mass spectrometer. Five SRM solution injections (EDF 9999 Cambridge Isotope Laboratories, Woburn, MA, USA) were used for quantitative determinations, and recoveries were always between 40% and 120% [45]. Toxic equivalent quantities (TEQs) were calculated using the International Toxic Equivalent Factors (I-TEFs) [49]. The composition of PCDD/F mixtures was calculated as the permil (‰) contribution of each homologue group with respect to the total.

Arsenic was analyzed on freeze dried samples by atomic absorption spectrometry (AAS) after total dissolution. Accuracy was within the uncertainties for SRM NIST 2710 (Montana Soil), and precision was 7% [28].

4. Results and Discussion

4.1. Assessment of Hg Delivery to the Augusta Harbor: Direct Inputs or Resuspension?

The chlor-alkali plants discharged without treatments in the harbor until the 1970s, when a demercurization plant and the consortial biological depurator for waste treatments became operative, in response to national laws. In spite of that, high Hg levels are still measured in surficial sediments [30,34–36] implying the possibility of a recent contribution. As written in Section 2.1, this possibility was the focal point to be verified by the historical and scientific reconstruction here summarized.

Sediment chronologies defined for the cores collected in 2003 in the Augusta Bay and Harbor were of great interest for the purpose of this scientific investigation. For most sites, they covered time intervals sufficiently extended to include the industrial setting, its development, and decline (i.e., early 1960s–late 1980s) up to 2003 (Figure 3a). $^{210}Pb_{ex}$ derived accumulation rates (S, cm y^{-1}) were calculated with the assumption of constant flux of the radiotracer and constant sedimentation (CF-CS), thus providing an average S useful for a first approximation dating, but probably overestimated. Associated errors for S ranged from 0.01 cm·year^{-1} at the core top to 0.05 cm·year^{-1} at the core bottom [27]. Since no evaluation of the effect of bioturbation could be achieved, surficial levels were excluded from its calculation. Despite these difficulties, in most cases ^{210}Pb derived chronologies were confirmed at deeper layers by ^{137}Cs results (see Figure 2b) and other parameters (e.g., magnetic susceptibility peaks, HCB production history) [27]. S values in the Augusta Bay ranged from 0.21 cm year^{-1} (site F) to 0.42 cm year^{-1} (site H), in agreement with box core data from the same area [50] and determined that 50 years of industrial history of the area were represented by 11–21 cm thick sediment layers. S values in the Augusta Harbor were higher (from 0.43 cm year^{-1} at site A to 1.12 cm year^{-1} in the sub-surficial sediment of site L) therefore the time intervals represented by these cores are shorter, starting from the mid-1940s or early 1950s.

As expected, site G proved to be very peculiar, with an unrealistic high rate of 1.08 cm·year^{-1}, most likely influenced by the massive sediment input generated by the undocumented illegal dumping of dredged material on the way to the disposal areas. Further investigation through core sampling at different locations throughout the Augusta Bay were necessary to verify the extent and pathways of this illegal habit (Figure 1a) but it was beyond the scope of the presented research. Chronologies were also confirmed by magnetic susceptibility measurements related to documented Etna eruptions in 1892 [51].

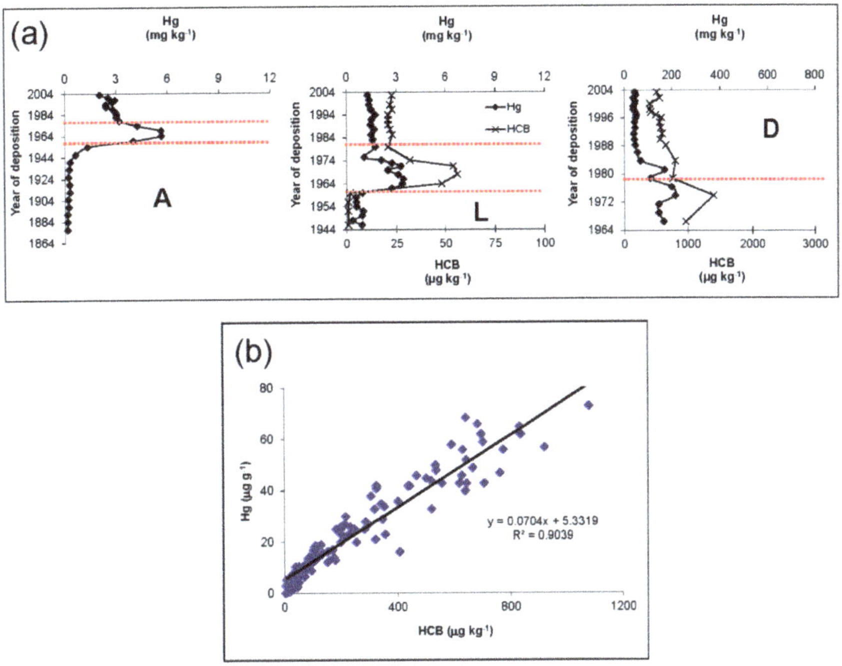

Figure 3. (**a**) Hg concentration vs. year of deposition in cores A, L, and D. HCB concentration profile for core L is also reported. The dotted lines encompass the period from the beginning of industrialization in the area (early 1960s) to the promotion of the first remediation measures (late 1970s); (**b**) Hg and HCB correlation in surficial samples collected in 2005 and in surficial layers of cores collected in 2003. The product moment and partial correlation coefficient R is 0.93, significant at $p < 0.0005$, and calculated with Statistica 7.0 StatSoft ©). Modified from [27].

Great variability was observed for Hg concentrations in sediments: values ranged from a minimum of 0.07 mg kg^{-1} to a maximum of 575 mg kg^{-1}; [27], due to the presence of point sources that strongly influenced nearby locations. Chronologies were reliable enough to allow the reconstruction of Hg historical trends in sediments of the Augusta Harbor. As an example, Figure 3a shows Hg temporal profiles in cores A, L, and D (Figure 1a). The sedimentary record of core A covered the entire 20th century and evidenced significant variations of Hg concentrations, passing from near background values in the pre-industrial period to ca. 6–8 mg kg^{-1} in the 1950s, at the peak of the industrial development of the area. Then, concentrations gradually and constantly decreased from the early 1970s, as the result of both technological improvements and remediation measures that came into use in that period [30–32,34–36]. However, the value measured at the sediment surface was still well above the natural background determined at the core bottom, as observed also in the box core sediments of the Augusta Bay [50]. In addition, such a surface value (0.25 mg kg^{-1}; Figure 3a) was still higher than the consensus-based threshold effect concentration of 0.18 mg kg^{-1} reported by [52], posing a serious threat to biota and people. The same trends and events were recorded with more detail at site L, as the last 50–60 years of industrial history in the area had been represented by a higher number of measurements. This chronological reconstruction was impossible for core D, due to a highly disturbed sedimentary record, but its proximity to the MCCAP in the southern sector of the Augusta Harbor explained the very high Hg values at depth, still high even at the surface (92 mg kg^{-1}; Figure 3a).

These high Hg values in surface sediments demanded further investigations because they could have been determined either by: (1) the presence of active industrial sources or

(2) the deposition of resuspended old, contaminated sediment layers. The first option is a serious violation of current regulations and, if verified, would have justified legal actions against the last managers of the industrial plants. Indeed, the Italian anti-pollution law 319 of 1976 came into effect in 1979 and from that moment graphite anodes present in Hg cells of the chloralkali plant were removed. This happened because the demercurization plant could not work with the sludge deriving from the graphite anodes [53]. No active source of Hg in the Augusta Harbor could then be tolerated from that moment on.

The second option was equally critical as it assigns the responsibility for the contamination of surficial sediments primarily to dredging activities and their management by the authorities in charge.

The new scientific challenge was then to establish which of the two was the process at work in the Augusta Harbor. In order to provide a reliable answer, an alternative tracer of industrial pollution with a perfectly time-constrained history in the study area had to be found. After further historical research, the choice fell on HCBs, whose input chronologies in the Augusta Bay were well known and documented, as these compounds are formed by graphite anodes in chloralkali plants, and as by-product of the manufacture of tetrachloroethylene (TCE) [54]. In the industrial area of Augusta, TCE production started in 1959 and ceased definitively in 1979, taking the following nine years for complete decommissioning. In addition, production wastes were discharged in the Augusta Harbor with no treatment or disposal procedures until 1976. In the last 40 years, no HCB sources have been active in the area [54].

The HCB temporal record for core L (Figure 3a) was consistent with the timing of TCE production in the Augusta Bay until the 1980s; values were close to detection limits in the preindustrial period, then increased from the early 1960s, reached the maximum around the beginning of the 1970s, and then abruptly decreased at the end of the 1970s. However, if sediments were to deposit undisturbed from that moment on, measured values had to be close to zero, since HCB inputs ceased definitively in the 1980s. On the contrary, measured values in the topmost section remained quite constant around 25 $\mu g\ kg^{-1}$ (Figure 3a). The highly significant correlation observed between Hg and HCB in surficial sediment samples (Figure 3b) was interpreted as evidence that the two contaminants were controlled in surface sediments by the same depositional process. This was likely linked to the mobilization of polluted sediments deposited in the period 1958–1980 by dredging and/or ship traffic. This process can also explain the lower values of ^{210}Pb observed in levels dated around the 1980s (Figure 2b), when dredging activities began [27]. For example, core L activities measured at 15 cm depth (dated with the CF-CS model around the 1980s; Figure 2b) are lower than those of the immediately deeper levels, suggesting the influence of relatively older sediments (and therefore with lower ^{210}Pb activities). These sediments add to those naturally settling in that area and have the effect of lowering the detected total activity. No evidence was found that could prove the existence of direct inputs from productions, at least in the northern sector of the Augusta Harbor at the time of sampling. The second option was then held responsible for the high Hg values at the sediment surface. Consequently, management recommendations derived by this research included the adoption of all possible measures to control the re-suspension and re-deposition of sediments whenever dredging activities were necessary or during ship transit [55,56]. Indeed, other research studies in the following years focused on the possibility that Hg discharged in the Augusta Harbor might be a primary source to the adjacent marine system through the outflow of bottom waters along its steep slope and gullies [57] and a secondary source through reworking, resuspension, and transport [58]. In addition, ref. [57] also recognized that a substantial contribution to Hg contamination in the Augusta Bay might result from past uncontrolled dredging activities that deposited part of the contaminated sediments on the slope system outside the harbor (Figure 1a). Additional risks posed by the high seismicity of the area were measured with the modelling of the sudden breaching and complete collapse of the system damming by a destructive earthquake [59]. The resulting dispersion of contaminants (using Hg as an example of those present in the study area) is expected to occur almost instantaneously at a large scale, suggesting that natural and artificial hazards are inextricably connected and demand innovative methodological approaches for planning appropriate risk reduction and policy management [59]. As the environmental persistence of Hg and HCB availability facilitates their bioaccumulation and affects the health status of marine organisms, the possible implications for environmental risk, pollutant transfer, and human health were investigated by other studies in the same years [35,60]. Specific analyses on Hg content and the Hg isotopic signatures of sediments, fish, and human hair shown as pelagic fish consumption is the principal transfer mechanism to the human population [35]. In addition, ref. [60] observed the progressive decrease of bioavailability and biological effects on local fauna over the last 20 years, is likely linked to the closure of many industrial facilities.

4.2. The origin of PCDD/F Contamination in the Industrial Area of Porto Marghera: A Compelling Scientific Investigation

The identification of specific industrial processes responsible for the presence of PCDD/Fs in the area of Porto Marghera (Venice) was the core of the trial against past and contemporary managers of the petrochemical plants during the late 1990s–early 2000s [6,28]. Research studies aiming at providing scientific evidences on this regard started from the observation of contaminant concentration distribution and the relative congener composition pattern (i.e., PCDD/F signatures or fingerprints) within industrial canals and lagoon environments. Two main PCDD/F fingerprints were identified (Figure 4a): (1) one characterized exclusively the core collected in the Lusore–Brentelle canal in the 2nd Industrial Area (core C9, Figure 1b), while (2) the other was found in the canals of the 1st Industrial Area (core C12 in the Brentella Canal, Figure 1b) and at sub-surficial depths in cores collected all over the lagoon (cores E1 and M3, Figure 1b). The first pattern was composed almost exclusively of octachlorodibenzofuran (OCDF) with no significant contribution by other furans and dioxins, while the second one was characterized by the presence also of other heavy furans (Figure 4a). In addition, samples collected in the 1st Industrial Area showed concentrations sensibly higher than those measured in the 2nd Industrial Area (core C9, Figure 1b). This first evidence testified the presence of two different sources of PCDD/Fs in the industrial canals of Porto Marghera, one in the 1st Industrial Area and another in the 2nd Industrial Area. Different types of industries have settled in the two industrial areas: chemical and electro-metallurgical industries were located mostly along the Nord Industrial Canal in the 1st Industrial Area, while the 2nd Industrial Area hosted petrochemical plants [61]. It was therefore legitimate to hypothesize that the two observed PCDD/F signatures were linked to these different industrial clusters. The questions arising from this first evidence were then as follows: (1) Were there in the 1st Industrial Area specific industrial metal processes that could have given levels of contamination and footprints of PCDD/Fs comparable to those measured in core C12 (Figure 4a)? and (2) Was it possible to recognize a similar link between industries located along the Lusore–Brentelle Canal (2nd Industrial Area) and the congener composition of dioxins and furans measured in its sediments (core C9; Figure 4a)? The results of the challenging literature search required to answer the first question are now summarized in the following part.

Ref. [62] tried to correlate the contamination of the Venice lagoon sediments to potential sources, analyzing literature data (i.e., contaminated emissions and sediments) through a statistical approach (principal component analysis, PCA). Lagoon sediments presented in [62] for the Venice urban area and at the boundary with Porto Marghera fell into the same group with the sediment of the Norwegian Frijefjorden contaminated by a magnesium processing plant [63]. This industrial process involved the chlorination of Mg under carbon presence at high temperatures (800–900 °C) and plant discharges were found to be enriched in OCDF, with PCDD/F emissions in water amounting to about 500 g TEQs per year in 1989 [63]. Indeed, the congener composition (as chlorination classes, Figure 4b) of the Norwegian fjord sample was characterized by the presence of heavy furans with a small contribution of octachlorodibenzodioxin (OCDD), a pattern similar to that observed in sediments of the 1st Industrial Area (C12, Figure 4a). In addition, total concentrations of

PCDD/Fs measured in the fjord and in the more contaminated canals of the 1st Industrial Area were of the same magnitude (hundreds of thousands of ng kg^{-1}, Figure 4a,b). A study conducted in 1997 [64] verified the level of sediment contamination about ten years after the emissions from the Mg production plant in the Frijefjorden had been drastically reduced. The results showed that pollutant levels in surface sediments were reduced by about two orders of magnitude (in the most contaminated area, where Mg processing slags were located, concentrations passed from about 37,000 to 3800 ng kg^{-1} TEQs), but the PCDD/F fingerprints were unchanged. This proved incontestably that the Mg industry was the major factor responsible for PCDD/F pollution at that site, with negligible contribution from other production processes.

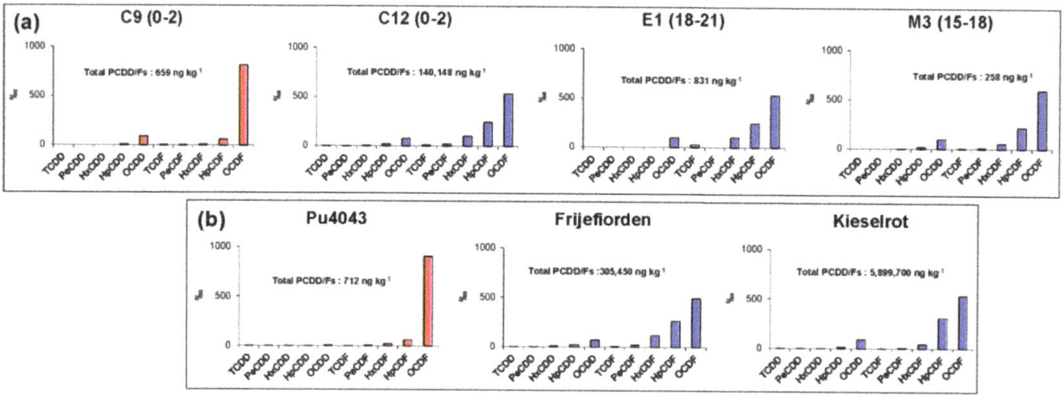

Figure 4. (**a**) PCDD/F congener composition (as chlorination classes) in the Venice Lagoon (E1 and M3) and in the industrial canals of Porto Marghera (C9 and C12). Total PCDD/F concentrations are also reported. Data are from [28,45,46]. (**b**) PCDD/F congener composition (as chlorination classes) retrieved in literature and relative to specific industrial processes. Total PCDD/F concentrations are also reported. Data are from [63,65,66].

As no Mg industry had ever been present in the Industrial Area of Porto Marghera, a new question arose: Was it possible to identify another industrial metal process that could explain the observed PCDD/F signatures in the 1st Industrial Area? At that point, it became necessary to verify whether the sediments of the most contaminated canals of Porto Marghera presented associations between PCDD/Fs and metals that could lead to specific industrial processes actually operating in the area. Using data from the survey carried out by the Venice Water Authority and the Port Authority [67], a good correlation (R^2: 0.742) was found between PCDD/Fs and As (Figure 5).

Since the formation of PCDD/Fs requires oxygen, a hydrocarbon compound, Cl, a catalyst (e.g., Cu or Fe), an ideal temperature (400–600 °C), and an adequate residence time [68], there was the necessity to identify the industrial processes operating under these conditions and simultaneously related to As pollution. One of these processes occurred regularly in the 1st Industrial Area and involved the extraction of Cu from pyrite ashes (very rich in As [69]), through the addition of 10% NaCl (chlorination) and the subsequent roasting at temperatures of about 450–550 °C with the formation of $CuCl_2$ and $FeCl_2$ [70]. Both compounds (and also NaCl) are regarded as catalysts for the formation of dioxins and especially furans [68,71], through chlorination and oxidation reactions of carbonaceous particles with degenerate graphitic structure, present in the combustion gases. Reactions take place on metal oxides in the flying ashes and are catalyzed by transition metal compounds. Chlorination takes place essentially by means of inorganic salts (KCl, CuCl, $CuCl_2$, NaCl) while the oxygen supply is given both by that present in the gaseous phase and by that existing in the aromatic nuclei present in the degenerate graphitic structures. Table 1 summarizes the main reactions that occur when roasting pyrite ashes. It is evident that a high amount of metal compounds potentially active as PCDD/F catalysts is produced. The residues of the roasting are then leached and metallic Cu, small amounts of Ag, Au, and also Zn are recovered from the resulting solution. Furthermore, the residual Fe oxides (called "purple ore") are agglomerated into tiles through a sintering process that, due to the presence of residual chlorine from the previous treatment, can produce dioxins and furans.

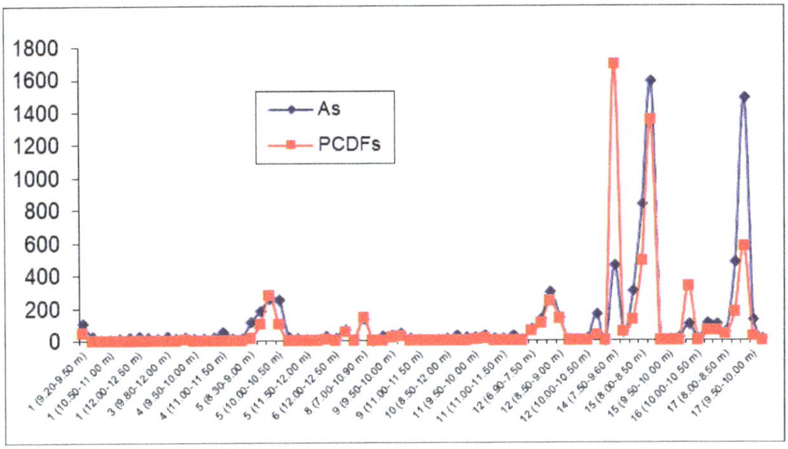

Figure 5. Association between As peaks (μg g^{-1}) and PCDD/Fs (μg kg^{-1}) in sediments of the industrial canals Nord and Brentella. Data are from [67].

Table 1. Main chemical reactions occurring during the roasting of pyrite ashes [67].

1. Double exchange between iron sulfate or copper sulfate with sodium chloride $FeSO_4 + NaCl \rightarrow Na_2SO_4 + FeCl_2$, $CuSO_4 + NaCl \rightarrow Na_2SO_4 + CuCl_2$,
2. Oxidation of iron (II) chloride to iron (III) chloride $12\ FeCl_2 + 3\ O_2 \rightarrow 8\ FeCl_3 + 2\ Fe_2O_3$
3. Reaction of iron (III) chloride with Cu sulfide $4\ FeCl_3 + 3\ Cu_2S + 6\ O_2 \rightarrow 6\ CuCl_2 + 2\ Fe_2O_3 + 3\ SO_2$

The industrial process able to produce PCDD/Fs as by-product was identified, but the so-called smoking gun had still to be found, i.e., a case study that linked undoubtedly the process of Cu extraction from pyrite ashes to PCDD/F production. A new literature search brought to light the "Kieselrot" case [59]. Kieselrot (German for red pebble) is the name given to the red siliceous slag obtained from Cu extraction processes which were carried out in Marsberg (North Rhine-Westphalia, Germany), until 1945. The extraction process involved the chlorination, roasting, and leaching of a "black siliceous mineral" containing Cu called Schwarzkiesel. During the 1950s–60s, the Kieselrot was used to build sport fields, recreational areas, roads, and paths. In the following years, this material was found to be highly contaminated by chlorinated aromatic hydrocarbons, in particular PCDD/F, HCB, polychlorinated biphenyls, polycyclic aromatic hydrocarbons, as well as by heavy metals [65]. The concentration of PCDD/Fs measured in a Kieselrot slag was extremely high (5,899,700 ng kg^{-1}, Figure 4b), comparable to the highest values found in sediments of the Brentella Canal (Figure 5). The relative signatures were the same, with OCDF as the most abundant congener, followed by heavy furans and a small amount of OCDD (Figure 4a,b). In addition, As was measured in Kieselrot slags with a good correlation with PCDD/Fs [65], as observed also in the industrial canals Nord and Brentella of the 1st Industrial Area of Porto Marghera (Figure 5). No further evidence was needed, and the PCDD/F contamination found in the Brentella Canal was then ascribed entirely to sources linked to Cu extraction from pyrite ashes. This industrial process had been active in the 1st Industrial Area since 1927 and, until the decommissioning in the early 1970s, produced about 130,000 units year of tiles and iron minerals [72].

The answer to the second question was more straightforward: the Lusore–Brentelle Canal was under the direct influence of a VCM stripping plant and the comparison of PCDD/F congener composition measured at site C9 with that observed in sediments from the Gulf of Finland under the direct influence of VCM stripping discharges (sample PU4043, [66], Figure 4b) left no doubt about the origin of dioxins and furans in the two areas. Therefore, it was concluded that VCM production in the 2nd Industrial Area affected only a restricted nearby area.

4.3. The History of PCDD/F Contamination in the Venice Lagoon

As already evidenced in Section 4.1, also in the Venice Lagoon the acquisition of reliable sediment chronologies was a mandatory prerequisite for the correct definition of PCDD/F temporal trends. Once sedimentary levels are correctly associated to a reasonable range of deposition years (i.e., an error confined below 5 years in the most recent levels, increasing to 20–30 years at higher depths) it is possible to "read" sediment profiles as historical documents.

Salt marshes are coastal wetlands that are regularly flooded and drained by seawater following the tidal cycle. They are found mostly at middle to high latitudes and are common habitats in the Venice Lagoon, where they are called "barene". Among their many ecosystem services [42], Venetian salt marshes represent an important historical and scientific archive that is able to record important environmental parameters over many decades [23,47,48], when their average accretion rates are comparable to the long-term average rate of sea level rise (SLR) in the lagoon (~0.25 cm year^{-1}; [23]).

Figure 6 shows the pattern of PCDD/F contamination recorded by salt marsh core M3 (Figure 1b). It is evident that PCDD/F concentrations reached the highest values between 1940 and 1980, and then decreased significantly. This $^{210}Pb_{ex}$ dating is highly reliable, since it was verified by Cs-137 peak activity at 4 cm depth, following the Chernobyl accident in 1986. The recent decreasing trend of PCDD/F concentrations in salt marshes was attributed to a strong reduction of inputs, due to the closure of the most polluting activities within the first industrial area, and to the promulgation of the first environmental regulations in Italy in the mid 1970s. In particular, the already cited Law 319/1976 regulated the composition of industrial effluents, and the Ministerial Decree 23 April 1998 ("Water Quality in the Venice Lagoon" published on the Official Gazette n. 140 and further revised in 1999) established very strict limits for industrial effluents to lagoon waters, banning the discharge of several toxic, persistent, and bioaccumulative contaminants (As, Cd, PCDD/Fs, PAHs, Hg, Pb, PCBs, chlorinated pesticides, cyanides, and tributyltin compounds, TBTs). During this period, several wastewater treatment plants became operative [28].

The definition of these contaminant input chronologies in salt marshes and lagoon sediments [73] has permitted assessment of the risks for both the environment and humans, with the use of models that were combined to create several past and future scenarios [74–76]. They have highlighted the predominant role of deep sediments as secondary sources of contamination through dredging [74,75] and the varying distribution pathways under different wind scenarios (i.e., bora and scirocco) [76]. Input chronologies were also used to implement a human intake and a physiologically-based pharmaco-kinetic model [77] used to simulate human internal exposure and provide a preliminary ecological and human health risk assessment for selected chemicals. Moreover, contaminant concentrations in sediments were at the base of a dynamic multimedia model applied to selected PCB and PCDD/F congeners, simulating the effects of climate change on their distribution and fluxes over the next 50 years in the Venice Lagoon (Italy) [78]. This model's results suggested that global warming would probably enhance PCB and PCDD/Fs mobility and hence their potential for long range atmospheric transport, reducing the environmental levels of these chemicals in the lagoon.

67. MAV Port Authority of Venice. *Rapporto Sullo Stato di Compromissione Delle Sponde e dei Canali di PORTO Marghera*; MAV Autorità Portuale: Venice, Italy, 1999. (In Italian)
68. Ryan, S.P.; Altwicker, E.R. The formation of polychlorinated dibenzo- p -dioxins/dibenzofurans from carbon model mixtures containing ferrous chloride. *Chemosphere* **2000**, *40*, 1009–1014. [CrossRef]
69. Oliveira, M.L.; Ward, C.R.; Izquierdo, M.; Sampaio, C.H.; de Brum, I.A.; Kautzmann, R.M.; Sabedot, S.; Querol, X.; Silva, L.F. Chemical composition and minerals in pyrite ash of an abandoned sulphuric acid production plant. *Sci. Total. Environ.* **2012**, *430*, 34–47. [CrossRef]
70. Mariani, M. *Chimica Applicata e Industriale*; Unione Tipografico-Editrice Torinese: Turin, Italy, 1972; Volume I, pp. 908–909. (In Italian)
71. Gianotti, G. Meccanismi di formazione delle diossine. In Proceedings of the Congress "Le Diossine–Esame a 360°", Milan, Italy, 19 November 1999. (In Italian)
72. *Elenco Delle Ditte e Degli Stabilimenti del Porto e Della Zona Industriale di Marghera*; Ente Della Zona Industriale di Porto Marghera: Venice, Italy, 1968. (In Italian)
73. Frignani, M.; Bellucci, L.G.; Favotto, M.; Albertazzi, S. Pollution historical trends as recorded by sediments at selected sites of the Venice Lagoon. *Environ. Int.* **2005**, *31*, 1011–1022. [CrossRef]
74. Sommerfreund, J.; Gandhi, N.; Diamond, M.; Mugnai, C.; Frignani, M.; Capodaglio, G.; Gerino, M.; Bellucci, L.; Giuliani, S. Contaminant fate and transport in the Venice Lagoon: Results from a multi-segment multimedia model. *Ecotoxicol. Environ. Saf.* **2010**, *73*, 222–230. [CrossRef]
75. Sommerfreund, J.; Arhonditsis, G.; Diamond, M.; Frignani, M.; Capodaglio, G.; Gerino, M.; Bellucci, L.; Giuliani, S.; Mugnai, C. Examination of the uncertainty in contaminant fate and transport modeling: A case study in the Venice Lagoon. *Ecotoxicol. Environ. Saf.* **2010**, *73*, 231–239. [CrossRef] [PubMed]
76. Mugnai, C.; Giuliani, S.; Bellucci, L.G.; Romano, S.; Frignani, M. Modelling Environmental Fate and Transport of POPs and Metals using the Fugacity/Aquivalence approach: Two Aquatic Environments as a Case Study. In *Marine Research at CNR*; Brugnoli, E., Cavarretta, G., Mazzola, S., Trincardi, F., Ravaioli, M., Santoleri, R., Eds.; Dipartimento Terra-Ambiente: Rome, Italy, 2011; Volume 6, pp. 965–978.
77. Giubilato, E.; Radomyski, A.; Critto, A.; Ciffroy, P.; Brochot, C.; Pizzol, L.; Marcomini, A. Modelling ecological and human exposure to POPs in Venice lagoon. Part I—Application of MERLIN-Expo tool for integrated exposure assessment. *Sci. Total. Environ.* **2016**, *565*, 961–976. [CrossRef] [PubMed]
78. Valle, M.D.; Codato, E.; Marcomini, A. Climate change influence on POPs distribution and fate: A case study. *Chemosphere* **2007**, *67*, 1287–1295. [CrossRef] [PubMed]

MDPI
St. Alban-Anlage 66
4052 Basel
Switzerland
www.mdpi.com

Applied Sciences Editorial Office
E-mail: applsci@mdpi.com
www.mdpi.com/journal/applsci

www.ingramcontent.com/pod-product-compliance
Lightning Source LLC
Chambersburg PA
CBHW061514210926
51850CB00008B/1913